9/30/81

D0887024

CHEMISTRY OF NATURAL WATERS

CHEMISTRY OF NATURAL WATERS

by
SAMUEL D. FAUST
OSMAN M. ALY

ANN ARBOR SCIENCE
PUBLISHERS INC / THE BUTTERWORTH GROUP

PREFACE

Increasing concern is being expressed about the quality of fresh ground and surface waters and oceanic waters in our environment. One of the highest priority uses is, of course, ingestion of water by humans. Consequently, we demand that potable water should have a quality to meet a rather rigid regimen of standards. On the other hand, it has been recognized that natural waters have uses other than drinking by man. For example, water quality is extremely important to freshwater and marine aquatic life, for industrial and agricultural uses, and for recreation and esthetic uses. These latter uses of water have moved the governmental and scientific communities to establish quality criteria for these various uses of natural waters.

There are many natural factors, physical, chemical and biological, that affect water quality. There are also many anthropogenic influences on water quality. This volume attempts to document the natural chemical factors. Wherever appropriate, however, some physical and biological aspects are cited.

It is hoped that this book will be utilized by the many scientific and engineering disciplines which encounter water quality problems in their professional endeavors. The authors have attempted to provide the essential chemical bases that control the many dissolved constituents in natural waters. Also, a considerable quantity of "raw" water quality data is provided that may be helpful in the management of lakes, reservoirs, streams, rivers, etc., and in the design, perhaps, of a potable water treatment plant. The authors have researched the scientific literature as thoroughly as possible on a particular water quality subject. Undoubtedly, a few papers have been overlooked. Also, we are as current as possible, but the reader must recognize that the "lag" time between manuscript preparation and ultimate publication will omit some recent water quality developments. The authors invite constructive criticism of the contents of this book that will assist in preparation of a second edition. We are optimistic about future needs for water quality information.

Samuel D. Faust
Osman M. Aly

ACKNOWLEDGMENTS

The authors have rather personal acknowledgments in the preparation of this book. First, we appreciate the patience of our families, who were neglected during the long hours of preparation. Second, we are indebted to Samantha's "Nanna," Ms. Mildred McHose, who typed the major share of the manuscript and who kept us "organized." Whenever "Nanna" was traveling, Ms. Eleanor E. Kover and Ms. Anne E. Faust willingly typed a page here and there. Third, we are delighted that someone, in the past, invented the photocopy process. Fourth, we are appreciative of our professional colleagues whose works are cited throughout the text. And last, but not least, we enjoyed splendid cooperation with the editors and publishers of Ann Arbor Science Publishers, Inc.

Samuel D. Faust is Professor of Environmental Science at Rutgers University. He received his PhD in environmental sciences and chemistry from Rutgers University and a BA in chemistry from Gettysburg College. Dr. Faust was a Research Fellow in applied chemistry during the 1966–1967 term at Harvard University.

Dr. Faust is the editor of three books on the chemistry of aquatic environments and has published 90 papers in research journals. He has served as chairman and organizer of several conferences. Dr. Faust is a member of several honorary professional societies and committees, which include Standard Methods for the Examination of Water and Wastewater.

Osman M. Aly is Manager of Environmental Quality at the corporate headquarters of Campbell Soup Company. He received his PhD in environmental sciences from Rutgers University, and held various research and teaching positions before joining the Campbell Soup Company in 1970. His major research interests include the chemistry of natural waters and environmental control.

Dr. Aly is the author of more than 30 publications on water and wastewater chemistry and treatment processes.

To
Our
Children

Samantha Anne Faust

and

Maggie, Sherry and Sherif Aly

CONTENTS

CHAPTER 1

CHEMICAL COMPOSITION OF NATURAL WATERS

INTRODUCTION

Natural ground, surface and atmospheric waters contain many chemical species in the dissolved state. The occurrence of these constituents results from many physical and chemical weathering processes on geologic formations and from many chemical reactions in the atmosphere. Furthermore, the nature of these constituents is a function of the type of geology, the distribution of this geology, and the physical and chemical constraints of the many weathering processes. (The anthropogenic sources of dissolved chemical species are not considered in this chapter, with the exception of acid rain.) Consequently, these natural waters achieve definite concentrations of the several dissolved substances that are controlled, in turn, by equilibrium conditions of the weathering processes. In other words, natural waters will acquire specific chemical characteristics. This is commonly given the description of "water quality." This loose and somewhat vague term has a different interpretation, which is dependent on utilization of the water.

Thus, there are several types of "water quality"—chemical, biological, physical, bacteriological, viral, sanitary, drinking water, etc., which suggests that there is no single, universal definition of water quality. This chapter considers the chemical constituents of natural waters, designates their principal geological sources, and defines their major water quality characteristics. Examples of chemical water quality of natural waters are cited where appropriate. This chapter does not present the analytical procedures for quantitative determination of the various chemical constituents in water. Instead, the reader is referred to the analytical work of the profession, *Standard Methods* [1], and to Hem's [2] excellent water supply paper for this information. In addition, the reader will find geographical differences in chemical water quality, for example, throughout the United States in various publications of the U.S. Geological Survey (USGS) [3].

GROUND AND SURFACE WATERS

Total Dissolved Solids

A general indication of chemical water quality is obtained from the total dissolved solids (TDS) content. TDS are defined analytically either by the total filterable residue (dried at 180°C) method [1] or by a conductivity technique [2]. Of course these procedures yield a gross indication of the dissolved cationic, anionic and neutral constituents found in natural waters. Normally, the analytical procedures for this water quality parameter yield only the inorganic constituents. There are occasions, however, when dissolved organic matter is included in the analysis, especially by the total filterable residue procedure.

The specific conductance and total residue on evaporation of natural waters have a wide range of values and are dependent on the weathering characteristics of the geologic formations and the degree of anthropogenic pollution, of course. Frequently, this parameter is employed for comparison of the water quality of different bodies of water and to compute the rates at which rivers transport dissolved solids to the oceans. For example, Leifeste [4] gave a brief report on the worldwide runoff of dissolved solids. This report also contains some hydrological and chemical data for the 27 major basins in the conterminous United States (Table I). The total annual yield of dissolved

Table I. Summary of Drainage Area, Water and Dissolved Solids Discharge [4]

Receiving Body	Drainage Area		Water Discharge		Dissolved Solids Discharge		
	Square Miles	Percentage of Total	Cubic Feet per Second	Percentage of Total	Thousands of Tons per Year	Percentage of Total	mg/l
Atlantic Ocean	285,900	10.7	359,400	20.6	37,500	14.2	105.0
Gulf of Mexico	1,750,500	65.6	887,400	50.8	183,000	69.3	210.0
Pacific Ocean	632,500	23.7	499,060	28.6	43,400	16.5	85.0
Total	2,667,900	100.0	1,745,860	100.0	264,000	100.0	–

solids is approximately 264 million tons and ranges from 0.492 to 157 million tons for individual basins.

Excessive quantities of dissolved solids in drinking water are objectionable because of possible physiological effects, unpalatable mineral tastes and higher costs from additional water treatment [5]. The physiological effects that are directly related to TDS include laxative effects primarily from sodium sulfate and magnesium sulfate. Other water quality problems caused by TDS are: corrosion and incrustation of metallic surfaces, agricultural uses of water for irrigation and adverse effects on fresh water species of aquatic life, especially fish. However, to date no water quality standard has been established for TDS.

Principal Cations

Calcium

Calcium is perhaps the most abundant dissolved cationic constituent of natural fresh waters. Consequently, this element is widely distributed in the minerals of rocks and soils. The geological sources of calcium are many: such feldspars as anorthite ($CaAl_2Si_2O_8$); from the igneous-rock group such precipitates as calcite and aragonite ($CaCO_3$), dolomite ($CaMg(CO_3)_2$), gypsum ($CaSO_4 \cdot 2H_2O$), anhydrite ($CaSO_4$), fluorite (CaF_2), fluorapatite ($Ca_5(PO_4)_3F$); and hydroxylapatite ($Ca_5(PO_4)_3OH$) from the sedimentary rocks. Calcium carbonate is frequently found as a cementing agent between mineral particles of sandstone and other detrital rocks.

Any calcium content of a natural water depends on the geological source and the extent of the chemical weathering process (see Chapter 5). Table II shows the analyses of several waters in the United States in which calcium is a major constituent. One general observation is made from these analyses: the calcium contents of waters from limestone areas usually are lower than those from gypsum areas. This may be seen from a comparison of analyses 1 and 2 with 3, for example. Analysis 6 shows the calcium content of a river that drains a generally humid climatic area. Here the calcium is the predominant cation. Analysis 7 shows the calcium content of a river that drains an arid climatic region in New Mexico where more soluble rock types occur.

Calcium is one of the constituents of "hard" water and is a scale former in hot water systems. Prevention of corrosion of cast iron water distribution systems may be obtained through controlled precipitation of calcium carbonate. Lime ($Ca(OH)_2$), limestone ($CaCO_3$) and dolomite ($CaMg(CO_3)_2$) are frequently employed as neutralizing agents in water and wastewater treatment.

Table II. Analyses of Waters in Which

Constituent	1 (mg/l)	2 (mg/l)	3 (mg/l)
Silica (SiO$_2$)	8.4	22	29
Iron (Fe)	0.04	–	–
Manganese (Mn)	–	–	–
Calcium (Ca)	46	144	636
Magnesium (Mg)	4.2	55	43
Sodium (Na)	1.5	} 29	} 17
Potassium (K)	0.8		
Bicarbonate (HCO$_3$)	146	622	143
Sulfate (SO$_4$)	4.0	60	1570
Chloride (Cl)	3.5	53	24
Fluoride (F)	0.0	0.4	–
Nitrate (NO$_3$)	7.3	0.3	18
Dissolved solids:			
Calculated	149	670	2410
Residue on evaporation	139	–	–
Hardness as CaCO$_3$	132	586	1760
Noncarbonate	12	76	1650
Specific conductance (micromhos at 25°C)	250	1120	2510
pH	7.0	–	–
Color	5	–	–

1. Big Spring, Huntsville, Alabama. Water-bearing formation, Tuscumbia Limestone.
2. Spring on Havasu Creek, near Grand Canyon, Arizona. Water-bearing formation,
3. Jumping Springs, SE¼ sec, 17, T. 26 S., R. 26 E., Eddy County, New Mexico. Water-
4. Rattlesnake Spring, sec. 25, T. 24 S., R. 23 E., Eddy County, New Mexico. Water-
5. Irrigation well NE¼ sec. 35, T. 1 S., R. 6 E., Maricopa County, Arizona. Water-bearing
6. Cumberland River at Smithland, Kentucky. Discharge, 17,100 cfs.
7. Pecos River near Artesia, New Mexico. Discharge-weighted average 1949 water year;
8. City well at Bushton, Rice County, Kansas. Depth, 99 ft. Water-bearing formation,
9. Industrial well, Williamanset, Massachusetts. Depth, 120 ft. Water-bearing formation,

Magnesium

Magnesium is also a major constituent of igneous and sedimentary rocks. The abundance of magnesium ions in fresh natural waters is, perhaps, second to calcium. Principal sources of magnesium from igneous rocks are: forsterite (Mg_2SiO_4), serpentine ($Mg_6(OH)_8Si_4O_{10}$), magnesioferrite ($MgFe_2O_4$), and cordierite ($Mg_2Al_3(AlSi_5O_{18})$). Major sources of magnesium from sedimentary rocks are: magnesite ($MgCO_3$), dolomite ($CaMg(CO_3)_2$), huntite ($Mg_3Ca(CO_3)_4$), and brucite ($Mg(OH)_2$).

It is somewhat rare to find a natural water in which magnesium is the predominant cation. It is more common to observe magnesium in occurrence with calcium. Table III shows the analyses of several waters in the United States where magnesium is a major constituent. Analyses 1, 2, 5, 6, 8 and 9

Calcium is a Major Constituent [2]

4 (mg/l)	5 (mg/l)	6 (mg/l)	7 (mg/l)	8 (mg/l)	9 (mg/l)
–	74	4.9	27	24	15
–	–	0.04	–	0.02	1.0
–	–	–	–	0.00	0.01
99	277	25	394	88	96
28	64	3.9	93	7.3	19
} 4.1	} 53	4.5	} 333	19	18
		1.4		2.8	1.5
287	85	90	157	320	133
120	113	12	1150	6.7	208
6	605	2.2	538	13	25
–	0.2	0.1	–	0.3	0.4
2.8	35	1.9	5.0	4.6	0.4
401	1260	100	2610	323	449
–	–	99	–	322	468
362	954	78	1370	250	318
127	884	5	1240	0	209
651	2340	172	3540	543	690
–	–	6.7	–	7.5	7.8
–	–	5	–	2	3

Temp, 16.1°C.
limestone in Supai Formation; 100 gpm. Temp, 19.4°C. Water deposits travertine.
bearing formation, gypsum in Castile Formation; 5 gpm.
bearing formation; alluvium, probably fed by Capitan Limestone; 2500 gpm.
formation; alluvium. Temp, 25.6°C.

mean discharge 298 cfs.
Dakota Sandstone.
Portland Arkose. Temp, 12.2°C.

have their origin in dolomitic or limestone geologic areas. Analysis 3 shows the magnesium content of waters that have passed through olivine deposits. Water in Analysis 4 has participated in reactions with magnesium silicates. The water in Analysis 5 was found in the Carlsbad Caverns of New Mexico, where much of the calcium was lost through the precipitation of calcite on the cave formations. This may account for the greater content of magnesium in this water. However, it is more nearly the case in most natural waters that the calcium content exceeds the magnesium content.

Magnesium is also one of the constituents of "hard" water and is also a scale former in hot water systems. The carbonates of magnesium may be used for neutralization in water and wastewater processes. Magnesium hydroxide has found use as a coagulant in water treatment.

Occasionally, attempts are made to establish a water quality standard for

Table III. Analyses of Waters in Which

Constituent	1 (mg/l)	2 (mg/l)	3 (mg/l)
Silica (SiO$_2$)	8.4	18	31
Aluminum (Al)	1.4	–	–
Iron (Fe)	0.24	1.4	–
Calcium (Ca)	40	94	20
Magnesium (Mg)	22	40	42
Sodium (Na)	0.4	17	} 19
Potassium (K)	1.2	2.2	
Carbonate (CO$_3$)	0	0	0
Bicarbonate (HCO$_3$)	213	471	279
Sulfate (SO$_4$)	4.9	49	22
Chloride (Cl)	2.0	9.0	7
Fluoride (F)	0.0	0.8	0.2
Nitrate (NO$_3$)	4.8	2.4	2.5
Dissolved solids			
Calculated	190	466	281
Residue on evaporation	180	527	–
Hardness as CaCO$_3$	190	400	222
Noncarbonate	16	13	0
Specific conductance (micromhos at 25°C)	326	764	458
pH	7.4	6.7	8.2
Color	5	5	–

1. Spring 2½ miles northwest of Jefferson City, Tennessee. Flow, 5000 gpm. From Knox
2. Well number 5, City of Sidney, Ohio. Depth, 231 ft. Water-bearing formation, Niagara
3. Spring in Buell Park, Navajo Indian Reservation, Arizona. Water-bearing formation,
4. Main spring at Siegler Hot Springs, NE¼ sec. 24, T. 12 N., R. 8 W., Lake County,
5. Green Lake in Carlsbad Caverns, New Mexico. Pool of ground water seepage.
6. Wisconsin River at Muscoda, Grant County, Wisconsin. Flows through area of
7. Spring, SW¼ sec. 26, T. 16 S., R. 7 E., Calhoun County, Alabama. Supplies city of
8. Oasis flowing well, SW¼ sec. 15, T. 11 S., R. 25 E. Chaves County, New Mexico.
  Limestone (limestone and dolomitic limestone with minor amounts sandstone,
9. Drilled well, NW¼ sec. 6, T. 6 N., R. 21 E., Milwaukee County, Wisconsin. Depth,

hardness constituents [5,6]. For example, as stated in the "Quality Criteria for Water" [5]: "Because hardness concentrations in water have not been proven health related, the final level achieved principally is a function of economics. Since hardness in water can be removed with treatment by such processes as lime-soda softening and zeolite or ion exchange systems, a criterion for raw waters used for public water supply is not practical."

Sodium

Almost all natural fresh waters contain sodium cations, but in concentrations less than calcium and magnesium. This is reversed, however, in saline and brackish waters. Sodium ions tend to remain in solution and very seldom enter into precipitation reactions that would control its solubility.

Magnesium is a Major Constituent [2]

4 (mg/l)	5 (mg/l)	6 (mg/l)	7 (mg/l)	8 (mg/l)	9 (mg/l)
175	11	4.2	12	13	18
–	–	–	1.0	–	0.2
–	0.04	0.13	0.03	0.01	0.39
34	16	18	23	140	35
242	71	8.5	12	43	33
184	} 7.8	2.5	1.7	} 21	28
18		2.4	1.6		1.3
0	21	0	0	0	0
1300	320	90	127	241	241
6.6	21	9.5	1.6	303	88
265	8	1.5	2.0	38	1.0
1.0	1.0	0.0	0.0	0.8	0.9
0.2	19	0.8	0.9	4.1	1.2
1580	333	92	119	682	326
–	321	106	108	701	329
1080	332	80	107	526	224
14	34	6	3	329	27
2500	570	165	197	997	511
6.5	8.3	7.0	7.6	7.4	8.2
–	–	35	5	–	10

Dolomite. Temp, 14.4°C.
Group. Temp, 11.7°C.
olivine tuff-breccia; 18 gpm. Temp, 12.2°C.
California. Water issues from contact with serpentine and sediments. Temp, 52.5°C.

magnesium limestone.
Anniston. Flow, 46 cfs. Water-bearing formation, quartzite. Temp, 17.8°C.
Depth, 843 ft. Flow over 9000 gpm when drilled. Water-bearing formation, San Andres gypsum and anhydrite).
500 ft. Water-bearing formation, Niagara Dolomite. Temp, 10.0°C.

Clay mineral surfaces adsorb sodium ions from solution where ion exchange reactions occur.

The feldspars of igneous rocks are good sources of sodium when weathered. Albite ($NaAlSi_3O_8$), for example, is chemically weathered to yield sodium ions and kaolinite, among other things. Such evaporites as halite (NaCl) and thenardite (Na_2SO_4) yield sodium ions on dissolution. However, their importance in fresh water may be insignificant compared with their contribution in saline waters.

The geological source of sodium is the principal factor that influences its ultimate concentration in water. Analyses 1, 2 and 3 in Table IV show three ground waters in which sodium is the principal cation that was derived from igneous terrains. Analysis 4 gives the sodium content of a surface water at low flow where the rock is fine grained and contains much soluble material.

Analysis 5 shows the very high sodium content that may result from the use and reuse of water in an irrigated area. Analysis 6 is a water that has been in contact with halite with the result of a sodium content of 121,000 mg/l. The effect of shale on sodium content of a ground water is seen in Analysis 7. It should be noted also that many of these waters in Table IV contain significant quantities of bicarbonate, which suggests a sodium carbonate (soda ash) origin.

In most fresh waters, sodium is not usually considered as a significant factor in water quality unless unusually high concentrations are encountered. When the latter occurs, a "salty" taste problem frequently results, which renders the water unpalatable. This results where the sodium chloride and sodium sulfate contents exceed their threshold taste levels. Sodium hydroxide (caustic soda) and sodium carbonate (soda ash) are employed frequently in water and wastewater treatment.

Table IV. Analyses of Waters in Which

Constituent	1 (mg/l)	2 (mg/l)
Silica (SiO_2)	22	16
Iron (Fe)	0.20	0.15
Calcium (Ca)	2.5	3.0
Magnesium (Mg)	2.1	7.4
Sodium (Na)	1,182	857
Potassium (K)		2.4
Carbonate (CO_3)	30	57
Bicarbonate (HCO_3)	412	2,080
Sulfate (SO_4)	3.5	1.6
Chloride (Cl)	9.5	71
Fluoride (F)	1.7	2.0
Nitrate (NO_3)	0.6	0.2
Boron (B)	–	0.40
Dissolved solids		
Calculated	457	2,060
Residue on evaporation	452	–
Hardness as $CaCO_3$	15	38
Noncarbonate	0	0
Specific Conductance (micromhos at 25°C)	718	2,960
pH	8.7	8.3
Color	1	–

1. Well at Raleigh-Durham airport, Wake County, North Carolina. Depth, 184 ft. Water-
2. Well, SE¼ NE¼ sec. 2, T. 22 N., R. 59 E., Richland County, Montana. Depth, 500 ft.
3. Irrigation well, SE¼ sec. 3, T. 1 N., R. 5 E., Maricopa County, Arizona. Depth, 500 ft.
4. Moreau River at Bixby, South Dakota; composite of two daily samples; mean
5. Gila River at Gillespie Dam, Arizona. Weighted average 1952 water year; mean
6. Test well 3, sec. 8, T. 24 S., R. 29 E., Eddy County, New Mexico. Depth, 292 ft.
7. Well in SW¼ sec. 7, T. 17 N., R. 26 E., San Miguel County, New Mexico. Depth, 50 ft.

Occasionally, concern is expressed by the medical profession about the sodium content of natural drinking waters in relation to cardiovascular diseases and to women with toxemia associated with pregnancy [6]. Although the study by the National Academy of Sciences did not propose any limit for sodium in drinking water, there was a recommendation that "knowledge of the sodium ion content of the water supply and maintenance of it at the lowest practicable concentration is clearly helpful in arranging diets with suitable sodium intake. In many diets, allowance is made for water to contain 100 mg/l of sodium" [6].

Potassium

The potassium content of natural waters is usually less than that of sodium, calcium and magnesium. The water chemistry of potassium is similar

Sodium is a Major Constituent [2]

3 (mg/l)	4 (mg/l)	5 (mg/l)	6 (mg/l)	7 (mg/l)
22	8.2	31	–	13
0.00	0.04	0.01	–	–
49	40	353	722	30
18	50	149	2,490	31
168	699	1,220	121,000	279
	16	9.8	3,700	
0	26	0	63	0
202	456	355	40	445
44	1,320	1,000	11,700	303
246	17	1,980	189,000	80
0.1	1.0	1.9	–	1.2
2.2	1.9	24	–	17
–	0.60	2.4	–	–
649	2,400	4,940	329,000	973
651	2,410	–	–	–
196	306	1,490	–	202
31	0	1,200	–	0
1,200	3,140	7,620	225,000	1,510
7.7	8.2	–	–	–
2	–	–	–	–

bearing formation, Coastal Plain sediments.
Water-bearing formation, Fort Union Formation (sandstone and shale).
Water-bearing formation, valley fill. Temp, 20.5°C.
discharge 1.7 cfs. Drains Pierre Shale, Fox Hills Sandstone, and Hell Creek Formation.
discharge 71.1 cfs.
Brine from Salado Formation and Rustler Formation.
Water-bearing formation shale of the Chinle Formation.

to sodium because it seldom enters into precipitation reactions but does undergo ion exchange reactions.

The principal sources of potassium are the feldspars orthoclase and microcline ($KAlSi_3O_8$) and the feldspathoid leucite ($KAlSi_2O_6$). These silicate minerals are very resistant to weathering processes but are altered at a very slow rate. Potassium is slightly less common than sodium in igneous rocks but is more abundant in sedimentary rocks. Dissolution of such evaporites as sylvite (KCl) and glaserite ($K_3Na(SO_4)_2$) account for the higher concentrations of potassium in brine waters.

The potassium content of natural fresh waters seldom exceeds the sodium content and rarely, if ever, is the principal cation. Frequently, the potassium and sodium contents have been reported together as sodium. Analyses 1, 6 and 7 of Table III show three waters that have nearly equal potassium and sodium contents. One is a ground water and one is a surface water from sedimentary rocks, whereas the third water is a spring from a quartzite formation. Brine waters such as Analysis 6 in Table IV may achieve unusually high potassium contents, 3700 mg/l.

Potassium has no special significance in natural water quality. At present, it is used as a tracer element and an indicator of the geological origin of the water. However, potassium is considered one of the four major cations that appears in natural waters and is included in the equivalence calculation.

Aluminum

Aluminum is the third most abundant element in the earth's outer crust and is widely distributed throughout rocks and soils. However, aluminum contents of natural fresh waters are rarely greater than 0.5 mg/l.

Many silicate rock minerals (feldspars, etc.) contain aluminum in coordination with oxygen. Gibbsite ($Al(OH)_3$) is a fairly common mineral. Alunite ($K_2Al_6(OH)_{12}(SO_4)_4$) is also a source of aluminum. Of course the various clay minerals contain aluminum, which enters into many weathering reactions, as noted in Chapter 5.

Very few data are available on the aluminum contents of natural fresh waters. Table V shows a water, Analysis 1, of unusually high aluminum content, 28 mg/l. This water has its origin in a shale, sand and marl geologic formation. The pH value of 4.0 is conducive to acidic weathering of this geology. Analyses 2, 3 and 5 show aluminum contents more nearly typical of natural fresh waters.

Despite the extremely low aluminum content of natural waters, this element plays a major role in many chemical weathering reactions. Also, aluminum salts are utilized in the chemical coagulation of waters and wastewaters. Aluminum may be toxic to certain plants under the appropriate soil conditions. For this reason, the recommended maximum concentrations are

Table V. Analyses of Waters High in Dissolved Aluminum or Manganese [2]

Constituent	1 (mg/l)	2 (mg/l)	3 (mg/l)	4 (mg/l)	5 (mg/l)
Silica (SiO_2)	98	10	92	9.7	31
Aluminum (Al)	28	0.1	0.35	3.5	0.2
Iron (Fe)	0.88	0.04	0.02	0.10	2.7
Manganese (Mn)	9.6	1.3	0.31	2.5	0.22
Calcium (Ca)	424	58	67	32	28
Magnesium (Mg)	194	13	0.0	11	1.9
Sodium (Na)	416	23	477	12	6.8
Potassium (K)	11	2.8	40	3.7	4.2
Hydrogen (H)	–	–	–	–	–
Carbonate (CO_3)	0	0	0	0	–
Bicarbonate (HCO_3)	0	101	1020	0	121
Sulfate (SO_4)	2420	116	169	171	1.4
Chloride (Cl)	380	39	206	5.0	1.0
Fluoride (F)	1.8	0.0	6.8	0.1	0.1
Nitrate (NO_3)	3.1	0.6	1.8	5.3	0.2
Orthophosphate (PO_4)	0.0	0.1	0.11	–	0.0
Boron (B)	–	–	2.8	–	–
Dissolved solids					
Calculated	3990	314	1570	256	137
Residue at 180°C	4190	338	1560	260	–
Hardness as $CaCO_3$	1860	198	168	125	78
Noncarbonate	1860	115	0	125	0
Specific Conductance					
(micromhos at 25°C)	4570	517	2430	507	192
pH	4.0	7.0	6.7	3.8	6.9

1. Well, 7 miles northeast of Montecello, Drew County, Arkansas. Depth, 22 ft. Water-bearing formation, shale, sand, and marl of the Jackson Group. Also contained radium (Ra), 1.7 μμc/l and uranium (U), 17 μg/l.
2. Composite from two radial collector wells at Parkersburg, Kanawha County, West Virginia. Depth, 52 ft. Water from sand and gravel. Also contained copper (Cu), 0.01 mg/l and zinc (Zn), 0.01 mg/l.
3. Wagon Wheel Gap hot spring, Mineral County, Colorado. Discharge, 20 gpm. Temp. 62.2°C. Associated with vein of Wagon Wheel Gap fluorite mine. Also contained 2.3 mg/l Li, 0.5 mg/l NH_4, 0.3 mg/l Br, and 0.3 mg/l I.
4. Kiskiminitas River at Leechburg (Vandergrift), Pennsylvania. Composite of nine daily samples. Mean discharge for period, 10,880 cfs.
5. Well, 167 ft. deep, Baltimore County, Maryland. Water-bearing formation, Port Deposit granitic gneiss. Also contained 0.01 mg/l copper (Cu).

5.0 mg/l aluminum for continuous use on all soils and 20 mg/l for use on fine-textured neutral to alkaline soils over a period of 20 years [7].

Iron

Iron is a widespread constituent of rocks and soils. It is the fourth most abundant, by weight, of the elements in the earth's crust. Almost all ground and surface waters contain significant quantities of dissolved or suspended iron.

In igneous rocks, iron may occur as fayalite (Fe_2SiO_4), or as hematite (Fe_2O_3) and magnetite (Fe_3O_4), which are essential components of the pyroxenes, amphiboles, biotite and olivine. The ferrous oxidation state occurs most frequently in these igneous rocks. In the sediments, iron may occur as pyrite or marcasite (FeS_2), siderite ($FeCO_3$), hematite and magnetite. The silicate forms of iron weather very slowly, whereas the oxides, sulfides and carbonates weather more quickly and are the principal sources of iron for ground water.

In a consideration of iron concentrations in natural waters, the source must be considered. For example, the water of a polluted stream or river does not usually contain uncomplexed and dissolved ferrous iron. In this situation, the iron probably would occur as suspended ferric hydroxide or in some form of an organic complex. The highly colored waters of some lakes and peat bogs usually have high iron contents. Here, the iron is complexed with naturally occurring organic matter (see below). An example is given in Table VI, Analysis 4, where the color value is 140. Surface waters that drain coal mine regions frequently have substantial iron contents and acidic pH values, as seen by Analysis 7. Ground waters with neutral pH values and with bicarbonate alkalinity exhibit the most frequently encountered iron concentrations in the 1 to 10 mg/l range. These may be seen in Analyses 1, 2, 3 and 8.

Iron in natural waters represents a nuisance type of constituent [5,8]. Frequently, the water initially contains dissolved ferrous iron, which is slowly oxidized to ferric hydroxide in the presence of dissolved oxygen. This precipitate imparts turbidity to the water, stains clothing in laundering processes, and interferes with the brewing of tea and coffee and the flavor of other beverages. Ferrous iron is also a product from the corrosion of pipes in water distribution systems. Iron also has some aquatic microbiological significance. The metabolic activities of the so-called "iron bacteria," *Crenothrix* and *Leptothrix,* may influence the iron content of water. Usually this involves the oxidation of ferrous iron to ferric hydroxide, as these are aerobic bacteria.

Manganese

Manganese and iron together have been considered as water quality parameters. However, their aquatic chemistries are considerably different and they should be evaluated and treated separately; however, manganese is similar to iron from its water quality effects [5,8].

Manganese is somewhat less abundant than iron in the earth's crust but is a common constituent of rocks and soils. Such ferromagnesian minerals as biotite and hornblende frequently contain some manganese. In sediments, manganese may be found as rhodochrosite ($MnCO_3$), rhodonite ($MnSiO_3$), hausmanite (Mn_3O_4), bixbyite (Mn_2O_3), pyrolusite (MnO_2) and manganosite (MnO). The oxides and hydroxides of manganese are the most commonly found constituents in rocks and soils. Since manganese is essential in plant metabolism, it frequently appears in surface waters as the result of decaying vegetation (leaves, aquatic plants, soil organic matter, etc.).

The concentrations of manganese in natural waters are usually less than 1.0 mg/l; however, there are some unusual waters in which the manganese contents are somewhat higher than 1.0 mg/l. Analysis 1, Table V, shows an abnormally high manganese content of 9.6 mg/l, which may be due, in part, to the acid pH value of 4.0. Acidic waters from coal mine drainage also contain manganese, such as Analysis 4, Table V. Frequently, iron and manganese occur together in natural waters at objectionable contents. An example of this is given in Analysis 5, Table V.

Manganese in natural waters represents a nuisance type of constituent [5,8]. Usually, it enters into the water as dissolved manganous (Mn(+II)) ions, which are slowly oxidized to black colloidal particles of MnO_2. This precipitate imparts turbidity to water, stains clothing in laundering processes, and interferes with the brewing of tea and coffee.

Principal Anions

Sulfate

Sulfate and other sulfur species are ubiquitous constituents of all natural waters. Sulfate may result from the chemical weathering of geologic formations or from biologically mediated oxidations of reduced sulfur species.

Sulfate occurs mainly in such evaporite sediments as gypsum ($CaSO_4 \cdot 2H_2O$), anhydrite ($CaSO_4$), epsomite ($MgSO_4 \cdot 7H_2O$) and mirabilite ($Na_2SO_4 \cdot 10H_2O$). An extremely significant source of sulfate is the chemical weathering of pyrite (FeS_2), which yields S_2^{2-} anions to the water phase. In turn, these reduced sulfur anions are oxidized catalytically in microbiological systems to sulfate (also see Chapter 4). Since hydrogen ions are produced in the course of this oxidation, waters with acidic pH values result.

The sulfate content of natural waters is, of course, variable and dependent on the geological source. Gypsum is reasonably soluble in water and can yield sulfate contents in excess of 1000 mg/l, as seen in Analyses 3 and 7, Table II. The influence of pyrite dissolution and subsequent oxidation on sulfate content may be seen in Analysis 4, Table V and Analysis 7, Table VI.

Sulfate is usually considered with the hardness constituents, calcium and magnesium. Gypsum scales can be formed in hot water systems. Sulfate may be used as a source of oxygen or, more appropriately, as an electron acceptor in microbiological processes. Odorous hydrogen sulfide usually evolves as the

Table VI. Analyses of

Constituent	1 (mg/l)	2 (mg/l)	3 (mg/l)
Silica (SiO_2)	20	12	26
Aluminum (Al)	–	1.2	1.2
Iron (Fe)	2.3	2.9	10
Manganese (Mn)	0.0	–	–
Calcium (Ca)	126	2.7	8.8
Magnesium (Mg)	43	2.0	8.4
Sodium (Na)	13	35	34
Potassium (K)	2.1	1.7	2.9
Bicarbonate (HCO_3)	440	100	65
Sulfate (SO_4)	139	5.6	71
Chloride (Cl)	8.0	2.0	2.0
Fluoride (F)	0.7	0.1	0.3
Nitrate (NO_3)	0.2	0.6	0.0
Dissolved solids			
Calculated	594	113	187
Residue on evaporation	571	101	180
Hardness as $CaCO_3$	490	15	56
Noncarbonate	131	0	3
Specific conductance (micromhos at 25°C)	885	162	264
pH	7.6	7.4	6.4
Color	1	23	7
Acidity (total) as H_2SO_4	–	–	–

1. Well 3, Nelson Rd , Water Works, Columbus, Ohio. Depth, 117 ft. Water from glacial
2. Well 79:8–50 Public supply, Memphis, Tennessee. Depth, 1,310. Water from sand
3. Well 5: 290–1, 6 miles southeast of Maryville, Blount County, Tennessee. Depth, 66
4. Partridge River near Aurora, Minnesota. Composite sample. Mean discharge, 30.8 cfs.
5. Brine produced with oil from well in NW¼ sec. 3, T. 11 N., R. 12 E., Okmulgee
   Formation.
6. Drainage at collar, drill hole 89, 7th level Mather A iron mine, Ishpeming, Michigan.
7. Shamokin Creek at Weighscale, Pennsylvania. Discharge, 64.2 cfs; affected by drainage
8. City well 4, Fulton, Mississippi. Depth, 210 ft. Water from the Tuscaloosa Formation.

result of these transformations. The aluminum and ferric salts of sulfate are employed as chemical coagulants in the treatment of water and wastewaters. Laxative effects are experienced when sulfate concentrations in drinking water are "high, circa 1000 mg/l" [7].

Chloride

In nonpolluted fresh waters, chloride may be considered as a trace constituent, that is, the concentrations are usually a few mg/l. However, where wastewater pollution occurs, the concentrations may be somewhat higher and on the order of tens of mg/l. In estuarine and oceanic waters, the chloride concentrations are several thousand mg/l.

Waters Containing Iron [2]

4 (mg/l)	5 (mg/l)	6 (mg/l)	7 (mg/l)	8 (mg/l)
11	9.1	8.1	21	7.9
–	–	–	29	0.6
1.4	32	0.31	15	11
–	–	0.34	10	0.32
18	7,470	264	119	8.4
8.0	1,330	17	68	1.5
} 9.3	43,800	52	} 17	1.5
	129	31		3.6
69	76	61	0	30
29	47	757	817	5.9
6.4	83,800	24	22	1.8
–	0	0.8	0.1	0.1
2.9	–	0.0	0.4	0.4
–	137,000	1,280	1,260	47
156	140,000	1,180	–	44
78	24,200	730	845	27
21	24,100	679	845	2
188	146,000	1,460	1,780	68.8
6.9	7.4	7.5	3.0	6.3
140	15	2	8	3
–	–	–	342	–

outwash. Temp, 13.3°C.
of the Wilcox Formation. Temp, 22.2°C.
ft. Water from Chattanooga Shale. Temp, 14.4°C.

County, Oklahoma. Depth, 2,394 ft. Water from the Gilcrease sand of drillers, Atoka

Temp, 15.1°C.
from coal mines.
Temp, 17.2°C.

The occurrence of chloride in rock formations is, perhaps, the lowest of any other major constituent because most chloride salts are extremely soluble in water. The rather complex feldspathoid sodalite ($Na_8Cl_2(AlSiO_4)_6$) occurs in igneous rocks. Frequently, chloride crystals may be found within the pores and cracks of igneous rocks. Perhaps the largest sources of chloride for fresh waters are the evaporites: halite ($NaCl$), sylvite (KCl), bischofite ($MgCl_2 \cdot 6H_2O$) and carnallite ($KMgCl_3 \cdot 6H_2O$).

Chloride is present in all fresh waters but the concentrations are usually lower than sulfate and bicarbonate. Analyses 1, 2, 3, 4, 8 and 9 of Table II show the more typical chloride concentrations of ground waters. The effect of irrigation practices may be seen in Analysis 5 of Table II, in which calcium and chloride are among the principal constituents. This water was drawn originally from the Salt River in Arizona, where the original calcium content was 48 mg/l and the chloride content was 50 mg/l. Ion exchange reactions in the soil increased the concentrations of these constituents. A more typical chloride water is one in which sodium is the major cationic constituent. These waters are usually brines, however, and cannot be considered fresh waters. Analysis 6 of Table IV is a brine with a chloride content of 189,000 mg/l and a sodium content of 121,000 mg/l.

In most natural, nonpolluted fresh waters, the chloride anion is relatively unimportant. Occasionally, chloride is used as a tracer constituent because it reacts conservatively in oxidation–reduction and ion exchange reactions. Where chloride concentrations exceed 400 mg/l, a "salty" taste may be imparted to the water [7]. Chloride is one of the reduction products from the use of chlorine as a disinfectant and oxidant in water and wastewater treatment.

Bicarbonate and Carbonate

These anionic constituents occur in all natural waters with various contents dependent on the geological source and pH value (Tables II and III). The full significance of bicarbonate (HCO_3^-) and carbonate (CO_3^{2-}) is discussed thoroughly in Chapter 3, where their role as proton acceptors in the determination of total alkalinity of natural waters is presented. Also, the roles of bicarbonate and carbonate in chemical weathering reactions and chemical equilibrium models are discussed in Chapters 5 and 6, respectively.

Minor Cations

There are many inorganic substances occurring in natural waters that have not been routinely determined until recent years. These substances usually are present in the waters at concentrations less than 1.0 mg/l and more than likely occur in the μg/l range. In this case, "trace" refers to the concentration at which these substances occur [2]. Frequently, the term "minor" is employed to denote the importance or significance of these substances as water quality constituents.

Alkali Metals

The elements in group I of the periodic table are the alkali metals: lithium, sodium, potassium, rubidium, cesium and francium. Of this group, lithium, rubidium and cesium are considered as the trace elements in natural waters. Since lithium is a potential toxicant to plants, a limit of 2.5 mg/l was established for continuous use of irrigation waters [7]. Where citrus fruits are grown, the lithium content is limited to 0.075 mg/l. Lithium and rubidium are found usually in the $\mu g/l$ concentration range. Very few data are available for cesium contents in natural waters.

Alkaline Earths

Group II of the periodic table contains the elements beryllium, magnesium, calcium, strontium, barium and radium. Of these six elements, only beryllium, strontium and barium will be considered here.

Beryllium. This element is usually found in association with silicates in igneous rocks since beryllium ions are small enough to replace silicon. The chemical properties of beryllium are such that only very low concentrations would occur in natural waters. Beryllium oxide and hydroxide species have very low solubilities, i.e., $10^{-7}-10^{-10}$ molar concentrations $(1.0-0.001\ \mu g/l)$ of beryllium at a pH value of 7.0. In a report on the water quality characteristics of the 100 largest cities in the United States, only one sample showed a detectable concentration of beryllium $(0.75\ \mu g/l)$ [9]. In a study of 1577 surface water samples collected at 130 sampling points in the United States, 85 samples had beryllium contents of $0.1-1.22\ \mu g/l$ [5]. Various toxic considerations led to the following water quality criteria for Be: for the protection of aquatic life, 11 $\mu g/l$, in soft fresh water and 1100 $\mu g/l$, in hard fresh water; for continuous irrigation on all soils, 100 $\mu g/l$, and for irrigation on neutral to alkaline fine-textured soils, 500 $\mu g/l$ [7].

Strontium. Strontium and calcium have similar chemical and geological properties. Strontium carbonate $(SrCO_3)$, strontianite and strontium sulfate $(SrSO_4)$, celestite, are commonly found in sediments. Strontium frequently replaces calcium or potassium in igneous rock minerals. A very high strontium content (52 mg/l) was reported for a public well supply in Waukesha, Wisconsin [2]. However, the more normal contents average <1.0 mg/l in ground and surface waters [10,11].

Barium. Barium may be found in trace quantities in the major groups of igneous and sedimentary rocks. Barium is somewhat more abundant in igneous rocks than strontium but is less abundant in carbonate rocks. The solubilities of barite $(BaSO_4)$ or witherite $(BaCO_3)$ perhaps control the concentration of barium in water wherever these solid phases occur. Durfor and Becker [9] reported a median concentration of 43 $\mu g/l$ for barium in public water supplies. Durum and Haffty [10] reported a median concentration of 45 $\mu g/l$ in some of the larger rivers of North América. Since barium is a muscle stimulant, especially of the heart, and also affects blood vessels and the nervous system, a 1.0 mg/l limit has been imposed for drinking water [7,8].

Table VII shows a summary of the results of a survey of 15 rivers in North America for the alkali metals and alkali earths by Durum and Haffty [10]. Spectrographic methods were employed. In all cases reported in this survey, the concentrations were <1.0 mg/l and fell within the range indicated in the table. A zero concentration was reported for several of the grab samples, which should be interpreted as less than the analytical sensitivity or that the element was not detected.

Other Trace Elements. Almost all metals may be found in trace concentrations in the surface waters and in selected ground waters in the United States [2,10,12,13]. Table VIII gives a summary of these concentrations, which are <1.0 mg/l. The spring water concentrations from California suggest the depth of origin of these elements. In general, ground water concentrations tend to be higher than surface waters. For the latter, it is difficult to indicate the source, whether it be natural or a wastewater discharge. (See Chapter 7 for the occurrence, distribution and chemistry of cadmium, chromium, lead, mercury and selenium in aquatic environments.)

Minor Anions

Arsenic

In natural waters, arsenical compounds would include the meta-arsenite anion (AsO_2^-), mono-ortho arsenite ($H_2AsO_3^-$), mono-ortho arsenate ($H_2AsO_4^-$), di-ortho arsenate ($HAsO_4^{2-}$) and tri-ortho arsenate (AsO_4^{3-}). Since the water chemistry of arsenic anions is similar to that of the phosphates, its significance in ground and surface waters may be similar.

Table VII. Concentrations of the Alkali Metals and Alkali Earths in Surface Waters of the United States [10]

Minor Constituents	Surface Water Concentrations— Rivers of North America (μg/l)
Alkali Metals	
Lithium	<0.075–37.0
Rubidium	0–7.4
Cesium	ND[a]
Alkali Earths	
Beryllium	0–<0.22
Strontium	6.3–802.0
Barium	9.0–152.0

[a]Not detected.

Table VIII. Concentrations of Some Trace Elements in Surface
and Ground Waters of the United States

Minor Constituents	Surface Water Concentrations— Rivers of North America (μg/l)	Spring Waters California[a] (μg/l)
Metals		
Titanium	$<0.80-107.0^b$	4.3
Vanadium	$0-6.7^b$	34.0
Chromium	$<0.72-84.0^b$	11.7
Molybdenum	$0-6.9^b$	67.0
Cobalt	$0-5.8^b$	0.71, 22.0
Nickel	$0-71.0^b$	11.0
Copper	$0.83-105.0^b$	8.6
Silver	$0-1.0^b$	—
Tin	$0-2.1^b$	—
Zinc	$0-<144.0^b$	397.0
Cadmium	$<1.0-130.0^c$	8.2
Mercury (total)	$<0.5-6.8^c$	—
Germanium	3.7^a	~50.0
Lead	$<1.0-890.0^c$	17.0

[a]From Silvey [13], average concentrations.
[b]From Durum and Haffty [10].
[c]From Durum et al. [12].

Mineral sources include: realgar (AsS), orpiment (As_2S_3) and arsenolite (As_2O_3). Arsenic is a minor constituent of almost all soil types. Pollutional sources of arsenic include: arsenical herbicides and insecticides; wastewaters from mining operations; dye, pigment, tanning, pottery and porcelain industries; metallurgical, pharmaceutical and textile industries; and laundry wastewaters.

The U.S. Public Health Service (USPHS) surveyed several drinking water supplies in 45 states in 1962-1963 for total arsenic content [14]. All contents were reported as <0.010 mg/l, with the exceptions of Idaho (<0.010–0.010 mg/l) and California (<0.010–0.020 mg/l). In 1970, the arsenic content of the Kansas River at Lawrence was reported to be 0.003 mg/l and, at Topeka, 0.008 mg/l. Kopp and Kroner [15] reported on the arsenic contents of several U.S. rivers that were sampled from 1962–1967. These contents ranged from a low (mean values) of 0.020 mg/l in the Great Basin to a high of 0.308 mg/l in the Lake Erie Basin. The Maumee River at Toledo, Ohio had an arsenic content of 0.336 mg/l. Durum et al. [12] reported a range of 10–1100 μg/l for the arsenic content of several surface waters in the United States in October, 1970.

Trivalent arsenic is the toxic form of arsenic and, if present in natural waters, anaerobic conditions may lead to the release of this constituent from bottom muds and sediments. Also, there is strong epidemiological evidence that relates the incidence of skin cancer to the arsenic content of drinking water [16]. There is also the possibility that arsenic may play a nutrient role in biological systems because of its similar chemistry to phosphorus. In natural waters, arsenic compounds should be considered as potentially hazardous substances. (See Chapter 7 for additional information on the significance of arsenic in aquatic environments.)

Boron

There is a considerable amount of information available about the boron content of natural waters, especially irrigational waters. In trace quantities, boron is an essential element to plant growth. Excessively higher amounts are harmful and even toxic to some plants. This element is included under anionic constituents because the ortho- and tetraboric acids protolyze into: $H_2BO_3^-$, HBO_3^{2-}, and BO_3^{3-} and $HB_4O_7^-$ and $B_4O_7^{2-}$, respectively.

In igneous rocks, the widely distributed mineral tourmaline contains boron. In the evaporite deposits of southeastern California and Nevada, boron occurs in colemanite $(Ca_2B_6O_{11} \cdot 5H_2O)$ and kernite $(Na_2B_4O_7 \cdot 4H_2O)$ which is a partially dehydrated form of borax. Domestic and industrial wastewaters may also be sources since sodium tetraborate is used as a cleansing agent.

The average concentration of boron in the surface waters of the United States may be 0.1 mg/l, ranging from a low of 0.02 mg/l in the Western Great Lakes Basin to an average of 0.3 mg/l in the Western Gulf Basin [15]. A few tenths of mg/l is more nearly typical of the ground and surface waters, as seen in Analyses 2 and 4 of Table IV. There are some thermal springs in Nevada and California with unusually high boron contents (see Analyses 8 and 9, Table IX).

The major significance of boron lies in its occurrence in irrigational waters in parts of the Western United States. Criteria have been proposed for the boron content of irrigation waters [7]: for use on sensitive crops the maximum boron content should be 0.75 mg/l; for neutral and alkaline fine-textured soils, it should be 2.0 mg/l over a 20-year period.

Fluoride

In terms of concentration, fluoride should be considered a minor constituent of natural fresh waters. This anion is, however, a significant factor in the human physiological aspects of water quality [7,8].

Minerals in which fluoride is a constituent usually are quite insoluble in water. Fluorite (CaF_2) is a common fluoride-bearing mineral with a rather low solubility, which occurs in igneous and sedimentary rocks. Other mineral sources of fluoride are fluorapatite ($Ca_5(PO_4)_3F$), sellaite (MgF_2), villiaumite (NaF) and cryolite (Na_3AlF_6).

In most natural fresh waters, fluoride concentrations range from a few tenths of a mg/l to one or two mg/l. Analyses 1, 2, 3, 4 and 7 of Table IV show these more typical fluoride contents of ground waters. There are some natural waters, however, that have unusually "high" fluoride concentrations. For example, Analysis 1, Table IX, is a ground water in southeastern Arizona that contains 32 mg/l F^-. A water from the Union of South Africa was reported to have a fluoride content of 67 mg/l [2].

The major significance of fluoride in natural waters is its role in the formation of human teeth and bones. Some natural waters contain the optimum concentration of 1.0 mg/l that is required for the prevention of dental caries. Concentrations of fluoride in excess of 4.0 mg/l may cause mottled or stained teeth [8]. Some communities in the United States permit their potable waters to be supplemented with fluoride salts for the control of dental caries.

Nitrate

Nitrate and other nitrogen species occur in natural waters mainly from biological systems and sources. Nitrate is usually considered the end product from a sequence of biologically mediated reactions in which organic nitrogen compounds are oxidized. Industrial and domestic wastewaters are excellent sources of nitrogen that eventually becomes nitrate in polluted surface waters. Nitrate is considered an essential element or nutrient in the biological process of eutrophication.

There are few, if any, mineral sources of nitrate for natural waters. Soda niter ($NaNO_3$) is the principal component of the famous nitrate deposit in northern Chile. In addition to wastewater sources, commercial fertilizers may be leached from soils into ground and surface waters. Residues from farm animals have considerable quantities of organic nitrogen compounds that are eventually oxidized to nitrate.

It is difficult to establish precisely a range of concentrations for nitrate in unpolluted natural waters. These contents are quite variable and are influenced by the sources cited above. Table X cites some nitrate contents of several ground and surface waters that may be more or less typical.

As cited above, nitrate is the end product in the biological decomposition of organic nitrogen compounds, whether they be natural or pollutional. If this decomposition occurs in a surface water, then an "oxygen" demand is created. Conversely, nitrate can be utilized as an electron acceptor by some microorganisms in their metabolic activities. Perhaps the most significant factor in water quality is the role of nitrate in the disease infantile methemoglobinemia [5,8].

Table IX. Analyses of Waters Containing Fluoride,

Constituent	1 (mg/l)	2 (mg/l)	3 (mg/l)
Silica (SiO_2)	–	23	27
Aluminum (Al)	–	–	–
Boron (B)	–	–	–
Iron (Fe)	–	–	0.28
Manganese (Mn)	–	–	–
Arsenic (As)	–	–	–
Strontium (Sr)	–	–	–
Calcium (Ca)	5.5	92	64
Magnesium (Mg)	4.4	38	19
Sodium (Na)	} 157	} 110	114
Potassium (K)			9.5
Lithium (Li)	–	–	–
Carbonate (CO_3)	58	0	0
Bicarbonate (HCO_3)	163	153	402
Sulfate (SO_4)	42	137	74
Chloride (Cl)	10	205	30
Fluoride (F)	32	0.6	0.1
Nitrate (NO_3)	–	83	60
Phosphate (PO_4)	–	–	–
Ammonium (NH_4^+)	–	–	–
Dissolved solids			
Calculated	389	764	596
Residue on evaporation	–	–	578
Hardness as $CaCO_3$	32	386	238
Noncarbonate	0	260	0
Specific Conductance (micromhos at 25°C)	660	1320	875
pH	–	–	7.4
Color	–	–	–

1. Flowing well NE¼ sec. 24, T. 13 S., R. 30 E., Chochise County, Arizona. Depth, 850
2. Irrigation well, SE¼ sec. 25, T. 2 N., R. 2 W. Maricopa County, Arizona. Depth, 275
3. Well, SE¼ sec. 21, T. 12 S., R. 10 W., Lincoln County, Kansas. Depth, 32 ft. Water-
4. Well, NW¼ sec. 2, T. 8 S., R. 5 W., Maricopa County, Arizona. Depth, 495 ft. Water-
5. Peace Creek at State Highway 17 bridge, Salfa Springs, Florida. 140 cfs.
6. Iowa River at Iowa City, Iowa. Discharge-weighted average of composites of daily
7. Powder River, 4.5 miles north of Baker, Baker County, Oregon.
8. Nevada Thermal well 4, Steamboat Springs, Washoe County, Nevada. Depth, 746 ft. mg/l.
9. Spring at Sulphur Bank, sec. 5, T. 13 N., R. 7 W., Lake County, California. Temp.,

Phosphates

In this case, phosphates include these ionic species: $H_2PO_4^-$, HPO_4^{2-} and PO_4^{3-}. The phosphates (or, more generally, phosphorus) represent an essential element in biological systems. Phosphorus is similar to nitrogen when pollutional

Nitrogen, Phosphorus or Boron in Unusual Amounts [2]

4 (mg/l)	5 (mg/l)	6 (mg/l)	7 (mg/l)	8 (mg/l)	9 (mg/l)
–	18	15	17	314	72
–	–	–	–	0.22	–
–	–	–	–	48	660
–	–	0.05	–	0.52	–
–	0.00	–	–	0.00	–
–	–	–	–	4.0	0.02
–	–	–	–	0.67	–
36	42	49	24	3.6	7.0
18	19	14	11	0.0	22
} 102	29	5.4	34	660	1100
	0.7	3.1	6.2	65	33
–	0.3	–	–	7.0	4.8
0	0	–	–	–	–
303	65	168	129	312	2960
34	114	40	32	108	454
32	13	4.3	18	874	690
0.4	5.0	0.2	0.4	2.6	1.0
68	0.3	14	13	2.7	0
–	30	–	14	0.24	–
–	–	–	0.1	0.0	476
–	303	–	236	2240	4990
440	318	251	–	2360	–
164	183	180	106	9.0	108
0	130	42	0	0	0
724	413	365	361	3430	7060
–	7.2	–	7.5	8.9	7.5
–	20	–	20	–	–

ft. Water-bearing formation, valley fill. Temp., 18.3°C.
ft. Water-bearing formation valley fill. Temp., 29.4°C.
bearing formation, alluvium. Temp., 14.4°C.
bearing formation, valley fill.

samples, October 1, 1950 to September 30, 1951; Mean discharge, 2543 sec-ft.

Bottom temperature, 186°C. Also contained bromide (Br) 1.5 mg/l and iodide (I) 0.6

77°C. Also contained H_2S, 3.6 mg/l; Br, 1.4 mg/?; I, 3.6 mg/l.

aspects are considered; however, it is dissimilar to nitrogen as there are naturally occurring mineral sources of phosphates.

The mineral sources of phosphates are: fluorapatite ($Ca_5(PO_4)_3F$); hydroxylapatite ($Ca_5(PO_4)_3OH$); strengite ($Fe(PO_4)\cdot2H_2O$); whitlockite ($Ca_3(PO_4)_2$); and berlinite ($AlPO_4$). These minerals may be found in igneous rocks and

Table X. Analyses of Waters Containing High Proportions of Silica [2]

Constituent	1 (mg/l)	2 (mg/l)	3 (mg/l)	4 (mg/l)
Silica (SiO_2)	99	103	363	49
Aluminum (Al)	–	–	0.2	–
Iron (Fe)	0.04	0.0	0.06	0.01
Manganese (Mn)	–	–	0.0	0.00
Calcium (Ca)	2.4	6.5	0.8	32
Magnesium (Mg)	1.4	1.1	0.0	12
Sodium (Na)	100	} 40	352	30
Potassium (K)	2.9		24	5.2
Carbonate (CO_3)	24	0		0
Bicarbonate (HCO_3)	111	77	–	220
Sulfate (SO_4)	30	15	23	11
Chloride (Cl)	10	17	405	7.9
Fluoride (F)	22	1.6	25	0.2
Nitrate (NO_3)	0.5	0.4	1.8	2.9
Boron (B)	0.61	0.0	4.4	0.08
Dissolved solids				
Calculated	348	222	1310	259
Residue on evaporation	–	–	–	257
Hardness as $CaCO_3$	12	20	2	129
Noncarbonate	0	0	0	0
Specific Conductance (micromhos at 25°C)	449	167	1790	358
pH	9.2	6.7	9.6	7.8
Color	–	–	–	10

1. Flowing well 7S 6E-9ba2. Owyhee County, Idaho. Depth, 800 ft. Temp., 50.0°C.
2. Spring on Rio San Antonio, SW¼ sec. 7, T. 20 N., R. 4 E. (unsurveyed) Sandoval County, New Mexico. Temp., 38.3°C. Flow, about 25 gpm. Water from rhyolite.
3. Spring, 650 ft. south of Three Sisters Springs in Upper Geyser Basin, Yellowstone National Park, Wyoming. Temp., 94°C. Also reported: 1.5 mg/l As, 5.2 mg/l Li, 1.5 mg/l Br. 0.3 mg/l I, 1.3 mg/l PO_4^{-3}, and 2.6 mg/l H_2S.
4. Drilled well, NW¼ sec. 10, T. 2 N., R. 32 E., Umatilla County, Oregon. Water from basalt of the Columbia River Group. Depth, 761 ft.

marine sediments; however, they are quite insoluble in water and may be responsible for controlling the phosphate content of natural waters. Pollutional sources would be the commercial fertilizers, domestic and industrial wastewaters and, to a lesser extent, decomposition of organic phosphorus compounds in biological systems.

It is somewhat difficult to establish a range of concentrations for phosphates in natural waters because the inputs from the many sources are quite variable. Until recent years the phosphate content was not a routine water analysis. Perhaps the more normally observed contents are a few tenths of mg/l, as seen in Analyses 2 and 3 of Table V. Unusually higher contents may be found in Analyses 5 and 7 of Table IX, where phosphate mining and wastewater disposal may be the responsible causative agents.

Phosphorus is, of course, an essential element for biological systems and transformations. It may or may not be the so-called limiting nutrient in eutrophication processes. Polyphosphates are used in mg/l concentrations in water distribution systems to inhibit calcium carbonate or iron hydroxide precipitation.

An attempt was made to establish a water quality criterion for phosphates in fresh waters [5]. It was suggested that the total phosphate content as P should not exceed 50 μg/l in any tributary to a lake or reservoir and should not exceed 25 μg/l within these waters. The rationale was to prevent the development of biological nuisances and to control eutrophic processes.

Selenium

In natural waters, selenium occurs either as a weak acid, selenious acid (H_2SeO_3) or selenic acid (H_2SeO_4). The accompanying anions are acid selenite ($HSeO_3^-$), selenite (SeO_3^{2-}), acid selenate ($HSeO_4^-$) or selenate (SeO_4^{2-}). Mineral sources of selenium are rarely found in nature alone but it is frequently associated with sulfur and sulfides. Traces of selenium are found in igneous rocks [17]. The selenium content of natural fresh waters has been reported to range from 0.114 to 0.348 μg/l for nine major rivers of the world, and from 1.0 to 400 μg/l for eight surface waters in Colorado [17]. Selenium is considered to be toxic to man [5] because of its symptoms similar to those of arsenic. For this reason primarily, a limiting concentration of 10 μg/l was established for drinking water [5,8]. (Additional information is given in Chapter 7 for Se in our environment.)

Nonionic Silica

Silicon is the second most abundant element in the earth's crust. The term "silica" refers to silicon in natural waters and is usually represented by the hydrated form of the oxide: H_4SiO_4 or $Si(OH)_4$, silicic acid. This is included as a nonionic species because it is a "weak" acid, which does not protolyze until the pH value of water exceeds 9.4.

There are innumerable mineral sources of silica for natural waters, but most are quite resistant to chemical weathering processes. Crystalline quartz (SiO_2) is a major component of many igneous rocks and the grains of sandstones. Perhaps the greatest sources of silica are the feldspars, micas and clay minerals, or the hydrous aluminum silicates. Many structural formulas may be cited but the reader is referred to Chapter 5 for details.

For the most part, the solubility of quartz or amorphous silica determine its concentration in water. At 25°C the solubility of quartz has been reported to be 6.0 mg/l, whereas the solubility of amorphous silica is 115 mg/l. These solubility values may be increased by increasing the water temperature or pH value. Silica concentrations of 1–30 mg/l are most frequently encountered in natural waters. Higher concentrations may be observed where the water temperatures are greater than 25°C or where alkaline pH values are observed. This is seen in Analyses 1 and 3 of Table X.

Dissolved silica enters into many chemical weathering reactions that occur in ground and surface waters. The major significance lies in the ability of silica to form "glassy" scales in steam-generating systems.

Dissolved Gases

Dissolved Oxygen

Oxygen is perhaps the most significant of all of the dissolved gases in natural waters. The role of oxygen in the transformation of organic matter in polluted surface waters is well documented. Therefore, the dissolved oxygen content is frequently employed as an indicator of the degree of pollution. Dissolved oxygen plays a significant role in the corrosion of iron pipes in water distribution systems. Of course oxygen is a by-product of photosynthetic activity of aquatic microorganisms. The solubilities of oxygen in fresh and saline waters, which are given in Table XI, are the quasi-legal values from *Standard Methods* [1]. The range of concentrations of oxygen in natural waters would be zero to the saturation values at a given temperature. The degree of undersaturation would be influenced by the extent and rate at which organic matter and other microbiological processes are occurring. Supersaturated oxygen contents can occur from extensive photosynthetic activities.

Nitrogen

This gas is soluble to the extent of 15.1 mg/l in water at 20°C in accord with the Henry's law constant. As a gas, nitrogen has no particular significance as a water quality factor. It may be significant, however, in eutrophic situations because some species of blue-green algae (*Anabaena, Nostoc*) can "fix" nitrogen to nitrates. This process, therefore, becomes a major source of one of the essential elements in eutrophication.

Hydrogen Sulfide and Methane

Trace concentrations of dissolved hydrogen sulfide and methane occur in ground and surface waters from the reduction of sulfate and carbon dioxide, respectively. Usually, this reduction occurs where a large amount of organic matter is found either from natural or anthropogenic sources. The microbially mediated oxidation of the organic matter may utilize the sulfate anion or the dissolved carbon dioxide as electron acceptors, which results in the reduced sulfur as H_2S and the reduced carbon as CH_4. These two gases have little significance as water quality constituents other than to serve as indicators of gross organic pollution. They may have some toxicity effects on aquatic life in surface waters; however, their aqueous solubilities are reduced considerably as these waters become aerated and turbulent.

Carbon Dioxide

This gaseous substance is discussed in Chapter 3 under acidity and alkalinity.

Radioactive Elements

The principal naturally occurring radioactive elements in water are uranium-238, thorium-232 and uranium-235. The decay of uranium-238 yields, among other things, radium-226, which has been detected in natural waters. In turn, radium-226 decays to a gaseous element, radon-222, which also has been found in ground waters. Thorium-232 yields radium-228 and radium-224 on disintegration. Radium concentrations greater than 3.3 pCi/l have been reported in waters from deep aquifers in Iowa, Illinois and Wisconsin [2]. Smith et al. [18] surveyed the ground water supplies in Maine and New Hampshire for naturally occurring ^{238}U, ^{226}Ra and ^{222}Rn. The latter two radioactive elements were the predominant ones found.

The reader is referred to the literature [2,10,18] for greater details on naturally occurring radioactive elements in ground and surface waters.

Naturally Occurring Organic Compounds

Organic Color

Some natural waters exhibit a yellow–brown color, which is quite common. These waters have various descriptors: "swamp water," "humus water" or "colored water." The last description should not be confused with colored waters arising from the discharge of industrial wastewaters.

Relatively little information is available about the broad classes or types of organics that occur naturally in surface waters. Organic matter in soils, aquatic vegetation and aquatic organisms would be the major sources. Christman and several co-workers (references cited below) have reported some of

Table XI. Solubility of Oxygen in
(reproduced from *Standard Methods* [1]

	Chloride Concentration in Water (mg/l)					Difference/100 mg Chloride
	0	5,000	10,000	15,000	20,000	
Temperature (°C)	Dissolved Oxygen (mg/l)					
0	14.6	13.8	13.0	12.1	11.3	0.017
1	14.2	13.4	12.6	11.8	11.0	0.016
2	13.8	13.1	12.3	11.5	10.8	0.015
3	13.5	12.7	12.0	11.2	10.5	0.015
4	13.1	12.4	11.7	11.0	10.3	0.014
5	12.8	12.1	11.4	10.7	10.0	0.014
6	12.5	11.8	11.1	10.5	9.8	0.014
7	12.2	11.5	10.9	10.2	9.6	0.013
8	11.9	11.2	10.6	10.0	9.4	0.013
9	11.6	11.0	10.4	9.8	9.2	0.012
10	11.3	10.7	10.1	9.6	9.0	0.012
11	11.1	10.5	9.9	9.4	8.8	0.011
12	10.8	10.3	9.7	9.2	8.6	0.011
13	10.6	10.1	9.5	9.0	8.5	0.011
14	10.4	9.9	9.3	8.8	8.3	0.010
15	10.2	9.7	9.1	8.6	8.1	0.010
16	10.0	9.5	9.0	8.5	8.0	0.010
17	9.7	9.3	8.8	8.3	7.8	0.010
18	9.5	9.1	8.6	8.2	7.7	0.009
19	9.4	8.9	8.5	8.0	7.6	0.009
20	9.2	8.7	8.3	7.9	7.4	0.009
21	9.0	8.6	8.1	7.7	7.3	0.009
22	8.8	8.4	8.0	7.6	7.1	0.008
23	8.7	8.3	7.9	7.4	7.0	0.008
24	8.5	8.1	7.7	7.3	6.9	0.008
25	8.4	8.0	7.6	7.2	6.7	0.008

[a]At a total pressure of 760 mm Hg. Under any other barometric pressure, P, the equation: $S' = S(P - p)/(760 - p)$, in which S is the solubility at 760 mm and p is the less than 1000 meters and temperatures below 25°C, p can be ignored. The equation oxygen, calculations made by Whipple and Whipple, *J. Am. Chem. Soc.* 33:362(1911).

the most significant information about the chemical nature of organic color in water that arises from: (1) the aqueous extraction of living woody substances, (2) the solution of degradation products in decaying wood, (3) the solution of soil organic matter, or (4) a combination of these processes.

Water Exposed to Water-Saturated Air
courtesy of the American Public Health Association)[a]

Temperature (°C)	Chloride Concentration in Water (mg/l)					Difference/100 mg Chloride
	0	5,000	10,000	15,000	20,000	
	Dissolved Oxygen (mg/l)					
26	8.2	7.8	7.4	7.0	6.6	0.008
27	8.1	7.7	7.3	6.9	6.5	0.008
28	7.9	7.5	7.1	6.8	6.4	0.008
29	7.8	7.4	7.0	6.6	6.3	0.008
30	7.6	7.3	6.9	6.5	6.1	0.008
31	7.5					
32	7.4					
33	7.3					
34	7.2					
35	7.1					
36	7.0					
37	6.9					
38	6.8					
39	6.7					
40	6.6					
41	6.5					
42	6.4					
43	6.3					
44	6.2					
45	6.1					
46	6.0					
47	5.9					
48	5.8					
49	5.7					
50	5.6					

solubility, S' (mg/l), can be obtained from the corresponding value in the table by the pressure (mm) of saturated water vapor at the temperature of the water. For elevations then becomes: $S' = S(P/760) = S(P'/29.92)$. Dry air is assumed to contain 20.90%

It is somewhat difficult to offer typical organic color contents of natural waters because of the variability of the sources. Perhaps a range of values may suffice. Lakes, streams, reservoirs, etc. may show color contents of 5 to 25 or so color "units." These waters would have a light yellow–brown color. On the other hand, the dark brown waters from peat bogs may have color units in excess of 1000.

Classification Schemes. Because of the relation of colored substances in water to organic matter in soils, almost all classification schemes are based on this area of research into "humus." The earliest schemes came from Oden [19], Ramann [20] and Waksman [21]. Essentially, the classifications are based on solubility of the organic matter under various conditions of acidity and alkalinity and in various solvents. Figure 1 shows a diagram of the fractionation procedure employed by most investigators of the nature of colored water. This is based on Oden's procedure [19], which was modified slightly by Page and Dutoit [23]. There are, of course, a multitude of names, etc., for the three fractions shown in Figure 1. The terminology employed in this chapter is, more or less, standard. The term, "humic acid," is preferred over "humus acid" [23].

There have been some recent modifications of the fractionation procedure shown in Figure 1. For example, Farrah et al. [24] concentrated humic acid from tap water after acidification by passage through a 0.25-μm epoxy–fiber-

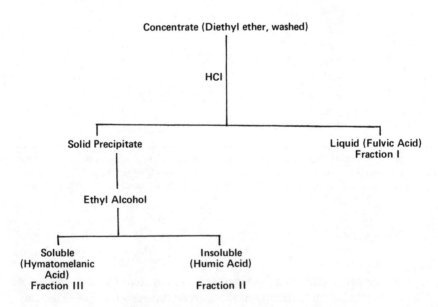

Figure 1.   Schematic diagram of fractionation procedure (reproduced from Black and Christman [22] courtesy of the American Water Works Association).

glass filter. As much as 30 mg/l of humic acid was detected in tap water! Weber and Wilson [25] extracted fulvic and humic acids from New Hampshire waters by passage first through anion exchange resins in the hydroxide form. The acids were eluted with 2 M NaCl. This strongly basic NaCl solution was adjusted to a pH value of 1, evaporated and centrifuged. Solid humic acid was formed. In another procedure, fulvic and humic acids were concentrated and were separated by passage of the acidified NaCl solution through an XAD-2 resin. This resin adsorbed the fulvic acid, whereas the insoluble humic acid remained at the top of the column. These two acids then were purified further. Apparently the technique of Weber and Wilson yields substantially more quantities of the two acids than the previous separation and fractionation technique of Oden [19], Page and Dutoit [23], etc.

Physical Characteristics. Much of the early research (1900-1940), reviewed by Black and Christman [26], was concerned with the physical characteristics of colored substances in water. It was learned very early, for example, that organic color in water is a negatively charged colloid [27].

In 1963, Black and Christman [26] reported on some of the fundamental physical properties of organic color in water. Samples were collected from seven states—Massachusetts, Connecticut, Virginia, North Carolina, Wisconsin, Florida and Georgia—in order to provide some diversity. One unique property of these natural colored waters is the variation of color intensity with $[H_3^+O]$. This is the so-called "indicator" effect. Figure 2 shows the effect of pH value on the color intensity (reported in color "units") of five waters. It appears that color intensity is increased as the $[H_3^+O]$ is decreased. However, this color change is not linear with pH value because alkaline waters are proportionally more intense than acid waters. This "pH-effect" was observed also by Shapiro [28]. For another physical property, particle size estimations were made through the use of membranes with pore sizes of 3.5, 4.8 and 10 mμ. The maximum amounts of color retained on these membranes were 91.0, 87.5 and 13.0%, respectively. This suggests that organic color has the particle size of colloids. Confirmation of this observation was made with light-scattering techniques and ultraviolet spectroscopy.

None of the colored waters investigated by Black and Christman [26] indicated any absorption in the visible region of the atomic spectrum. All waters showed only end adsorption (approaching 200 nm) in the ultraviolet region (Figure 3). Furthermore, the spectra were continuous and did not show any point of maximum adsorption. Dilution had no effect on the shape of the adsorption curve other than to decrease the absorbance value. Apparently the yellow–orange color to the eye comes from light scattering since no absorption is observed in the 450–550 nm region of the atomic spectrum. Changing the pH value had no effect on shape of the ultraviolet spectrum. Similar ob-

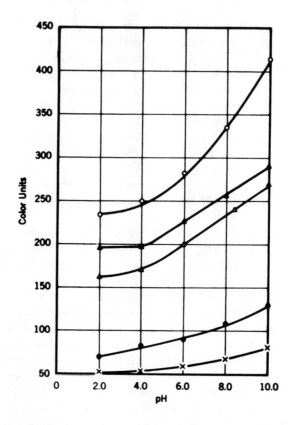

Figure 2.   Effect of pH on color of samples A, B, D, E and G (reproduced from Black and Christman [26] courtesy of the American Water Works Association). Key: A △; B ○; D ●; E ✕; and G ▲.

servations had been reported by Shapiro [28] on the colored waters of Linsley Pond in Connecticut.

Chemical Characteristics. Considerable research has been conducted by numerous investigators into the chemical nature of organic color. Yet, the precise structure of color molecules remains unsolved. An early effort was given by Black and Christman [22]. One of the controversies is the elemental analysis of organic color. The data of Black and Christman [22] in Table XII clearly show the percentages of C, H, N and O in the three fractions (see Figure 1 for the key to I, II and III). The point of disagreement lies with

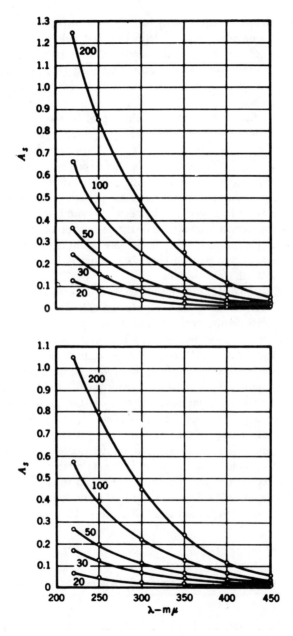

Figure 3.   Ultraviolet absorption of two samples as a function of color value (reproduced from Black and Christman [26] courtesy of the American Water Works Association). The upper graph is for Sample B; the lower for Sample I. Each curve represents the results obtained with the sample diluted with pH 8.0 buffer to the color unit value indicated by the number near the curve.

presence of nitrogen. Shapiro [28] had reported earlier that any "nitrogen was believed to be an impurity." Midwood and Felbeck [29] support the data of Black and Christman with a report of nitrogen in colored waters of Rhode Island. In fact, these investigators identified 14 amino acids. This information is suspect also since these amino acids may have originated from bacterial contamination.

Nature of the functional groups on the organic color molecule is another area of disagreement. Black and Christman [22] offer infrared data that show the presence of hydroxy carboxylic acids and "strong indications of unsaturation and aromaticity." The latter may be due to phenyl groups. On the other hand, Shapiro [28] states: "the possibility of unsaturation or aromatic rings is not completely ruled out, but if these are present they are probably relatively unimportant." Midwood and Felbeck [29] and Ungar [30], using infrared spectroscopy, report aromaticity in the organic color molecule. Wershaw and Bohner [31] employed pyrolysis and gas–liquid chromatography (GLC) to identify aromatic and polysaccharide structures in humic and fulvic acids isolated from a North Carolina soil.

Black and Christman [22] attempted to determine the number of phenolic hydroxyl and carboxylic groups on the organic color molecule through techniques of methylation and titration in nonaqueous solvent. This should yield, in acidic terms, an equivalent weight of color compounds. For 10 different fulvic acid fractions (I), equivalent weights ranged from 89 to 138. These values are in disagreement with the average equivalent weight of 228 reported by Shapiro [28]. The latter investigator titrated in an aqueous system and computed the equivalent weight from the quantity of NaOH required to reach a pH value of 7.0. This is not an acceptable calculation in view of the "weak" nature of the acidic groups on the organic color molecule. The titration, methylation and infrared spectroscopic investigations of the color molecule by Black and Christman [22] have established the presence of phenolic hydroxyl and carboxyl groups. Weber and Wilson [25] reported the organic oxygen-containing functional group analyses for fulvic and humic samples from a variety of water and soil sources. These analyses are seen in

Table XII. Elemental Analysis of Fractions of Sample A [22]

Fraction	C	H (% of total)	N	O
I	41.5	5.72	1.98	50.8
II	29.3	5.94	1.85	62.91
III	49.3	5.11	1.24	44.36

Table XIII. Oxygen-Containing Functional Groups in Fulvic
and Humic Acid Samples (meq/g) [25][a]

Sample	Total Acidity	Carboxyl	Phenol OH[b]	Carbonyl
S-HA[c]	7.9	3.7	4.2	3.1
S-FA[c]	12.8	8.9	3.9	2.0
S-FA(C)[d]	13.4	8.2	5.2	3.5
OR-HA	8.2	4.5	3.7	4.3
JP-HA	7.1	4.9	2.2	5.1
OR-FA	10.6	6.8	4.3	4.3
SR-OM[e]	13.7	8.8	5.0	

[a]S, soil; HA, humic acid; FA, fulvic acid; OE, Oyster River; JP, Jewell Pond; C, Conway, New Hampshire; SR, Satilla River; OM, organic matter.
[b]Difference between total acidity and carboxyl values.
[c]Literature values.
[d]Soil fulvic acid from Conway, New Hampshire (this work).
[e]Average value for all organic matter from waters of the Satilla River system in southeastern Georgia.

Table XIII where, for example, the total acidity and carboxyl values are higher in fulvic acid than in humic acid, regardless of their source. Additional information about these functional groups is found in an excellent review by Steelink [32].

In an effort to deduce the structure of a "color" macromolecule, Christman and Ghassemi [33] employed a degradative technique with alkaline-CuO on materials isolated from natural waters. Seven products were identified: vanillin, vanillic acid, syringic acid, catechol, resorcinol, protocatechuic acid and 3,5-dihydroxybenzoic acid (Figure 4). The most significant point is the presence of phenolic nuclei in the color macromolecule. There is some concern about environmental contamination with phenolic compounds. These products provide conclusive evidence for the presence of aromatic nuclei and phenolic groups in the color macromolecule. Figure 4 also shows a proposed structure for the color molecule by Christman and Ghassemi. A similar structure has been proposed by Schnitzer [34] for soil humic and fulvic acids.

LaMar and Goerlitz [35,36] provided some additional information about the chemical nature of the organic "color" molecule. These investigators maintain that "most of the organic matter in naturally colored surface waters consists of a mixture of carboxylic acids or salts of these acids." Water samples were collected from unpolluted streams in California, Hawaii and Washington. The organic acids were recovered by two techniques: continuous liquid–liquid extraction with n-butanol and vacuum evaporation at 50°C. Attempts at identification were conducted by chromatographic techniques of

(a) Identified Degradation Products of Color Solids

(b) Proposed Structure of Color Macromolecule

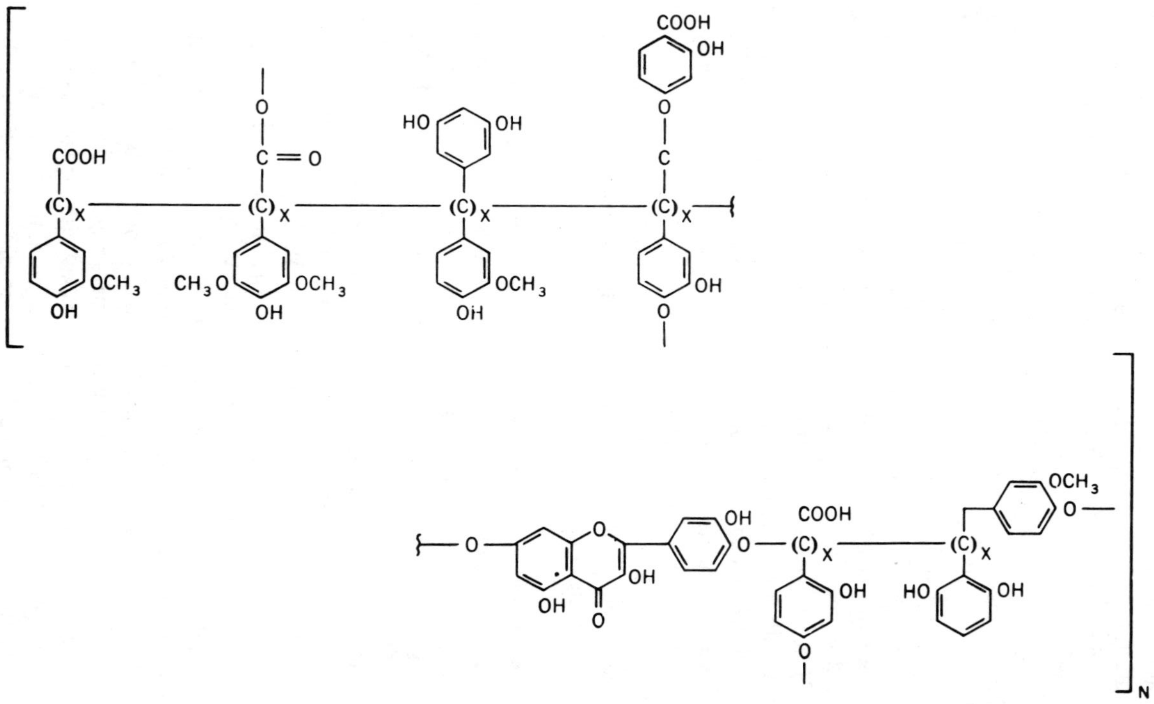

Figure 4. Degradation products and proposed structure of color macromolecules (reproduced from Christman and Ghassemi [33] courtesy of the American Water Works Association).

gas–liquid, paper and column, and by infrared spectroscopy. Approximately 30 carboxylic acids were isolated but only 10 were identified: butyric, valeric, caproic, lactic, oxalic, malonic, succinic, maleic, fumaric and adipic. The authors also indicated that a predominant part of the organic acids is nonvolatile for gas chromatographic purposes. Infrared analysis of these non-volatiles indicated polymeric hydroxyl carboxylic acids with aromatic and olefinic unsaturation.

Some interesting data came from the studies of Day and Feldbeck [37] on the yellow-colored, water-soluble organic exudate from the aquatic fungus *Aureobasidium pullulans*. This fungal was isolated from a yellow-colored fresh water pond in Rhode Island. It is one of nine significant fungi present in sewage, and strains of this species are also common soil organisms. Apparently this fungal has the ability to decompose large amounts of organic matter and exude yellow-colored substances in fresh waters. Day and Feldbeck [37] have analyzed this fungal exudate, which shows characteristics consistent with previous research on color in fresh water [29] and in humic substances [38]. A summation of this research is given in Table XIV, where the essential fragments are given: amino acids, hexosamines, polycyclic aromatics, carboxylic acid, phenols and methoxyl groups. Most of this material was found in the fulvic acid fraction. These data do, indeed, offer the intriguing possibility that this fungi may produce a major portion of naturally occurring color in fresh water.

Molecular Weight Characteristics. Several attempts have been made to separate the colored organic matter in natural waters into apparent molecular weight fractions. These attempts have effected the separations on columns of Sephadex (gel filtration). This is a type of molecular sieving in which a small volume of the concentrate ($<3\%$ of the column's volume) is placed on top of the column. Molecules larger than the pore size cannot penetrate the Sephadex particles. Consequently, they move rapidly down the column with the eluant. Smaller molecules penetrate into the gel particles to various extents that depend on their shape and size. The various types and fractionation ranges of Sephadex gels are given in Table XV.

Gjessing and Lee [39] reported a feasibility study on the fractionation of some moderately colored surface waters from Wisconsin. The Black Earth Creek (BEC) was a typical example whose color content was 80 Pt units. It is seen in Table XVI that the soluble organic matter in a 400-fold concentrate from BEC was separated into at least 10 fractions with the attendant molecular weight ranges. This feasibility study suggests that organics in natural waters are fractionable and "probably do not consist of a continuous distribution of all sizes of molecules." For the BEC organic matter, the greater percentage of molecular weights ranged from 100 to 30,000. Also, there is a significant number of fractions with molecular weights above 30,000. These

Table XIV. Summation of Fragments Found in Exudate and Humic Acid[a]

Fragment	Assumed Average Structure	Elemental Analysis				Percentage of Exudate	Percentage of Humic Acid
		C	H	O	N		
Amino Acids	CH_3CHNH_2COOH (alanine)	40	8	36	16	2.5	10.5
Hexosamines	$C_6H_{13}O_5N$ (glucosamine)	40.2	7.3	44.6	7.9	1.2	2.5
Polycyclic Aromatics	$C_{14}H_{10}$ (anthracene)	94.5	5.5			17.0	10.0
Carboxylic Acid	COOH	27.0	2.0	71.0		4.0	11.2
Phenolic OH	OH		6.0	94.0		11.4	11.1
Methoxyl	OCH_3	39.0	10.0	51.0		0.5	2.0
Total						36.6	47.5

[a]Reproduced from Day and Feldbeck [37] courtesy of the American Water Works Association.

Table XV. Sephadex Types and Fractionation Range

Type	Approximate Limit for Complete Exclusion (mol wt)	Fractionation Range (mol wt)
G-10	700	0–700
G-15	1,500	0–1,500
G-25	5,000	100–5,000
G-50	10,000	500–10,000
G-75	50,000	1,000–50,000
G-100	100,000	5,000–100,000
G-150	150,000	5,000–150,000
G-200	200,000	5,000–200,000

data are not in accordance with the value of 456 reported earlier by Shapiro [28]. This value was retracted later by Shapiro [40].

Ghassemi and Christman [41] fractionated several colored waters (from Washington, British Columbia and Alaska) on Sephadex gel columns. With the exception of one fraction having a molecular weight in excess of 50,000, color molecules in these waters fell into the 700 to 10,000 molecular weight range. These authors cited some uncertainties in the fractionation of colored waters by Sephadex columns: (1) possible interaction of color acids with the gels, (2) effect of sample concentration techniques, and (3) lack of confirmation of molecular weights by an independent method. This fractionation technique was employed also to study the effect of $[H_3^+O]$ on molecular size. In the acidic (pH = 3.5) to a slightly alkaline (pH = 8.0) range, the size of the color molecule was increased slightly. This may be attributed to an increase in the molecular radius due to greater dissociation of the functional groups or to a decrease in the extent of adsorption on Sephadex or, perhaps, both mechanisms.

Additional fractionation studies on Sephadex are those by Brodsky et al. [42] on peat extracts, Shapiro [40] on colored waters, Wershaw et al. [43] on sodium humates and Christman and Minear [44] on colored waters and soluble, naturally occurring organophosphorus compounds.

There is an additional source of organic matter that ultimately forms colored water, namely, dead leaves from various forms of riparian vegetation. The mechanisms by which this significant source operates are described by Slack and Feltz [45], Lush and Hynes [46] and Novak et al. [47].

Polynuclear Aromatic Hydrocarbons (PAH)

These compounds are considered as naturally occurring because their origin is usually in petroleum. They are released into the environment through a

Table XVI. Fractionation of Black Earth Creek Sample[a]

Peak	Grade	Molecular Weight Range	% COD	Molecular Size Range (radius cm $\times 10^{-8}$)
Ia$_1$	G-200	>200,000	6	100
Ia$_2$	G-200	100,000–200,000	4	10–50
Ib	G-100	50,000–100,000	6	5–30
II	G-75	20,000–30,000	11	10–25
IIIa	G-25	3,000–5,000	12	20
IIIb	G-25		2 ⎫	5–15
IIIc$_1$	G-25	100–5,000	13 ⎭	4
IIId$_1$	G-10		2	
IIId$_2$	G-10		4 ⎫	0–4
IIId$_3$	G-10	<700	6 ⎭	
Precipitate not soluble in H$_2$SO$_4$			9	
Precipitate soluble in H$_2$SO$_4$			9	
Loss on G-75			10	
Distillate			2	
Total			96	

[a]Reproduced from Gjessing and Lee [39] courtesy of the American Chemical Society.

pyrolytic industrial operation of some sort. Examples are the preparation of acetylene from natural gas; the hydrolysis of kerosene to benzene, toluene and other organic solvents; the pyrolysis of wood, etc. There is a legitimate concern about the occurrence of these compounds in man's environment because of the carcinogenic properties exhibited to laboratory animals by some PAH. For example, 3,4-benzpyrene (BP) is, perhaps, the most carcinogenic of all PAH.

Andelman and Snodgrass [48], Andelman and Suess [49] and Harrison et al. [50] have published excellent review articles about the incidence and significance of PAH in the water environment. Typical PAH may be seen in Figure 5. The ubiquitous distribution of 3,4-benzpyrene throughout man's environment is claimed [48]. This should be questioned in view of the analytical uncertainties surrounding detection of organic molecules in environmental samples. Undoubtedly, BP is widespread, but the ubiquitous claim should be questioned until more precise analytical data are available.

Origin and Sources of PAH. Industrial operations engaged in the pyrolytic processing of such organic raw materials as coal and petroleum form (i.e., origin) PAH whenever the process is conducted at temperatures circa 700°C. Also, the consumption of pyrolytic products (coal tar, coal tar pitch, gasoline, etc.) can produce PAH. Wastewater effluents from these industrial operations conduct PAH into receiving waters because existing methods of treatment do not reduce their content to any significant extent. Effluents from the thermal processing of mineral fuels are one of the largest sources of PAH contamination of surface waters [48]. The PAH contents of typical effluents are summarized in Table XVII. Domestic wastewaters may also contain significant quantities of PAH. Borneff and Kunte [52,53] cite a concentration range of 0.8 to 87.5 μg/l for "total" PAH in several European wastewater effluents. Also, a municipal incinerator has been cited as a source of PAH [54].

Solubility and Stability of PAH in Water. Two of the most important properties influencing the incidence of PAH in water are solubility and stability. For example, the solubility of 3,4-benzpyrene in clean river water or tap water was observed to be approximately 0.01 μg/l [55]. Il'nitskii et al. [56] performed several experiments on the stability of BP in pond and tap water at concentrations of 0.01 and 10 μg/l. The latter concentration was assured through the use of an acetone–water mixture. In general, undecomposed BP could be detected after 35-40 days to the extent of 5–20% of the initial concentration. Only 10–15% of the 10 μg/l BP was lost "in the first few days." On the other hand, almost 50% of the 0.01 μg/l BP was apparently decomposed within 24 hours. Both solutions were stored in the dark. No suggestions were offered for the disappearance of the BP from the water.

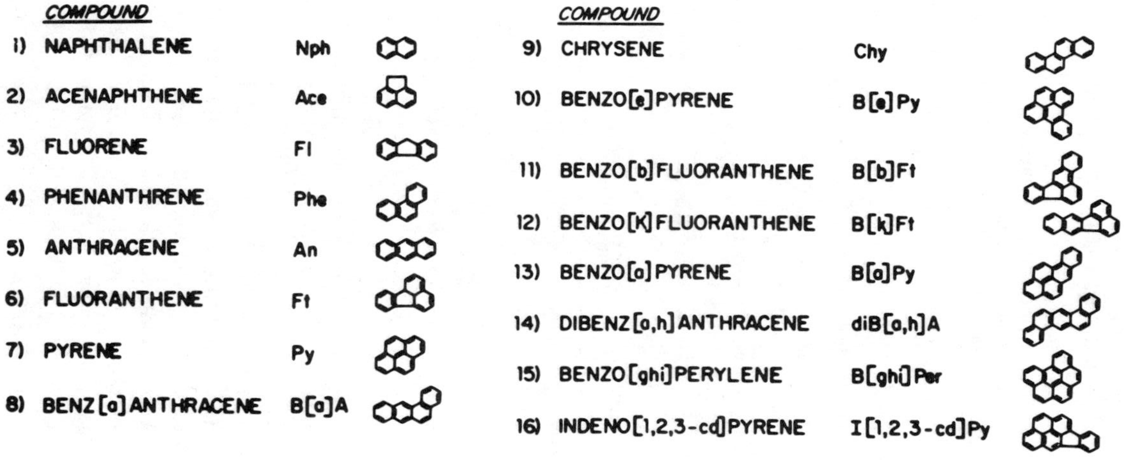

	COMPOUND			COMPOUND	
1)	NAPHTHALENE	Nph	9)	CHRYSENE	Chy
2)	ACENAPHTHENE	Ace	10)	BENZO[a]PYRENE	B[a]Py
3)	FLUORENE	Fl	11)	BENZO[b]FLUORANTHENE	B[b]Ft
4)	PHENANTHRENE	Phe	12)	BENZO[K]FLUORANTHENE	B[k]Ft
5)	ANTHRACENE	An	13)	BENZO[a]PYRENE	B[a]Py
6)	FLUORANTHENE	Ft	14)	DIBENZ[a,h]ANTHRACENE	diB[a,h]A
7)	PYRENE	Py	15)	BENZO[ghi]PERYLENE	B[ghi]Per
8)	BENZ[a]ANTHRACENE	B[a]A	16)	INDENO[1,2,3-cd]PYRENE	I[1,2,3-cd]Py

Figure 5.   Identification, structures and abbreviations of PAH. All but B[e]Py are on the U.S. Environmental Protection Agency's (EPA) priority pollutant list. Reproduced from Ogan et al. [51] courtesy of the American Chemical Society.

Table XVII. PAH Concentration in Industrial Effluents[a]

Industry	Source of Wastewater	BP Concentration (μg/l)
Shale Oil	After treatment for dephenolization	2–320
Coke By-products	Not indicated	Present
	After biochemical treatment	12–16
	After oil separation (5 samples)	6.5, 130, 250 290 and "big" quantity
	Spent gas liquor	Very small quantity
Coke or Oil–Gas Works	Before discharge to sewer	Not indicated
	Before discharge to sewer (2 plants)	1000 and 340
Oil–Gas Works	After oil separation (3 samples)	3, 6 and 30
Oil Refinery	After oil separation (3 samples)	None detected
Tar Paper	Not indicated	Present
Acetylene	Not indicated	15–100
Ammonium Sulfate	After cooling and settling	About 10[b]

[a]After Andelman and Suess [49].
[b]Other PAH present.

Presumably, microbiological action or adsorption on container walls was responsible.

PAH in Fresh Surface Waters. There is evidence that PAH are found in almost all aquatic environments. This is especially true in fresh surface waters receiving wastewaters that are known sources of PAH and downstream from petroleum refineries. Some typical concentrations of 3,4-BP and PAH are reported in Table XVIII from various Russian, German and American rivers [48]. These contents result from a special effort to sample the rivers below known sources of PAH. Basu and Saxena [57] reported "total" PAH concentrations of 0.6636 and 0.3518 μg/l for the raw drinking water supplies of Pittsburgh and Philadelphia, respectively. The finished drinking waters contained "total" PAH values of 0.0028 and 0.0149 μg/l, respectively. Herbes [58] investigated the partitioning of PAH between dissolved and particulate phases in natural waters. A "significant" fraction, 15-65% of anthracene, for example, would be associated with detrital and living organic matter in natural waters. Borneff and Kunte [59] have found PAH concentrations up to 0.10 μg/l in drinking water. These researchers have concluded that

Table XVIII. Typical Concentration Ranges of 3,4-BP
and PAH in Fresh Surface Waters [48]

Source	3, 4-BP (μg/l)	PAH[a] (μg/l)	Total PAH (μg/l)
Rhine River	0.05−0.11	0.01−0.73	0.73−1.50
German Rivers	0.001−0.04	0.04−1.3	0.12−3.1
A U.S. River	0.078−0.150	−	−
Rivers[b]	0.0001−12	−	−
Moscow Reservoirs	4−13	−	−
Volga River	0.0001	−	−
Sunzha River	0.05−3.5	−	−
Oyster River, Connecticut	0.078−0.150	−	−

[a]May be carcinogenic.
[b]Receiving wastewaters that are sources of PAH.

PAH contents between 0.15 and 0.20 μg/l in drinking water constitute questionable safety. In 1970, the World Health Organization (WHO) adopted a 0.20-μg/l concentration as a standard for European drinking water [60].

ATMOSPHERIC WATERS

Atmospheric waters (rain water and snow) have a complex chemical composition that varies considerably from location to location as well as from storm event to storm event and from season to season. Rain water, for example, contains constituents of local origin and some that have been imported by atmospheric currents from elsewhere. Occasionally there is "dry" precipitation of mineral and organic dusts that would add substances to ground and surface waters. Also, there is the universal problem of acid rain.

The chemical analysis of atmospheric waters is worldwide. A systematic examination of rain water in northern Europe was started in 1955 by the International Meteorological Institute, Stockholm [61]. A network of sampling stations was established in Scandinavia, Great Britain and northern Europe to collect and analyze on a monthly basis. Rain water has been collected and analyzed, of course, in the United States, England and Australia [61].

Rain Water

"Normal" Rain

The principal chemical constituents of rain water are the major ions: sodium, potassium, magnesium, calcium, chloride, bicarbonate and sulfate.

Minor constituents would be ammonia, nitrate, nitrite, nitrogen, carbon dioxide, iodine, bromine, boron, iron, alumina and silica. Sources of these substances would be evaporation and aerosols from saline and fresh waters, dusts from land masses, effluents from industrial stacks and, on occasion, volcanic emanations.

In Table XIX, the average chemical compositions of several rain waters are given for northern Europe and southeastern Australia. These two geographic locations receive similar amounts of rain and are near the oceans. There is very little difference in the major cationic and chloride contents of the two areas. Any apparent differences may be due to analytical error, atmospheric currents and distribution of the sampling stations. In any event, these concentrations should be considered as "low." The source of these cations and anions is rather obvious, namely, the salts are absorbed by winds over the open ocean and are eluted by the rainfall. The chemical composition (i.e., $Cl:Na$, $K:Na$, $SO_4:Cl$ ratios, etc.) of rain deposited near coastal areas is similar to that of diluted sea water.

In Table XX, the chemical composition of rain water from several locations in the conterminous United States is given. These locations represent such diverse areas as near the oceans or the Great Lakes and in inland areas. That there are considerable differences in chemical composition of the various rainfalls is obvious from the data. Location is, perhaps, the principal factor establishing these differences but the data may not always be consistent. For example, the coastal locations of Brownsville, Texas, Cape Hatteras, North Carolina and San Diego, California show considerable differences in chemical composition.

In a rather comprehensive study, Pearson and Fisher [62] examined the chemical composition of atmospheric precipitation in the northeastern United States. Monthly samples were collected from 18 sites for 12–36 months in New York, Pennsylvania and New England. As seen in Table XXI, the analytical data were combined with rainfall amounts to obtain annual average loads of the various ionic species expressed as $ton/day/mi^2 \times 10^{-3}$ The loads of certain species did not vary systematically with either area or annual amount of precipitation: calcium averaged 5.3×10^{-3}; magnesium averaged 1.0×10^{-3}; potassium averaged 1.4×10^{-3}; total nitrogen as N averaged 2.2×10^{-3}; and phosphate averaged 0.37×10^{-3} (9 sites for 12 months). Sodium and chloride loads (2.4×10^{-3} and 4.0×10^{-3}, respectively) did not vary with precipitation among the inland stations, but did vary at the coastal stations. Sodium loads ranged from 4.6×10^{-3} to 13×10^{-3} and chloride from 5.2×10^{-3} to 25×10^{-3} at 25 and 55 in./yr precipitation, respectively, for the latter stations. Sulfate loads varied with precipitation at both inland and coastal areas: 25×10^{-3} to 45×10^{-3} for 25 and 55 in./yr precipitation, respectively. This study highlights an important point often overlooked in stream pollution assessment. That is, atmospheric precipitation

Table XIX. Composition of Average Rain Water in Northern Europe and in Southeastern Australia [61]

Locality	Average Annual Rainfall (mm)	pH	Constituents (upper numbers, average; lower numbers, range)							
			Sodium (Na) (mg/l)	Potassium (K) (mg/l)	Calcium (Ca) (mg/l)	Magnesium (Mg) (mg/l)	Chloride (Cl) (mg/l)	Sulfate (SO_4) (mg/l)	Nitrate (NO_3) (mg/l)	Ammonia (NH_4) (mg/l)
Northern Europe (62 stations, 30 months)	560	5.47 3.9–7.7	2.05 0.6–63.2	0.35 0–11.2	1.42 0.20–25.5	0.39 0.12–2.93	3.47 0.06–64.0	2.19 0.18–6.52	0.27 0–1.6	0.41 0–8.7
Southeastern Australia (28 stations, 36 months)	590		2.46 0–82.8	0.37 0.04–6.6	1.20 0–20.0	0.50 0–27.6	4.43 0–138.5	Trace Trace		

Table XX. Chemical Composition of Rain Water from Several Localities in the Conterminous United States [61]

Locality	Distance from Sea (mi)	Average Annual Rainfall (mm)	Constituents						
			Sodium (Na) (mg/l)	Potassium (K) (mg/l)	Calcium (Ca) (mg/l)	Chloride (Cl) (mg/l)	Sulfate (SO₄) (mg/l)	Nitrate (NO₃) (mg/l)	Ammonium (NH₄) (mg/l)
Cape Hatteras, North Carolina	0	1370	4.49	0.24	0.44	6.50	0.88	1.03	0.11
San Diego, California	0	266	2.17	0.21	0.67	3.31	1.66	3.13	1.15
Brownsville, Texas	1	635	22.30	1.00	6.50	21.96	5.34	1.76	0.28
Akron, Ohio	27[a]	889	0.10	0.10	0.69	0.17	1.62	4.68	0.38
Tallahassee, Florida	37	1397	0.53	0.13	0.43	0.66	0.48	0.72	0.07
Greenville, North Carolina	50	1194	0.18	0.07	0.31	0.13	0.57	2.97	0.14
Tacoma, Washington	75	2032	14.50	0.59	0.73	22.58	1.69	0.99	0.05
Urbana, Illinois	85[a]	940	0.90	0.07		0.69	1.20	1.27	0.09
Washington, DC	85	1052	0.23	0.18	0.23	0.35	1.33	2.14	0.43
Fresno, California	112	240	0.30	1.11	0.37	0.35	0.54	2.94	2.21
Indianapolis, Indiana	128[a]	995	0.26	0.12	0.69	0.18	4.00	2.06	0.27
Albany, New York	150	914	0.21	0.09	0.43	0.23	0.10	4.05	0.21
Roanoke, Virginia	200	1270	0.22	0.13	0.32	0.23	1.33	3.12	0.24
Ely, Nevada	410	381	0.69	0.14	3.79	0.30	1.05	0.81	0.35
Amarillo, Texas	540	534	0.22	0.23	2.17	0.14	0.03	1.64	0.28
Glasgow, Montana	625	380	0.40	0.26	1.72	0.17	1.30	1.82	0.75
Grand Junction, Colorado	650	226	0.69	0.17	3.41	0.28	2.37	2.63	0.33
Columbia, Missouri	650	1016	0.33	0.31	2.18	0.15	1.20	3.81	0.44

[a]Distance from fresh water lake system.

is an extremely significant source of these chemical substances, especially the nitrogen species and phosphates that are essential elements in eutrophic processes.

Pearson and Fisher [62] cited several sources for the dissolved constituents in the atmospheric waters shown above. The sodium and chloride are, of course, derived from the dissolution of sea salt aerosols over the oceans. Atmospheric turbulence carries these aerosols to coastal land and adjacent inland areas. Soil dust is apparently the principal source for the potassium, calcium and magnesium ions in natural precipitation. Sulfate arises from the solution of sulfate aerosols and from the oxidation of $H_2S_{(g)}$ and $SO_{2(g)}$, which are released mainly from the combustion of fossil fuels. Hydrogen ions originate from two sources: (1) from the oxidation of $SO_{2(g)}$ in the atmosphere, and (2) from the dissolution of $CO_{2(g)}$, which yields $H_2CO_{3(aq)}$ in the precipitation. Bicarbonate anions may come from the dissolution of $CO_{2(g)}$ and from $CaCO_{3(s)}$ dusts. It is difficult to account for the sources of the NH_4^+ and NO_3^-. Vague reference is frequently given to "gaseous" sources for NH_4^+ and to "dust" sources for NO_3^-. Gambell and Fisher [63] offer evidence that the role of lightning is unimportant in the occurrence of NO_3^- in rain water from thunderstorms. Phosphates originate, in all probability, from atmospheric dusts.

"Acid" Rain

Considerable concern has been expressed in recent years about the acidic nature of precipitation in North America [64] and in Europe [65]. That is, pH values are less than the equilibrium value of 5.65 for "pure" rain water saturated with $CO_{2(g)}$. There is considerable evidence that there has been a significant change in the pH value of precipitation from the mid-1950s to date in northeastern United States and Canada. This acidity is due mainly to sulfuric acid, 60–70%, and to nitric acid, 30–40%. Cooper et al. [66] proposed the chemical mechanisms (Table XXII) whereby the two acids are formed. The major constituents from anthropic sources are the sulfur oxides emitted from combustion of coals or oil. Sulfur in fossil fuels is converted to SO_2 during combustion which, in turn, is oxidized to H_2SO_4 either in the gaseous or liquid phase. Also, nitrogen oxides can occur either from stationary combustion processes or from the emissions of automotive exhausts. At elevated temperatures, nitric oxide (NO) is formed from N_2 and O_2. Then the NO is oxidized to NO_2, which reacts with water to form nitrous (HNO_2) and nitric (HNO_3) acids. There may be some additional acids resulting from combustion processes: HCl, HF, H_3PO_4, organic acids (RCOOH), metallic ions (i.e., Fe^{3+}, Al^{3+}), phenols and the ammonium ion [67].

Geographically, acid rainfall is widespread. In the United States, however, the northeastern portion has been affected most severely over the years.

There is, however, some recent evidence that this problem is spreading to the Southeast and Midwest and to the large cities of Los Angeles, San Francisco and Seattle [64]. Likens et al. [68] described this spread of acid rain in eastern North America and northwestern Europe through the use of iso-plethic maps.

Typical ionic concentrations and pH values for acid rain in the Eastern United States are given in Table XXIII by Cogbill and Likens [69]. The principal cations are NH_4^+, Na^+, Ca^{2+}, Mg^{2+}, K^+ and H^+, whereas the principal anions are SO_4^{2-}, NO_3^- and Cl^-. Since the ionic equivalents were balanced, the H^+ ion is associated with the three principal anions which, in turn, indicates

Table XXI. Annual Loads and Amounts of Precipitation

Station	Year Ending	Precipitation (in./yr)	Calcium	Magnesium	Sodium
					Rural Inland
Houlton, Maine	8–66	30.0	8.36	0.77	1.51
Canton, New York	9–66	30.8	5.46	1.16	1.28
	9–67	25.7	4.88	1.10	1.79
	9–68	28.1	4.26	1.08	2.14
St. Albans Bay, Vermont	8–66	35.1	15.15	1.29	3.40
Pittsburg, New Hampshire	8–66	41.1	4.38	0.68	3.84
Corinna, Maine	8–66	36.4	4.58	0.74	2.71
	8–67	38.8	2.30	0.50	2.21
Mays Point, New York	9–66	30.8	5.58	1.11	1.46
	9–67	27.7	5.80	0.96	2.35
	9–68	33.0	14.29	2.27	1.85
Hinckley, New York	9–66	42.4	7.04	0.92	1.42
	9–67	41.8	8.34	1.46	2.69
	9–68	46.2	5.04	0.87	2.26
					Coastal
Taunton, Massachusetts	8–66	29.3	1.84	0.88	5.15
Upton, New York	9–66	38.0	3.99	1.63	9.16
	9–67	54.3	3.49	1.34	10.80
	9–68	40.5	2.62	1.43	9.88
Mineola, New York	9–66	33.5	7.92	2.66	7.14
	9–67	40.1	7.66	2.61	7.33
	9–68	38.2	7.48	2.54	9.43
					Urban
Albany, New York	9–66	31.9	17.53	1.00	3.20
	9–67	30.3	20.01	1.42	2.66
	9–68	30.1	16.59	1.41	2.58

the three major acidic constituents. Liljestrand and Morgan [70] reported the chemical composition of acid rain in Pasadena, California. Although the data in Table XXIV were obtained in a short time span (February 1976–September 1977), the principal constituents are typical and a report of the minor ions is noteworthy. Wolff et al. [71] published a unique study whereby the acid rain in the New York Metropolitan area was examined with its relationship to meteorological factors. This study was conducted at eight sites around New York City over the time span of 1975–1977. A seasonal effect was observed with the lowest mean pH value of 4.12 occurring during the summer. Rainfall events associated with cold fronts and air mass type of showers and

at Sampling Sites in the Northeastern United States [62]

Annual Load (10^{-3} ton/day/mi^2)					
Potassium	Ammonium	Hydrogen Ion	Bicarbonate	Sulfate	Chloride
Stations					
1.84	1.12	0.052	16.68	19.26	3.01
1.45	1.42	0.200	1.54	31.48	2.15
0.85	1.29	0.231	0.00	27.04	2.71
0.85	2.54	0.147	2.30	24.64	2.22
2.33	4.11	0.093	15.81	37.73	5.40
1.10	2.05	0.167	4.16	29.56	3.75
2.79	1.40	0.175	7.15	23.78	4.74
1.45	1.40	0.196	3.81	23.37	4.54
0.96	2.00	0.273	0.00	32.97	2.90
0.60	0.72	0.280	0.00	30.43	2.78
1.08	2.06	0.107	13.60	32.73	5.59
1.69	3.32	0.390	0.00	42.04	3.14
0.91	0.78	0.297	1.22	36.60	3.37
1.46	3.19	0.388	1.34	35.52	2.61
Stations					
1.29	1.48	0.095	0.00	22.82	8.93
1.27	2.15	0.303	0.65	31.55	15.94
0.93	1.30	0.377	0.00	37.33	24.65
9.83	2.60	0.372	0.00	29.80	16.24
0.67	2.35	0.203	0.55	31.31	12.65
0.33	1.37	0.300	0.78	43.02	12.45
2.52	3.29	0.260	0.00	43.84	14.76
Station					
1.26	1.16	0.134	4.91	45.63	6.59
1.05	2.14	0.078	1.65	51.73	5.57
1.02	1.22	0.143	3.71	44.24	4.22

Table XXI, continued

Station	Year Ending	Precipitation (in./yr)	Annual Load (10⁻³ ton/day/mi²)		
			Nitrate	Total Nitrogen as N	Phosphate
Rural Inland Stations					
Houlton, Maine	8–66	30.0	1.45	1.20	–
Canton, New York	9–66	30.8	1.55	1.45	–
	9–67	25.7	2.38	1.54	–
	9–68	28.1	2.05	2.43	0.23
St. Albans Bay, Vermont	8–66	35.1	12.57	6.03	–
Pittsburg, New Hampshire	8–66	41.1	3.51	2.38	–
Corinna, Maine	8–66	36.4	2.25	1.59	–
	8–67	38.8	7.83	2.86	–
Mays Point, New York	9–66	30.8	1.30	1.84	–
	9–67	27.7	2.28	1.07	–
	9–68	33.0	6.82	3.14	0.28
Hinckley, New York	9–66	42.4	2.16	3.06	–
	9–67	41.8	2.07	1.07	–
	9–68	46.2	1.22	2.75	0.50
Coastal Stations					
Taunton, Massachusetts	8–66	29.3	1.40	1.46	–
Upton, New York	9–66	38.0	3.87	2.54	–
	9–67	54.3	3.27	1.74	–
	9–68	40.5	5.22	3.20	0.17
Mineola, New York	9–66	33.5	7.22	3.46	–
	9–67	40.1	4.22	2.02	–
	9–68	38.2	4.69	3.61	0.49
Urban Station					
Albany, New York	9–66	31.9	8.41	2.80	–
	9–67	30.3	12.34	4.45	–
	9–68	30.1	8.32	2.83	0.23

thundershowers provided the lowest pH values of 4.17 and 3.91, respectively. These type of storms are associated generally with the back side of a polluted high-pressure system.

Numerous effects of acid rainfall on the environment may be cited [64,68]. The most significant of these are acidification of streams, lakes and soils; damage to such aquatic life as fish and to forest ecology; direct damage to the foliar surface of crop plants; and reduction of crop productivity, etc. Each of these effects has been documented or is under study [64,68]. Some

Table XXII. Chemical Mechanisms of Acid Rain Formation in the Atmosphere [66]

Sulfur Oxides

$$S + O_2 \longrightarrow SO_2$$

$$SO_2 + H_2O \longrightarrow H_2SO_3$$

$$H_2SO_3 + \frac{1}{2}O_2 \xrightarrow{\text{Catalyst}} H_2SO_4$$

$$2SO_2 + O_2 \xrightarrow{\text{Catalyst}} 2SO_3$$

$$SO_3 + H_2O \longrightarrow H_2SO_4$$

Nitrogen Oxides

$$N_2 + O_2 \longrightarrow 2NO$$

$$2NO + O_2 \longrightarrow 2NO_2$$

$$2NO_2 + H_2O \longrightarrow HNO_2 + HNO_3$$

of the more recent reports are those of Schindler et al. [72] on the acidification of lakes in Ontario; Schindler et al. [73] on the effects of mobilization of heavy metals and radionuclides from the sediments of fresh water lakes; and Dillon et al. [74] on the acidification of lakes in south-central Ontario. The reader is referred to two excellent review articles by Glass et al. [64] and Likens et al. [68] for additional information on the characteristics and effects of acid precipitation.

Snow

Melting snow provides a large portion of the water supply in the western portion of the United States. Consequently, knowledge of the chemical quality of snow would be especially important because the waters have domestic, industrial and agricultural uses. Most of our knowledge about the chemical composition of natural, unpolluted snow comes from a study reported by Feth and co-workers [75]. Data were collected from the northern Sierra Nevada, Utah and Colorado. A summary of the chemical data appears in Table XXV. It appears that all of the major chemical constituents usually found in surface and ground waters are found also in snow melt. The dissolved solids content is extremely low with a minimum value of 0.7 mg/l and a maximum value of 18 mg/l reported in this study. There is some diversity in chemical quality of the snow melt as seen by the variation in dissolved solids content from source to source. The Utah snows exhibited a preponderance of calcium, magnesium and bicarbonate. Continental atmospheric dust

Table XXIII. Weighted Precipitation Chemistry and pH for Locations in the Eastern United States during 1972-1973[a]

Location	Period	Water (cm/area)	Ca^{2+} (mg/l)	SO$_4^{2-}$ (mg/l)	NO$_3^-$ (mg/l)	Cl$^-$ (mg/l)	NH$_4^+$ (mg/l)	Na$^+$ (mg/l)	K$^+$ (mg/l)	Mg^{2+} (mg/l)	pH
Ithaca, New York	September 1972–August 1973	90.04	0.83	4.96	2.88	0.47	0.32	0.15	0.09	0.08	4.05
Aurora, New York	September 1972–August 1973	98.07	0.45	4.51	2.72	0.47	0.40	0.08	0.07	0.07	4.05
Geneva, New York	September 1972–August 1973	91.34	0.73	4.58	3.27	0.37	0.42	0.10	0.09	0.13	4.09
Hubbard Brook, New Hampshire	June 1973–August 1973	7.95	0.24	4.77	1.92	0.15	0.37	0.07	0.07	0.05	4.05
Gatlinburg, Tennessee	June 1973–August 1973	38.15	0.20	3.19	1.24	0.15	0.19	0.05	0.07	0.03	4.19

[a]Reproduced from Cogbill and Likens [69] courtesy of the American Geophysical Union.

Table XXIV. Rain Water Concentrations in Pasadena, California[a]

	Precipitation Weighted Mean Concentration	Maximum Concentration	Minimum Concentration
H^+	$8.7 \times 10^{-5} M$	$1.6 \times 10^{-3} M$	$4.7 \times 10^{-6} M$
NH_4^+	$3.3 \times 10^{-5} M$	$8.7 \times 10^{-4} M$	$5.2 \times 10^{-6} M$
Na^+	$2.5 \times 10^{-5} M$	$3.9 \times 10^{-4} M$	$9.5 \times 10^{-7} M$
K^+	$2.1 \times 10^{-6} M$	$4.7 \times 10^{-5} M$	$1.7 \times 10^{-7} M$
Ca^{2+}	$4.8 \times 10^{-6} M$	$4.2 \times 10^{-5} M$	$2.5 \times 10^{-7} M$
Mg^{2+}	$3.3 \times 10^{-6} M$	$2.9 \times 10^{-5} M$	$1.5 \times 10^{-7} M$
Fe	$3.6 \times 10^{-7} M$	$2.3 \times 10^{-6} M$	$2.5 \times 10^{-8} M$
Mn	$3.4 \times 10^{-8} M$	$3.9 \times 10^{-7} M$	$<2.0 \times 10^{-9} M$
Al	$2.8 \times 10^{-7} M$	$1.8 \times 10^{-6} M$	$<3.7 \times 10^{-8} M$
Pb	$3.6 \times 10^{-7} M$	$3.2 \times 10^{-6} M$	$1.8 \times 10^{-8} M$
Cl^-	$2.9 \times 10^{-5} M$	$4.6 \times 10^{-4} M$	$1.1 \times 10^{-6} M$
NO_3^-	$7.5 \times 10^{-5} M$	$1.9 \times 10^{-3} M$	$1.1 \times 10^{-5} M$
SO_4^{2-}	$3.0 \times 10^{-5} M$	$4.3 \times 10^{-4} M$	$7.5 \times 10^{-6} M$
Br^-	$1.5 \times 10^{-7} M$	$1.4 \times 10^{-6} M$	$<1.0 \times 10^{-7} M$
HCO_3^-	$5.8 \times 10^{-8} M$	$1.1 \times 10^{-6} M$	$2.8 \times 10^{-9} M$
$Si(OH)_4$	$9.0 \times 10^{-7} M$	$9.3 \times 10^{-6} M$	$8.6 \times 10^{-8} M$
Organic Carbon	0.65 mg/l	5.1 mg/l	<0.1 mg/l
Filterable Residue	2.3 mg/l	9.24 mg/l	<0.01 mg/l
Total Residue	10.8 mg/l	27.8 mg/l	2.1 mg/l

[a]Reproduced from Liljestrand and Morgan [70] courtesy of the American Chemical Society.

was assigned the source for the cations, whereas the dissolution of $CO_{2(g)}$ was indicated as the source of the bicarbonate. The Sierra Nevada snows showed sodium and chloride as major constituents, which undoubtedly originated from the atmospheric pollutant $SO_{2(g)}$. The range of pH values for the three sites suggests that the snow was relatively uncontaminated from anthropic sources. Feth et al. [75] indicate that the bicarbonate constituent controls these pH values.

REFERENCES

1. *Standard Methods for the Examination of Water and Wastewater*, 14th ed. (New York: American Public Health Association, 1976).
2. Hem, J. D. "Study and Interpretation of the Chemical Characteristics of Natural Water," *Geol. Survey Water Supply Paper 1473*, U.S. Government Printing Office, Washington, DC (1970).
3. "Quality of the Surface Waters of the United States," *Geol. Survey Water Supply Papers* 2091, 2152, 2093, 2094, 2095, 2096, 2097, 2015 and 2016, U.S. Government Printing Office, Washington, DC.

Table XXV. Comparison of Median and Mean Concentrations of Selected
Constituents in Snow from the Sierra Nevada,
Utah and Denver, Colorado [75]

Constituent (mg/l)	Sierra Nevada		Utah		Denver, Colorado	
	Median	Mean	Median	Mean	Median	Mean
Silica (SiO$_2$)	0.0	0.16	–	–	–	–
Calcium (Ca)	0.2	0.39	2.0	2.23	–	–
Magnesium (Mg)	0.0	0.16	0.35	0.33	–	–
Sodium (Na)	0.2	0.46	0.55	0.60	1.1	1.24
Potassium (K)	0.2	0.31	0.35	0.47	–	–
Bicarbonate (HCO$_3$)	2.0	2.90	5.0	6.29	11	10
Sulfate (SO$_4$)	0.65	0.93	2.2	2.25	1.5	2.88
Chloride (Cl)	0.4	0.50	0.7	0.97	1.0	1.60
Dissolved solids (calculated)	4.0	4.70	9.1	10.58	–	–
pH (range)	4.2–8.3		6.2–7.2		6.0–7.1	

4. Leifeste, D. K. "Dissolved-Solids Discharge to the Oceans from the Conterminous United States," *Geol. Survey Circ. 685,* Washington, DC
   (1974).
5. U.S. Environmental Protection Agency. "Quality Criteria for Water,"
   EPA-440/9-76-023, Washington, DC (1976).
6. National Academy of Sciences. "Drinking Water and Health," *Federal
   Register 42(132)*:35764 (July 11, 1977).
7. U.S. Environmental Protection Agency. "Water Quality Criteria, 1972,"
   EPA R3 73-033, Washington, DC (1973).
8. Tate, C. N., and R. R. Trussell. *J. Am. Water Works Assoc. 69:* 486
   (1977).
9. Durfor, C. N., and E. Becker. "Public Water Supplies of the 100 Largest
   Cities in the United States, 1962," *Geol. Survey Water Supply Paper
   1812,* Washington, DC (1964).
10. Durum, W. H., and J. Haffty. "Occurrence of Minor Elements in Water,"
    Geol. Survey Circ. 445, Washington, DC (1961).
11. Skougstadt, M. W., and C. A. Horr. "Occurrence of Strontium in Natural
    Water," *Geol. Survey Circ. 420,* Washington, DC (1960).
12. Durum, W. H., et al., "Reconnaissance of Selected Minor Elements in
    Surface Waters of the United States, October, 1970," *Geol. Survey Circ.
    643,* Washington, DC (1971).
13. Silvey, W. D. "Occurrence of Selected Minor Elements in the Waters of
    California," *Geol. Survey Water Supply Paper 1535-L,* U.S. Government
    Printing Office, Washington, DC (1967).
14. U.S. Environmental Protection Agency, *Water Quality Criteria Data
    Book,* Vol. 2, Inorganic Chemical Pollution of Fresh Water, Pub.
    #18010DPY, Washington, DC (1971).
15. Kopp, J. F., and R. C. Kroner. "Trace Metals in Waters of the United
    States," U.S. Department of the Interior, Federal Water Pollution Control
    Administration, Washington, DC (1967).

16. Chen, K. P., and H. Y. Wu. *J. Formosan Med. Assoc.* 61:611 (1962).
17. Lakin, H. W. "Selenium in Our Environment," in *Trace Elements in the Environment,* ACS 123, Washington, DC, American Chemical Society (1973).
18. Smith, B., et al. *J. Am. Water Works Assoc.* 53:75 (1961).
19. Oden, S. *Kolloidchem. Beihefte* 11:75 (1919).
20. Ramann, E. *Soil Science,* 3rd ed., Berlin, Germany (1911).
21. Waksman, S. A. *Humus,* 2nd ed. (Baltimore: Williams & Wilkins, Inc., (1938).
22. Black, A. P., and R. A. Christman. *J. Am. Water Works Assoc.* 55:897 (1963).
23. Page, H. J., and M. M. S. Dutoit. *J. Agric. Sci.* 20:478 (1930).
24. Farrah, S. R., et al. *Water Res.* 12:303 (1978).
25. Weber, J. H., and S. A. Wilson. *Water Res.* 9:1079 (1975).
26. Black, A. P., and R. F. Christman. *J. Am. Water Works Assoc.* 55:753 (1963).
27. Saville, T. *J. New England Water Works Assoc.* 31:79 (1917).
28. Shapiro, J. *Limnol. Oceanog.* 2:161 (1957).
29. Midwood, R. B., and G. T. Felbeck. *J. Am. Water Works Assoc.* 60:357 (1968).
30. Ungar, J. Research Report R106, Department of Mines and Technical Services, Mines Branch, Ottawa, Canada (November 1962).
31. Wershaw, R. L., and G. E. Bohner. *Geochim. Cosmochim. Acta* 33:757 (1969).
32. Steelink, C. *J. Chem. Ed.* 54:599 (1977).
33. Christman, R. F., and M. Ghassemi. *J. Am. Water Works Assoc.* 58:723 (1966).
34. Schnitzer, M., and S. U. Khan. *Humic Substances in the Environment* (New York: Marcel Dekker, Inc., 1972).
35. LaMar, W. L., and D. F. Goerlitz. "Organic Acids in Naturally Colored Surface Waters," *Geol. Survey Water Supply Paper 1817-A,* U.S. Department of the Interior, Washington, DC (1966).
36. LeMar, W. L., and D. F. Goerlitz. *J. Am. Water Works Assoc.* 55:797 (1963).
37. Day, H. R., and G. T. Felbeck, Jr. *J. Am. Water Works Assoc.* 66:484 (1974).
38. Felbeck, G. T., Jr. *Soil Sci.* 111:42 (1971).
39. Gjessing, E., and G. F. Lee. *Environ. Sci. Technol.* 1:631 (1967).
40. Shapiro, J. *Proc. Symp. of the Hungarian Hydrological Society,* Budapest-Tihany (September 1966).
41. Ghassemi, J., and R. F. Christman. *Limnol. Oceanog.* 13:583 (1968).
42. Brodsky, A., et al. *J. Am. Water Works Assoc.* 62:386 (1970).
43. Wershaw, R. L., et al. *Adv. X-ray Anal.* 13:609 (1970).
44. Christman, R. F., and R. A. Minear, In: *Organic Compounds in Aquatic Environments,* S. D. Faust and J. V. Hunter, Eds. (New York: Marcel Dekker, Inc., 1971).
45. Slack, K. V., and H. R. Feltz. *Environ. Sci. Technol.* 2:126 (1968).
46. Lush, D. L., and H. B. N. Hynes. *Limnol. Oceanog.* 18:968 (1973).
47. Novak, J. T. *J. Am. Water Works Assoc.* 67:134 (1975).
48. Andelman, J. B., and J. E. Snodgrass. *Crit. Rev. Environ. Control* 4(1):69 (1974).

49. Andelman, J. B., and M. J. Suess. *Bull. World Health Org.* 43:479 (1970).
50. Harrison, R. M., et al. *Water Res.* 9:331 (1975).
51. Ogan, K., et al. *Anal. Chem.* 51:1315 (1979).
52. Borneff, J., and H. Kunte. *Arch. Hyg. (Berlin)* 148:585 (1964).
53. Borneff, J., and H. Kunte. *Arch. Hyg. (Berlin)* 149:226 (1965).
54. Davies, I. W., et al. *Environ. Sci. Technol.* 10:451 (1976).
55. Il'nitskii, A. P., and K. P. Ershova. *Vop. Onkol.* 16(8): (1970) (Russian); *Chem. Abstr.* 74:79382K (1971).
56. Il'nitskii, A. P., et al. *Gig. Sanit.* 36(4):8 (1971) (Russian).
57. Basu, D. K., and J. Saxena. *Environ. Sci. Technol.* 12:795 (1978).
58. Herbes, S. E. *Water Res.* 11:493 (1977).
59. Borneff, J., and H. Kunte. *Arch. Hyg. (Berlin).* 153(3):220 (1969) (German); *Chem. Abstr.* 71:73890r (1969).
60. "European Standards for Drinking Water," 2nd ed. (Geneva, Switzerland: World Health Organization, 1970).
61. Carroll, D. "Rainwater as a Chemical Agent of Geologic Processes—A Review," *Geol. Survey Water Supply Paper 1535-G,* Washington, DC (1962).
62. Pearson, D. J., Jr., and D. W. Fisher. "Chemical Composition of Atmospheric Precipitation in the Northeastern United States," *Geol. Survey Water Supply Paper 1535-P,* Washington, DC (1971).
63. Gambell, A. W., and D. W. Fisher. *J. Geophys. Res.* 69:4203 (1964).
64. Glass, N. R., et al. *Environ. Sci. Technol.* 13:1350 (1979).
65. Barett, E., and G. Brodin. *Tellus* 7:251 (1955).
66. Cooper, H. B. H., et al. *Water Air Soil Poll.* 6:351 (1976).
67. Galloway, J. N., et al. *Water Air Soil Poll.* 6:423 (1976).
68. Likens, G. E., et al. *Scientific Am.* 241:43 (1979).
69. Cogbill, C. V., and G. E. Likens. *Water Resources Res.* 10:1133 (1974).
70. Liljestrand, H. M., and J. J. Morgan. *Environ. Sci. Technol.* 12:1271 (1978).
71. Wolff, G. T., et al. *Environ. Sci. Technol.* 13:209 (1979).
72. Schindler, D. W., et al. *Can. J. Fish. Aquat. Sci.* 37:342 (1980).
73. Schindler, D. W., et al. *Can. J. Fish. Aquat. Sci.* 37:373 (1980).
74. Dillon, P. J., et al. *J. Fish. Res. Bd. Can.* 35:809 (1978).
75. Feth, J. H., et al. "Chemical Composition of Snow in the Northern Sierra Nevada and Other Areas," *Geol. Survey Water Supply Paper 1535-J,* Washington, DC (1964).

CHAPTER 2

CHEMICAL EQUILIBRIA AND THERMODYNAMICS

INTRODUCTION

Water chemistry is concerned with the definition of chemical systems and chemical reactions that occur in aquatic environments. In turn, aquatic environments may be natural, i.e., a lake, stream, pond, river, etc., or they may be artificially created, i.e., a water or wastewater treatment plant. Some of these aquatic environments may be composed of none, one or more solid phases in contact with the water phase, which may or may not be open to the atmosphere. On the other hand, many chemical reactions may occur in the aqueous phase, where solid and gaseous phases have little or no influence. Such chemistry may be encountered in water and wastewater treatment plants.

This chapter presents elements of thermodynamics that affect chemical reactions in aquatic environments. It is not our intent to systematically develop the fundamentals of chemical thermodynamics as there are several excellent textbooks on this subject [1-3]. Rather, we will summarize some of the basic definitions of thermodynamics that govern aqueous reactions and systems.

Many water chemistry reactions may be fitted into a generalized model:

$$aA + bB = cC + dD$$

This reaction reads, from left to right, that a moles of A react with b moles of B to yield c moles of C and d moles of D. In turn, A may represent an impurity or pollutant that must be removed from the water phase by constituent B. C represents the reaction product of A which, in many cases, is a solid phase that is removed by sedimentation or filtration. D usually represents a by-product of the reaction. Another interpretation of this generalized model may be that A represents an ionic constituent of water and B is some sort of a solid phase. In this case, A and B react, perhaps by chemical weathering, to yield another ionic constituent, C, and another solid phase. The original water quality characteristics are altered by this process.

An important aspect may be noted from the generalized reaction model. This is the use of the symbol =, which denotes a condition or state of equilibrium. A frequent interpretation of equilibrium is one in which the forward reaction (left to right), A + B, and the backward reaction (right to left), C + D, are occurring simultaneously with equal rates. That is, there is a dynamic balance between these opposing reactions where no net change in the composition of the system is observed. If this system reaches such a state, then a mathematical expression may be formulated:

$$K_{eq} = \frac{[C]^c[D]^d}{[A]^a[B]^b} \tag{1}$$

Thus, if one multiplies the molar concentrations of the products (raised to the power of their molar coefficients) and then divides by the product of the molar concentrations of the reactants (raised to the power of their molar coefficients), a numerical value results that is called the equilibrium constant. That is, K_{eq} values are unique and constant for reactions that may be fitted to A + B = C + D.

A kinetic approach to equilibrium may be more pertinent to aquatic chemistry. Often, reactions are brought from a nonequilibrium position to an equilibrium position. That is, B is added to water to react with A (in a treatment plant, perhaps) to produce C and D. We are concerned now with the velocity of the forward reaction for which may be formulated:

$$v_f = k_f[A]^a[B]^b \tag{2}$$

where v_f is the velocity of the forward reaction and k_f is a proportionality constant. The product, $[A]^a[B]^b$, now acquires a different connotation. It is the driving force for the forward reaction. As [A] and [B] decrease with time, the v_f decreases and [C] and [D] increase. Thus, the driving force for the backward reaction increases with time for which the velocity may be formulated:

$$v_b = k_b[C]^c[D]^d \tag{3}$$

where v_b is the velocity of the backward reaction and k_b is a proportionality constant. Very rarely are v_f and v_b measured separately, but rather a v_{net} value is determined:

$$v_{net} = v_f - v_b \qquad (4)$$

The graphic relationships of v_{net}, v_f and v_b are seen in Figure 1. Eventually, $v_{net} = 0$, which means that $v_f = v_b$, whereupon a condition of equilibrium is reached from

$$k_f[A]^a[B]^b = k_b[C]^c[D]^d \qquad (5)$$

It follows that

$$K_{eq} = \frac{k_f}{k_b} = \frac{[C]^c[D]^d}{[A]^a[B]^b} \qquad (6)$$

The equilibrium constant becomes the ratio of the forward and backward proportionality constants. Under these conditions, the concentrations of

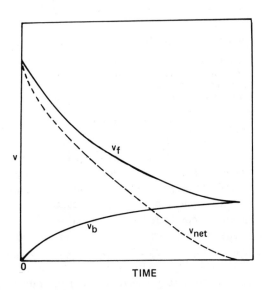

Figure 1.   Kinetic approach to equilibrium of a chemical reaction.

A, B, C and D do not change analytically with time. The essential points of water chemistry are the conditions of chemical equilibrium, the equilibrium constant and the kinetics of the reaction.

The generalized kinetic approach to equilibrium described above infers that reaction rate expressions for a net chemical reaction follow the reactant and product concentrations raised to exponents that are identical to the stoichiometric coefficients. This is simply not the case experimentally, where rate laws are often complex that reflect various mechanistic pathways and, perhaps, several rate-limiting steps. These exponents must be determined experimentally. In any event, an equilibrium constant expression can be derived from the ratio of the specific rate constants for the forward and backward reactions. Despite the complication of the reaction mechanism, the properties of the system at equilibrium are independent of the path taken from reactants to products. Consequently, the equilibrium constant expression can be written on the basis of the overall reaction.

THERMODYNAMIC VARIABLES AND FUNCTIONS

General Statement

The thermodynamic concepts underlying chemical equilibria are utilized in this book mainly to denote the spontaneity of a reaction to occur in water and to describe the natural water quality composition of ground and surface waters.

Several criteria may be cited for a system in a condition of equilibrium:

1. The mechanical properties must be uniform and constant, that is, there must be no unbalanced forces acting on or within the system.
2. The chemical composition of a system at equilibrium must be uniform and there must be no net change in the concentrations of the reactants and products.
3. The temperature of the system must be uniform and must be the same as the temperature of the surroundings.

As noted above, the equilibrium constant, K_{eq}, is a measure of the completeness to which the reaction occurs from *left* to *right*. The larger the value of K_{eq}, the farther will the reaction proceed from left to right. It is prudent, on occasion, to calculate the tendency for a reaction to occur before a laboratory experiment is conducted or before a water treatment process is installed. The reader should have three concerns about aquatic chemistry reactions: (1) the direction (i.e., left or right) in which a reaction tends to proceed under the applicable environmental conditions, (2) the extent to which the reaction tends to proceed, and (3) the rate with which the reaction occurs. Chemical thermodynamics answers the first two questions, whereas chemical kinetics answers the third. Chemical thermodynamics is that branch of science concerned primarily with the quantities of *energy* and *entropy* associated with physical and chemical processes of gaseous, solid and liquid systems or combinations of these three [2].

Some Basic Thermodynamic Definitions and Relationships

A summary of some of the basic definitions and relationships of chemical thermodynamics is given in Table I. We are concerned only with the relationship between a system and its surroundings. Herein the internal properties of the system (i.e., Helmholtz free energy) are neglected. We present only the thermodynamics of a chemical system that indicate a spontaneous or feasible change of a left to right reaction. Note from Table I that an intensive state variable (or property) of a system is independent of

Table I. Some Basic Thermodynamic Definitions and Relationships

Thermodynamic Variables or Functions of a System

T = absolute temperature, an intensive state variable
P = pressure, an intensive state variable
V = volume, an extensive state variable
E = internal energy, an extensive state function
H = enthalpy, an extensive state function
G = Gibbs free energy, an extensive state function
S = entropy, an extensive state property

Transfer of Energy between a System and its Surroundings

q = heat absorbed by a system from its surroundings
w = work performed by surroundings on a system

The First and Second Laws of Thermodynamics

$$\text{First law:} \quad dE = q + w \tag{7}$$

$$\text{Second law:} \quad dS_{total} = dS_{sys} + dS_{sur} \tag{8}$$

The total entropy change of system and its surroundings is the sum of the entropy flow between the entropy change in the system + the entropy change in the surroundings. For reversible, spontaneous processes:

$$dS_{total} = 0 \tag{9}$$

For the irreversible or natural process:

Table I, continued

$$dS_{total} > 0 \tag{10}$$

Entropy of a system is defined:

$$dS_{sys} = \frac{q_r}{T} \tag{11}$$

where q_r = heat evolved from a reversible process.

Entropy of the surroundings is defined:

$$dS_{sur} = \frac{\bar{q}}{T} \tag{12}$$

where $\bar{q}$ = heat actually absorbed by surroundings. Combining,

$$dS_{total} = \frac{q_r}{T} + \frac{\bar{q}}{T} \geqslant 0 \tag{13}$$

where dS_{total} is $\geqslant 0$ for spontaneous processes occurring at isothermal conditions.

Definitions of Thermodynamic State Functions

Enthalpy $H = E + PV$ $\tag{14}$

Gibbs Free Energy $G = H - TS$ $\tag{15}$

Further Relationships Between Thermodynamic Properties

(Closed Systems of Fixed Composition with $P - V$ work only)

$$dE = TdS - PdV \tag{16}$$
$$dH = TdS + VdP \tag{17}$$
$$dG = -SdT + VdP \tag{18}$$

Furthermore:

from (16): $\left(\dfrac{\partial E}{\partial S}\right)_v = T$ $\quad(19)\quad$ and $\left(\dfrac{\partial E}{\partial V}\right)_s = -P$ $\tag{20}$

from (17): $\left(\dfrac{\partial H}{\partial S}\right)_p = T$ $\quad(21)\quad$ and $\left(\dfrac{\partial H}{\partial P}\right)_s = V$ $\tag{22}$

from (18): $\left(\dfrac{\partial G}{\partial P}\right)_T = V$ $\quad(23)\quad$ and $\left(\dfrac{\partial G}{\partial T}\right)_p = -S$ $\tag{24}$

the amount of material involved, whereas an extensive variable is proportional to the quantity of material present.

CHEMICAL EQUILIBRIUM

Reaction Spontaneity

In aquatic chemistry we are concerned with the assemblage of various reactants from a nonequilibrium state to a state of equilibrium. First, however, one should determine the spontaneity of the reaction (Strong and Stratton [4] argue for use of the word feasibility instead of spontaneity). That is, can the left to right reaction between A and B occur? The spontaneity of a proposed chemical reaction may be calculated from

$$\Delta G = \Delta H - T\Delta S \qquad (25)$$

where, for a given system, ΔG = the change in the Gibbs free energy
ΔH = the enthalpy change
ΔS = the entropy change
T = the absolute temperature.

This expression emerges from the second law of thermodynamics and is applicable for systems under conditions of constant temperature and pressure and where only pressure–volume work is involved. The values of ΔG represent the difference between the two driving forces of a chemical reaction: ΔH and ΔS. For finite changes of ΔG, three cases may be distinguished:

1. when equilibrium exists:

$$\Delta G = 0 \qquad (26)$$

2. for a spontaneous reaction:

$$\Delta G < 0 \qquad (27)$$

3. for a nonspontaneous reaction:

$$\Delta G > 0 \qquad (28)$$

When $\Delta G = 0$ or when $\Delta H = T\Delta S$, there is no change in free energy of the reaction and a condition of equilibrium exists. When $\Delta G < 0$, a negative free energy change occurs and the left to right reaction is spontaneous and may occur. When $\Delta G > 0$, a positive free energy change occurs and the left to right reaction is not spontaneous and should not occur. A $+ \Delta G$ value forbids the occurrence of a reaction. There are four combinations of ΔH and $T\Delta S$ leading to ΔG values [1]:

ΔH Sign	ΔS Sign	ΔG Sign	Results
−	+	−	Spontaneous
−	−	$\begin{cases} - \text{ at low T} \\ + \text{ at high T} \end{cases}$	Spontaneous Not spontaneous
+	+	$\begin{cases} + \text{ at low T} \\ - \text{ at high T} \end{cases}$	Not spontaneous Spontaneous
+	−	+	Not spontaneous

For reactions occurring at a constant temperature, the term $T\Delta S$ is small in comparison to ΔH. Therefore, the sign of ΔG is usually the same as ΔH. That is, if ΔH is sufficiently large, say greater than 10 kcal, then the sign of ΔH will be the same as ΔG [3].

Standard Free Energy

To calculate free energy changes of a reaction on a uniform basis, reference conditions of pressure and absolute temperature for chemical systems are assigned. These conditions, called the standard state, are a pressure of one atmosphere and a temperature of $298.15°K$ ($25°C$). Furthermore, every chemical element is assigned a zero free energy of formation. In other words, elements are considered to be in their most stable physical form at 1 atm and $298.15°K$. For compounds, standard molar free energies of formation are employed. That is, whenever a compound is formed from an element, the standard free energy change accompanying the formation of *1 mole* of compound is the standard molar free energy of formation. Therefore, every compound will have a ΔG^0 (some texts use ΔG_f^0) value for 1 mole, 1 atm and $298.15°K$. The standard free energy change of any reaction may be computed from the expression

$$\Delta G^0_{Rex} = \Sigma \Delta G^0_{products} - \Sigma \Delta G^0_{reactants} \qquad (29)$$

The sign of ΔG^0_{Rex} now indicates whether the process:

REACTANTS (1 atm, 1 mol, 298.15°K) → PRODUCTS (1 atm, 1 mol, 298.15°K)

is spontaneous. If ΔG^0_{Rex} is −, then reactants in their standard state should be converted to products in their standard state. If ΔG^0_{Rex} is +, then reactants in their standard state should not be converted to products in their standard state. If ΔG^0_{Rex} is 0, then the reactants and products are already in equilibrium in their standard states.

Chemical Potential

Any chemical species may be assigned an intensive quantity called chemical potential. This characterizes the thermodynamic behavior of this species when reacting with other chemical species, both as to the direction in which reactions are permitted to occur and the positions of chemical equilibrium. Chemical potentials account for changes in energy of the system resulting from a change in its chemical content. This concept applies not only to homogeneous reactions (occurring in a single phase), but also to reactions at the interface between different phases, namely, heterogeneous reactions. Chemical potentials apply also to phase transformations and equilibria, regardless of whether they occur in pure substances or in mixtures. This latter point is especially important in water quality chemistry.

A mixture of n_1 moles of species 1, n_2 moles of species 2, etc., may be considered. The chemical potential, u_i, of each species i has the property that the free energy of the mixture is the sum of the u_i multiplied by the number of moles, n_i:

$$G = \sum_i n_i u_i^-$$ (30)

Chemical potentials depend on the composition of the mixture, the temperature, T and the pressure, P.

In a mixture of perfect gases, 1, 2, 3, etc., the chemical potential of the species i is

$$u_i = u_i^0(T) + RT \ln P_i$$ (31)

where P_i = partial pressure, atm, of the species, i, $u_i^0(T)$ = chemical potential of the species i at a standard pressure of 1 atm and at temperature, T, which is 298°K unless another T is chosen. ($u_i^0(T)$ may be set equal to $G^0(T)$, the standard molar free energy of formation for the pure gas i.)

For *dilute* solutions, the chemical potential of the solute species is

$$u_{is} = u_{is}^0 + RT \ln c_i \tag{32}$$

where c_i = concentration in mol/l of the species, i, and u_{is}^0 = the chemical potential of the species i in a solution at standard state conditions (u_{is}^0 may also be = $G^0(T)$, the standard molar free energy of formation for the particular species, i). u_{is} is dependent on T and the total pressure, P.

It should be noted that Equations 31 and 32 are given for perfect gases and dilute solutions. For solutions greater than about $0.01 M$ [1], Equation 32 should have the c_i term replaced by a_i, activities that represent effective concentrations and are calculated by multiplying c_i by an appropriate activity coefficient, γ_i. Similarly, for mixtures of real gases, Equation 31 remains valid if P_i is replaced by a corrected partial pressure called fugacity.

Chemical Potential and Chemical Equilibrium

Gas Reactions

If this reaction is considered,

$$3H_{2(g)} + N_{2(g)} = 2NH_{3(g)} \tag{33}$$

algebraically, this reaction may be rearranged as follows:

$$0 = 2NH_{3(g)} - 3H_{2(g)} - N_{2(g)} \tag{34}$$

whereupon a general, balanced equation of any chemical reaction may be written as follows:

$$0 = \sum v_i A_i \tag{35}$$

where the v_i are the coefficients of the species A_i. They are, by definition, always negative for the reactants and positive for the products. This is similar to Equation 30.

At the beginning of any reaction, the number of moles of the various species A_i are n_i^0, where i covers over all reactants and products. (The case of $n_i^0 = 0$ for all products is neglected.) It is now convenient to define a parameter called the *extent of reaction* or *reaction variable*. It is given the symbol ξ. It is defined by the observation that at any point in the reaction the change in the moles of species A_i has been ξv_i. That is, the species with positive v_i (products), ξv_i moles have been formed, whereas the negative quantities (reactants) $\xi|v_i|$ moles have disappeared. The unit of the ξ is the mole, since the v_i is a number. At any point in the reaction the number of moles of A_i is

$$n_i = n_i^0 + v_i\xi \tag{36}$$

Of course ξ is zero at the start of any reaction and reaches 1.0 when many moles of reactants as indicated by their coefficients $|v_i|$ have reacted to form products. The *extent of the reaction* may be smaller or larger than 1, which depends, of course, on the numbers of moles of reactants and products. More important, however, is that an increase of ξ means progress of the reaction to the right, whereas a decrease indicates that the reaction is proceeding to the left with the moles of reactants increasing at the expense of the products. An infinitesimal change of ξ corresponds to the change of moles:

$$dn_i = v_i d\xi \tag{37}$$

The free energy of a system of reactants and products (from Equations 30 and 36) at constant temperature and pressure is

$$G = \sum_i n_i u_i = \sum_i (n_i^0 + v_i\xi)u_i \tag{38}$$

When ξ is changed by $d\xi$, the change in G is

$$dG = \sum_i v_i u_i d\xi \tag{39}$$

Specifically, dG must be zero for *equilibrium* when P and T are constant, which requires

$$\sum_i v_i u_i = 0 \tag{40}$$

since $d\xi$ is not zero, by definition. This is the important *general equilibrium condition*, at constant P and T, for chemical reactions. Equation 40 can be expanded to a general case:

$$\sum v_i u_i = \Delta G = \sum G_{PROD} - \sum G_{REACT} \qquad (41)$$

whereupon the condition that $\Delta G = 0$ the reaction is at equilibrium conditions.

Since all of the species, A_i, in Equation 33 are gases, the u_i values are given in Equation 31 and

$$\Delta G = \sum_i v_i u_i = \sum_i v_i u^0 + RT \sum_i v_i \ln P_i \qquad (42)$$

Consequently, the first term on the right-hand side of Equation 42 becomes ΔG^0, the increase in free energy that would be incurred if the reactants at *standard conditions* were changed into products at *standard conditions*. This increase is to an extent indicated by the coefficients of the reaction in Equation 33. The change in the number of moles of the species A_i would be v_i. This ΔG^0 is denoted also the *standard free energy of the reaction*.

At any given point in a reaction, a mass-action quotient, Q, may be written. This Q value has the same mathematical form as K_{eq}, but is not limited to the condition of equilibrium. It reflects the composition of the reaction system. For the ammonia reaction (Equation 33), the Q would be

$$P_1^{v_1} \cdot P_2^{v_2} \ldots = \frac{P_{NH_3}^2}{P_{H_2}^3 P_{N_2}} \qquad (43)$$

whereupon Equation 42 becomes

$$\Delta G = \sum_i v_i u_i = \Delta G^0 + RT \ln Q \qquad (44)$$

From Equation 40, this becomes zero at equilibrium, so that

$$RT \ln Q = -\Delta G^0(T) \qquad (45)$$

At a constant T, the right-hand size is constant as well as RT. Therefore:

$$Q = K_{eq}(T) \tag{46}$$

Since this is the mass-action law and $K_{eq}(T)$ is the mass-action constant, Equation 45 may be stated as follows:

$$\Delta G^0(T) = -RT \ln K_{eq}(T) \tag{47}$$

or

$$K_{eq}(T) = \exp[-\Delta G^0(T)/RT] \tag{48}$$

Consequently, Equations 47 and 48 are two important relationships between K_{eq} and *the standard free energy of the reaction.*

Figure 2 [1] shows the typical behavior of the free energy of a reaction

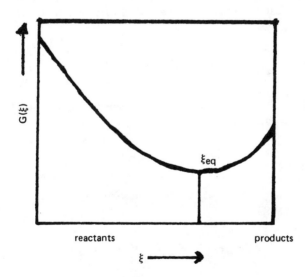

Figure 2.   The free energy of a reaction mixture [1]. As the reaction proceeds from left to right, the concentrations of the reagents change in such a way that the free energy of the system is lowered. Equilibrium is established when the free energy has reached its minimum value. Starting with the products, the reaction would proceed toward the left until the same minimum free energy has been reached. Reproduced from Waser [1] courtesy of W. A. Benjamin, Inc.

mixture in dependence of the extent of the reaction. The free energy of the system assumes a minimum value for equilibrium, ξ_{eq}.

General Solution Reactions

The above discussion applies to any type of chemical equilibrium so far as this general condition exists:

$$\Delta G = \sum_i v_i u_i = 0 \tag{49}$$

as well as Equations 47 and 48. An excellent example for water chemistry may be cited:

$$Mn^{2+} + Cl_{2(g)} + 2H_2O = 4H^+ + MnO_{2(s)} + 2Cl^- \tag{50}$$

Equation 50 involves ions, a gas, a solid and a liquid (H_2O), all of which are in equilibrium with the solution. For the ions, the chemical potentials depend on their concentrations (or activities), as shown by Equation 32. For H_2O, the chemical potential may be assumed to be approximately constant and equal to pure water:

$$uH_2O \approx const \approx u^oH_2O = G_f^o H_2O$$

At equilibrium, the chemical potential of dissolved Cl_2 is equal to that of Cl_2 in the gas phase, whereupon

$$uCl_2 = u^oCl_2 + RT \ln P_{Cl_2}$$

Finally, the chemical potential of $MnO_{2(s)}$ in solution, at equilibrium, is equal to "pure" solid MnO_2:

$$uMnO_{2(s)} = u^o MnO_{2(s)} = G^o MnO_{2(s)}$$

Combining the chemical potentials in Equation 50,

$$\Delta G = \sum_i v_i u_i = 4uH^+ + uMnO_{2(s)} + 2uCl^- - uMn^{2+} - uCl_{2(g)} - 2uH_2O \tag{51}$$

or,

$$\Delta G = 4u^o H^+ + u^o MnO_{2(s)} + 2uCl^- - u^o Mn^{2+} - u^o Cl_{2(g)} - 2u^o H_2O$$
$$+ RT(4 \ln [H^+] + 2 \ln [Cl^-] - \ln [Mn^{2+}] - \ln P_{Cl_2}). \tag{52}$$

Equation 52 is equal to Equation 44, so that $\Delta G = 0$ leads to the mass-action law expressed in Equations 46, 47 and 48, whereupon

$$Q = \frac{[H^+]^4 [Cl^-]^2}{[Mn^{2+}] PCl_2} \tag{53}$$

Once more, the equilibrium condition $\Delta G = 0$ leads to the mass-action law, $Q = K_{eq}(T)$, and to $\Delta G^0(T) = -RT \ln K_{eq}(T)$.

Equation 53 suggests several conventions adopted for the mass-action expression, of which some have been expressed in Equation 1. Other conventions are:

1. The $Cl_{2(g)}$ term must be expressed in its partial pressure because the standard state refers to 1 atm.
2. The concentration must be used for the dissolved species because the standard state refers to a concentration of 1 mol.
3. Solids do not appear in Q or K_{eq} because their chemical potentials are constant.
4. H_2O does not appear in Q for dilute solutions because its concentration and chemical potential remain essentially constant in the considered reaction.

Example calculations of u^0, ΔG^0 and $K_{eq}(T)$ are given below; however, it should be indicated that $K_{eq}(T)$ does, indeed, carry a mathematical unit of expression. For Equation 53, if $Q = K_{eq}(T)$, the unit is $(mol/l)^5/atm$. Other examples are given throughout this text.

It must be noted that the exposition in the above two sections are not derivations of the mass-action law from first principles but rather depend on the specific form of the chemical potentials given in Equations 31 and 32. Although the equilibrium condition expressed in Equation 40 is exact, the expressions used for the chemical potentials are valid only for perfect gases and dilute solutions. Nonetheless, these expressions can be made exact by replacing concentrations by activities and pressures by fugacities. The reader is referred to the literature [1,2] for a fuller development of these two points.

Standard Free Energy and the Equilibrium Constant

The relationship between ΔG^0_{Rex} and K_{eq} is

$$\Delta G^0 = -RT \ln K_{eq} = -1.364 \log K_{eq} \tag{47}$$

Interpretation of Equation 47 is:

1. When ΔG^0 is $-$, then $K_{eq} > 1$ and the left to right reaction is spontaneous. It follows that the larger the K_{eq} value, the more negative is the ΔG^0 value.
2. When ΔG^0 is $+$, then $K_{eq} < 1$ and the left to right reaction is not spontaneous. Another interpretation is that the right to left reaction is preferred and is spontaneous. It follows that the smaller the K_{eq} value, the more positive is the ΔG^0 value.
3. When ΔG^0 is 0, then $K_{eq} = 1$ and the reactants and products are in equilibrium under standard-state conditions. Consequently, if the ΔG^0 values are known for the reactants and products, it becomes a simple mathematical exercise to calculate the K_{eq} value.

Standard-state conditions are stressed in this paragraph; however, the reader should be aware that Equation 47 may be employed to calculate ΔG^0 from K_{eq} values obtained from nonstandard-state conditions, provided that the equilibrium concentrations or partial pressures of the reactants and products are rigorously known.

Two additional points should be made about the fundamental relations expressed in Equations 29 and 47. Firstly, all the coefficients, whether whole or fractional numbers of the reactants and products, do not change Equation 47. This is indicated in Equation 41. Secondly, if a reaction mixture is at equilibrium with respect to several reactions, it is at equilibrium also with respect to all reactions. This has the ramification that when the balanced equations of any two reactions, for example, are multiplied by coefficients and added, the respective ΔG^0 values combine in the same manner. As a consequence of Equation 47, the logarithms of the K_{eq} values also combine in the same manner. Consequently, the equilibrium constant of the combined reaction is the product of the constant of the original reactions raised to the appropriate powers. For example (HAc = acetic acid):

$$HAc = H^+ + Ac^- \qquad\qquad K_{eq1} = 10^{-4.75} \qquad\qquad (54)$$

$$\underline{H_2O + H^+ = H_3O^+ \qquad\qquad K_{eq2} = 1 \qquad\qquad\qquad (55)}$$

$$HAc + H_2O = H_3O^+ + Ac^- \qquad K_{eq3} = ? \qquad\qquad\quad (56)$$

Since $K_{eq3} = K_{eq1} \cdot K_{eq2} = 10^{-4.75} \times 1 = 10^{-4.75}$.

Important by-products have emerged from the thermodynamic development of the mass-action law from expressions for the chemical potentials. Equations 47 and 48 are significant for three reasons:

1. They permit derivation of the temperature dependence of K_{eq} (discussed in the next section).
2. They indicate that ΔG^0 values may be determined experimentally by measuring equilibrium constants.
3. Equilibrium constants and/or ΔG^0 values may be calculated for innumerable reactions to denote the left to right spontaneity, which is one of our concerns in water chemistry.

Effect of Temperature

Water chemistry reactions in nature rarely, if ever, occur at 25°C. For example, ground water temperatures are recorded frequently at 15°C. In computation of the spontaneity of a reaction at nonstandard-state temperatures, it is necessary to use the appropriate equilibrium constants and ΔG^0 values. The van't Hoff relation relates temperature and the equilibrium constant. If one combines Equations 25 and 47 under standard-state conditions, the following expression is obtained:

$$\ln K_{eq} = \frac{-\Delta H^0}{RT} + \frac{\Delta S^0}{R} \tag{57}$$

In application of Equation 57, it is assumed that ΔH^0 and ΔS^0 values remain constant and are independent of temperature over the range of interest. It follows, then, that a plot of $\ln K_{eq}$ vs $1/T$ should be linear whose slope is ΔH^0. If the left-hand side of Equation 57 is integrated between K_1 and K_2 and the right-hand side is integrated between T_1 and T_2, one obtains

$$\log \frac{K_{eq2}}{K_{eq1}} = \frac{-\Delta H^0}{2.303\ R} \left(\frac{T_1 - T_2}{T_1 T_2} \right) \tag{58}$$

Therefore, if we wish to calculate, for example, K_{eq2} at T_2, it is a simple matter provided, of course, that ΔH^0_{Rex} is known. It follows also that ΔG^0_{Rex} at T_2 may be calculated from Equation 47. A more frequent use of Equation 58 is to calculate ΔH^0_{Rex} after measurements of K_{eq1} and K_{eq2}.

Temperature also affects the value of ΔG. The Gibbs–Helmholtz equation permits calculation of ΔG at any temperature provided, of course, that it is known at one temperature. From Equation 25, the following expression may be derived (at constant pressure):

$$\Delta G = \Delta H + T \frac{\partial \left(\frac{\Delta G}{T} \right)_P}{\partial T} \tag{59}$$

(A more thorough discussion of the Gibbs–Helmholtz equation is given by Everdell [2].) A more convenient form of Equation 59 may be obtained by rearrangement and dividing through by T^2:

$$\frac{\partial \left(\frac{\Delta G}{T} \right)_P}{\partial T} = -\frac{\Delta H}{T^2} \tag{60}$$

From Equation 60 it follows that

$$\int d \left(\frac{\Delta G}{T}\right)_P = -\int \frac{\Delta H}{T^2} \, dT \qquad (61)$$

If ΔH is assumed constant over the temperature range of interest, one obtains, on integration:

$$\Delta G \simeq \Delta H + CT \qquad (62)$$

where C is the integration constant. Therefore, if ΔG is known at any one temperature, C may be evaluated so that ΔG at any other temperature may be calculated. Equation 62 must be regarded as an approximation, and the temperature range over which it holds is limited where ΔH and C do not change appreciably. If one substitutes Equation 47 into Equation 62, it follows that

$$\ln K_{eq} = -\frac{\Delta H}{RT} + \frac{C}{R} \qquad (63)$$

Note the similarity between Equations 57 and 63. An approximation of C may be obtained from ΔS.

Effect of Pressure

Consider this system in equilibrium:

$$CaCO_{3(s)} = CaO_{(s)} + CO_{2(g)} \qquad (64)$$

At equilibrium, the chemical potentials are

$$uCaCO_{3eq} = uCaO_{eq} + uCO_{2eq} \qquad (65)$$

It has been noted above that the chemical potential of a pure solid is dependent on the temperature and pressure but is independent of the quantity present. Consequently, $uCaCO_{3eq} = u^0CaCO_3$, $uCaO_{eq} = u^0CaO$ and, if the behavior of $CO_{2(g)}$ is assumed to be ideal (from Equation 31),

$$uCO_{2eq} = u^0CO_2 + RT \ln P_{CO_2} \qquad (66)$$

At equilibrium,

$$RT \ln P_{CO_{2(g)}} = u^0 CaCO_3 - u^0 CaO - u^0 CO_2 \tag{67}$$

whereupon

$$P_{CO_{2(g)}} = K_{eqp}(T,P) \tag{68}$$

Consequently, the K_{eqp} is shown to be a function of pressure as well as temperature. Everdell [2] states that this dependence on pressure is of relatively little importance because there are small differences between the sum of the partial molar volumes of reactants and products of condensed systems at higher pressures. Stumm and Morgan [5] present data to suggest that there are pressure effects on chemical equilibria in deep oceans and aquifers.

Relationship Between K_p and K_c

An equilibrium constant, K_c, can be derived in terms of molar concentrations for ideal gaseous systems. The ideal gas law is

$$PV = nRT \tag{69}$$

where n = the number of moles and R = the gas law constant. This equation may be rewritten as follows:

$$P = \left(\frac{n}{V}\right) RT = CRT \tag{70}$$

where C = the concentration in mol/l. Therefore,

$$P_i = C_i RT \tag{71}$$

for i number of gases.
From Equation 46 it may be stated:

$$RT \ln P_{eq} = RT \ln K_p \tag{72}$$

where P_{eq} = the equilibrium partial pressures of the ideal gases. Whereupon

$$\ln P_{eq} = \ln K_p \tag{73}$$

Substituting,

$$\ln(C_i RT) = \ln K_p \tag{74}$$

or

$$\ln C_i + \ln RT = \ln K_p \tag{75}$$

Rearranging and letting C_i come to their equilibrium values,

$$\ln C_i = \ln K_p - \ln RT = \ln K_c \tag{76}$$

or

$$\ln K_c = \Sigma v_i \ln C_i \tag{77}$$

Expanding Equation 76 to include v_i of gases,

$$\ln K_p = \Sigma v_i \ln C_i + \Sigma v_i \ln(RT) \tag{78}$$

or

$$K_p = K_c(RT)^{\Sigma v_i} \tag{79}$$

Equation 77 is valid for dilute solutions where ideal behavior is observed and the conditions of Equations 45 and 46 are met.

APPLICATIONS OF THERMODYNAMICS

Example Calculations

Values of ΔH^0, ΔS^0 and ΔG^0 are available in most thermodynamic textbooks and in the literature references [1,2,6,7]. A source of special note is

U.S. Geological Survey Bulletins 1259 and 1452, which contain the thermodynamic properties of minerals and related substances [6,7]. Other special sources are those of Latimer [8], Krauskopf [9] and Rossini et al. [10]. A few values are given in Table II.

Since the above thermodynamic equations are rather easy to apply, only a couple of examples are cited here. For Equation 29 we may use

$$Ca^{2+}_{(aq)} + CO^{2-}_{3(aq)} = CaCO_{3(s)} \tag{80}$$

ΔG^0 values: -132.18 -126.22 -269.78

$\Delta G^0_{Rex} = -269.78 - (-132.18 \ -126.22) = -11.38 \ kcal/mol$

Since the ΔG^0_{Rex} is negative, we can state that the left to right reaction is spontaneous, that is, Ca^{2+} and CO_3^{2-} ions in aqueous solutions combine to form $CaCO_3$ in a solid phase. Another interpretation of this reaction would be that the right to left reaction would yield a positive ΔG^0 value and that this reaction would not be spontaneous. However, the right to left reaction occurs to some extent as witnessed from the slight solubility of $CaCO_3$ in water.

For Equation 47 we may cite the following:

$$HOCl + H_2O = H_3^+O + OCl^- \tag{81}$$

ΔG^0 -19.110 -56.69 -56.69 -8.9

$\Delta G^0_{Rex} = (-56.69 \ -8.9) - (-19.110 \ -56.69) = +10.20 \ kcal/mol$

$$\log K_{eq} = \frac{10.20}{-1.364} = -7.48; K_{eq} = 10^{-7.48}$$

Thus the protolysis constant for hypochlorous acid has been calculated from the free energy values of the individual constituents in Equation 81. Since K_{eq} is less than 1, the protolysis of HOCl is not spontaneous and occurs only to a small extent.

The Phase Rule of Gibbs

J. Willard Gibbs [11] deduced the Phase Rule to define the conditions of equilibrium for heterogeneous systems. Gibbs regarded a chemical system as possessing only three independent variables: temperature, pressure and concentrations of the components. The Phase Rule attempts to relate the number of phases and the components of a system through the following mathematical relation:

Table II. Values of ΔH^0, ΔG^0 and ΔS^0 of Some Substances of Interest
in Water Chemistry[a]

Substance	ΔH^0 (kcal/mol)	ΔG^0 (kcal/mol)	ΔS^0 (cal/mol·deg)
$Br^-_{(aq)}$	−28.90	−24.57	19.29
$Br_{2(l)}$	0.00	0.00	36.4
$CH_{4(g)}$	−17.89	−12.14	44.50
$CO_{2(g)}$	−94.05	−94.26	51.06
$CO_{3(aq)}^{2-}$	−161.63	−126.22	−12.7
$H_2CO_{3(aq)}$	−167.0	−149.0	45.7
$HCO_{3(aq)}^-$	−165.18	−140.31	22.7
$HCOOH_{(aq)}$	−98.0	−85.1	39.1
$HCOO^-_{(aq)}$	−98.0	−80.0	21.9
$Ca^{2+}_{(aq)}$	−129.77	−132.18	−13.2
$CaCO_3$ (calcite)	−288.45	−269.78	22.2
$CaO_{(s)}$	−151.9	−144.4	9.5
$Ca(OH)_{2(s)}$	−235.80	−214.33	18.2
$Cl^-_{(aq)}$	−40.02	−31.35	13.17
$Cl_{2(g)}$	0.00	0.00	53.29
$Cu^{2+}_{(aq)}$	15.39	15.53	−23.6
$Fe^{2+}_{(aq)}$	−21.0	−20.30	−27.1
$Fe^{3+}_{(aq)}$	−11.4	−2.52	−70.1
$H_2O_{(g)}$	−57.80	−54.65	45.11
$H_2O_{(l)}$	−68.32	−56.69	16.72
$HS^-_{(aq)}$	−4.22	3.01	14.6
$S^{2-}_{(aq)}$	10.0	20.0	5.3
$I^-_{(aq)}$	−13.37	−12.35	26.14
$I_{2(g)}$	14.88	4.63	62.28
$K^+_{(aq)}$	−60.04	−67.47	24.5
$Mg^{2+}_{(aq)}$	−110.41	−108.99	−28.2
$Mn^{2+}_{(aq)}$	−52.3	−53.4	−20.0
$MnO_{4(aq)}^-$	−129.7	−107.4	45.4
$NH^+_{4(aq)}$	−31.74	−19.00	26.97
$Na^+_{(aq)}$	−57.28	−62.59	14.4
$OH^-_{(aq)}$	−54.96	−37.60	−2.52
$SO^{2-}_{4(aq)}$	−216.90	−177.34	4.1

[a]From Waser [1].

$$F = C - P + 2 \qquad\qquad (82)$$

where F = the number of degrees of freedom or the number of variable factors that must be arbitrarily fixed to determine a third and, therefore, to define the system

C = the number of components

P = the number of phases, and the integer, 2, is added to the equation for the number of independent variables of the system that are fixed or varied as necessary.

To define a system completely, it is necessary to have as many equations as variables. If this cannot be achieved, one or more of the variables will have an undefined value. The number of these undefined values also gives the degrees of freedom or the F-value of the system.

To apply the Phase Rule to a heterogeneous system, the number of phases and components must be selected. Let us take the easy one first: the number of phases. Those homogeneous, physically distinct and mechanically separable portions of a system are called phases. For example, three phases may be distinguished for the same substance—water, namely, ice, water and vapor. Several phases may coexist and may vary greatly in different systems. In general, it is a relatively simple matter to distinguish the number of phases in an aqueous chemical system.

However, it is often more difficult to distinguish the number of components of a system. Components are not synonymous with chemical elements or compounds present, although they may be. The choice of the number of components is restricted to those constituents that can undergo independent variation in concentration in the different phases. An additional restriction is placed, namely, there must be chosen the least number of independently variable constituents by which the composition of each phase can be expressed in the form of a chemical equation. For example, $CaCO_{3(s)} = CaO_{(s)} + CO_{2(g)}$. There are two phases: solid and gaseous. How many components? There are two and not three. Why? $CaCO_{3(s)}$ and $CaO_{(s)}$ are constituents of the solid phase, whereas $CO_{2(g)}$ is the gaseous phase. The concentration of $CaCO_{3(s)}$ in the solid phase depends on the concentration of $CaO_{(s)}$. On the other hand, $CaO_{(s)}$ can vary independently of $CaCO_{3(s)}$ in the solid phase. In the gaseous phase, the concentration of $CO_{2(g)}$ varies independently of both $CaCO_{3(s)}$ and $CaO_{(s)}$. Consequently, $CaO_{(s)}$ and $CO_{2(g)}$ are chosen as the number of components of the system.

Two examples of the phase rule may be cited:

Case 1 $CO_{2(g)} - H_2O_{(l)}$

$CO_{2(g)} + H_2O_{(l)} = H_2CO_{3(aq)}$

$C = 2, CO_{2(g)}, H_2O_{(l)}$

P = 2, gas, liquid

F = 2 - 2 + 2 = 2

There are two degrees of freedom, but there are three internal variables in the system: temperature, pressure and $H_2CO_{3(aq)}$ concentration. If temperature and pressure of $CO_{2(g)}$ are chosen then the aqueous concentration of H_2CO_3 is established and cannot vary.

Case 2 $CaO_{(s)} - CO_{2(g)} - H_2O_{(l)}$

$CaCO_{3(s)} + CO_{2(g)} + H_2O = Ca^{2+}_{(aq)} + 2HCO^-_{3(aq)}$

$C = 3$, CaO, CO_2, H_2O

$P = 3$, solid, liquid, gas

$F = 3 - 3 + 2 = 2$

There are two degrees of freedom, but there are several internal variables: temperature, pressure, $[Ca^{2+}]$, $[HCO^-_3]$, $[H^+]$, $[CO^{2-}_3]$, $[H_2CO_3]$ and $[OH^-]$. If temperature and pressure of $CO_{2(g)}$ are chosen, then the aqueous concentrations of the dissolved constituents are fixed and cannot vary. In this example, more degrees of freedom may be achieved if the three components are considered in only one phase. Consequently, $F = 3 - 1 + 2 = 4$. Now there are four internal variables. If pressure and temperature are selected as two, then variation is permitted in two of the remaining variables. It is the general observation that as the number of phases is increased with a fixed number of components, the degrees of freedom decrease and more constraints are placed on the system.

Greater detail about the Phase Rule may be obtained from Alexander Findlay's book *The Phase Rule and Its Applications* [12].

Solubility of Gases in Water

There are several of the atmospheric gases that play rather significant roles in water quality. Among these gases are oxygen, carbon dioxide, hydrogen sulfide, nitrogen, methane and hydrogen. The mean composition of the atmosphere is seen in Table III. The quantitative relation between solubility of a gas in water and pressure was stated by Henry in 1803: the mass of gas dissolved by a given volume of solvent at constant temperature is proportional to the pressure of the gas with which it is in equilibrium. In mathematical form, Henry's law is

$$m = kP \qquad (83)$$

where m is the mass of gas dissolved by unit volume of solvent at the equilibrium pressure P, and k is a constant. If the pressure of the gas is increased,

Table III. Mean Composition of the Atmosphere

Gas	Percentage by Volume	Partial Pressure (atm)
N_2	78.1	0.781
O_2	20.9	0.209
Ar	0.93	0.0093
H_2O	0.1–2.8	0.001–0.028
CO_2	0.03	0.0003
Ne	1.8×10^{-3}	1.8×10^{-5}
He	5.2×10^{-4}	5.2×10^{-6}
CH_4	1.5×10^{-4}	1.5×10^{-6}
Kr	1.1×10^{-4}	1.1×10^{-6}
CO	$(0.06-1) \times 10^{-4}$	$(0.06-1) \times 10^{-6}$
SO_2	1×10^{-4}	1×10^{-6}
N_2O	5×10^{-5}	5×10^{-7}
H_2	$\sim 5 \times 10^{-5}$	$\sim 5 \times 10^{-7}$
O_3	$(0.1-1.0) \times 10^{-5}$	$(0.1-1.0) \times 10^{-7}$
Xe	8.7×10^{-6}	8.7×10^{-8}
NO_2	$(0.05-2) \times 10^{-6}$	$(0.05-2) \times 10^{-8}$
Rn	6×10^{-18}	6×10^{-20}

the mass dissolved will increase in the same proportion. Henry's law may be expressed in another manner. The mass of gas dissolved per unit volume of water may be expressed in concentration units of moles per liter. Also, the pressure, P, of the gas is proportional to the number of molecules in the gas space which, in turn, is also proportional to the concentration (mol/l) of gas phase. Hence:

$$\frac{\text{Concentration of gas in water}}{\text{Concentration of gas in gaseous phase}} = \text{Constant} \qquad (84)$$

at a definite temperature. This constant is independent of pressure. Henry's law generally is obeyed by gases of low solubility if the pressures are not excessively high or the temperatures too low. This is generally the situation for the gaseous systems of concern in water quality chemistry. There are some deviations from Henry's law in aqueous systems due to such factors as ionization and compound formation between solute and water. Ammonia is a classic example:

$$NH_3 + H_2O = NH_4^+ + OH^- \qquad (85)$$

Frequently, Henry's law is expressed in terms of mole fractions:

$$\frac{N_2}{N_1 + N_2} = XH_a = P \tag{86}$$

where N_1 = moles of solvent (or water)
   N_2 = moles of solute
   X = mole fraction of the solute in the liquid phase
   H_a = the Henry law constant expressed in units of atm/mol fraction
   P = the pressure in atmospheres.

Table IV gives some of the Henry law constants for several gases for temperatures 0–100°C.

Another form of Henry's law may be employed. The concentration of a gas in a gaseous phase may be expressed:

$$[Gas_{(g)}] = P_{gas}/RT \tag{87}$$

where P_{gas} is the partial pressure of the gas, T is the temperature in degrees Kelvin and R is the ideal gas law constant expressed in liter atm/deg/mol. Substituting into Equation 83,

$$[Gas_{(aq)}] = \frac{Constant}{RT}P_{gas} \tag{88}$$

where now Constant/RT = K_H which is expressed in mol/l/atm. The Constant from Equation 88 is dimensionless and is frequently called the distribution constant. Conversion from H_a to K_H values may be accomplished by

Table IV. Henry's Law Constants for Various Gases in Water

T (°C)	$H_a \times 10^{-4}$, atm/mol fraction									
	Air	CO_2	CO	C_2H_6	H_2	H_2S	CH_4	NO	N_2	O_2
0	4.32	0.0728	3.52	1.26	5.79	0.0268	2.24	1.69	5.29	2.55
10	5.49	0.104	4.42	1.89	6.36	0.0367	2.97	2.18	6.68	3.27
20	6.64	0.142	5.36	2.63	6.83	0.0483	3.76	2.64	8.04	4.01
30	7.71	0.186	6.20	3.42	7.29	0.0609	4.49	3.10	9.24	4.75
40	8.70	0.233	6.96	4.23	7.51	0.0745	5.20	3.52	10.4	5.35
50	9.46	0.283	7.61	5.00	7.65	0.0884	5.77	3.90	11.3	5.88
60	10.1	0.341	8.21	5.65	7.65	0.103	6.26	4.18	12.0	6.29
70	10.5		8.45	6.23	7.61	0.119	6.66	4.38	12.5	6.63
80	10.7		8.45	6.61	7.55	0.135	6.82	4.48	12.6	6.87
90	10.8		8.46	6.87	7.51	0.144	6.92	4.52	12.6	6.99
100	10.7		8.46	6.92	7.45	0.148	7.01	4.54	12.6	7.01

$$K_H = \frac{[Gas_{(aq)}]^*}{XH_a} = \frac{mol/l}{\frac{mol\ fraction \times atm}{mole\ fraction}}$$

(89)

*Saturation concentration

K_H values will be given through this volume wherever appropriate.

Ionic Strength and Dissolved Solids

The active masses of cations and anions in solution are somewhat less than the analytically measured and reported concentrations. This ionic activity is due to electrostatic effects of ions repelling and attracting each other and of the dielectric effects of the solvent which, in this case, is water. Consequently, ions participate in chemical reactions in proportion to their activities, which are computed from concentrations by multiplication with a correction factor, the so-called activity coefficient (given the symbol γ). In turn, the activity coefficient is computed from the ionic strength (a measure of the strength of the electrostatic field caused by the ions).

The computation from the mathematical formulas of activity coefficients and ionic strengths may be tedious and time-consuming. Consequently, nomographic and graphic solutions have been devised and reported by Hem [13]. These solutions are less accurate than direct computation from the mathematical formulas and should be employed only where approximations are required.

The nomographic plate (Figure 3) of Hem is used in the following manner for computation of ionic strength. The right- and left-hand scales are identical and represent the individual ionic increments of ionic strength. A transparent straight edge is laid on the nomograph horizontally, which intersects the concentration value of the given ion. The individual ionic contribution to the total ionic strength is read on either the right- or left-hand scales. This value is recorded. This procedure is repeated for the other ions that affect the total ionic strength significantly. Then the total ionic strength is determined simply by adding the increments together. The results are accurate to about two significant figures. The appropriate activity coefficient is calculated now from the graph given in Figure 4.

THERMODYNAMICS AND NATURAL SYSTEMS

Conceptual versus Natural Systems

A chemical system is that portion of the "real" world under thermodynamic inspection and description. Imaginary boundaries separate the

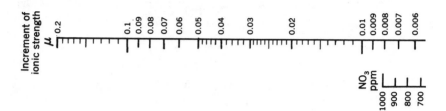

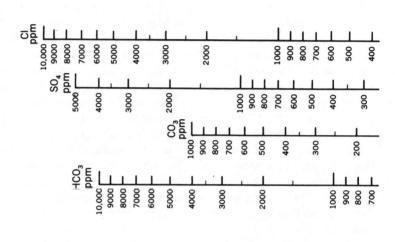

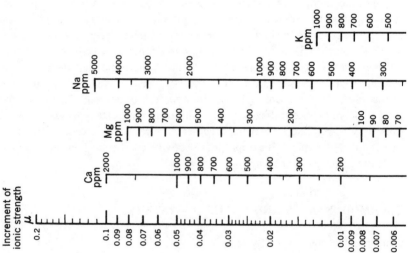

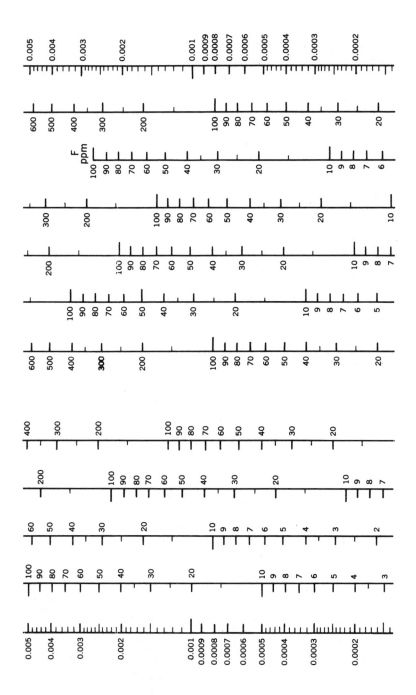

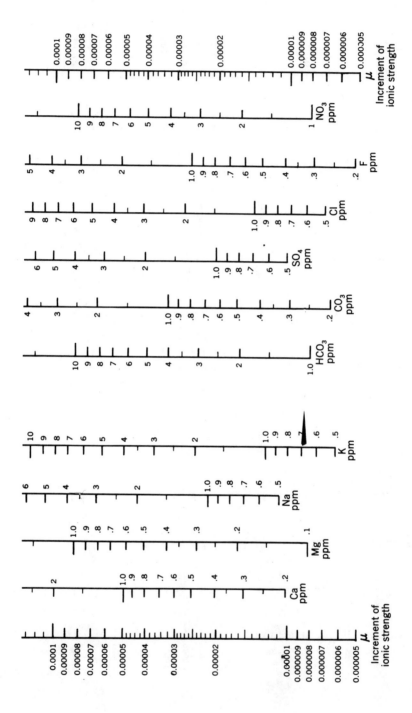

Figure 3.   Nomograph for computing ionic strength of natural waters [13].

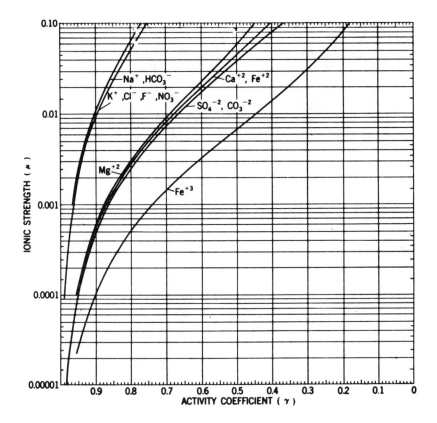

Figure 4.   Relation of activity coefficients for important ions in natural water to ionic strength of solutions [13].

system from its surroundings. This is important because it leads to the designation of three types of systems: isolated, closed and open. Thermodynamically, isolated systems are not permitted to exchange heat, work or matter with their surroundings. Closed systems are permitted to exchange energy, but not matter. Open systems are permitted to exchange energy and matter with their surroundings. There are no limitations of size to systems. They may be small and confined to 25-ml reaction vessels or they may be as large as a lake or an ocean.

Difficulties are encountered here with the application of thermodynamic equilibrium concepts to natural aquatic environments. The relation between the equilibrium constant and the standard free energy change cited above in Equation 47 was derived for constant temperature and pressure conditions of a closed system. One question immediately arises: do natural aquatic systems meet the thermodynamic criteria of a closed system? The answer is simply no. Natural systems are very seldom closed in the strict thermodynamic sense. No doubt there are some natural systems that are sufficiently self-contained and separated from other parts of the world that could be considered closed. For the most part, however, energy and matter are constantly being transported across the boundaries of the natural system in and out of its surroundings. In this case, the "open" designation should be invoked and dynamic models should be employed to describe natural systems.

Phases

To describe a thermodynamic system rigorously, the composition of the three phases of nature must be known. Generally it is not difficult to obtain accurate information about the gaseous, aqueous and solid phases of a conceptual or carefully prepared synthetic thermodynamic system. In many cases it is extremely difficult to distinguish between several solid phases in natural aquatic systems. For example, ground water systems or the bottoms of lakes may have several kinds of solid phases that would, in turn, affect the chemical composition of the aqueous phase. Furthermore, it may be difficult to separate and identify solid phases to assign standard free energies of formation to them. Also, it may be difficult to justify that certain geologic formations are separate and discrete solid phases. The inability to account for the composition of a natural system will lead, of course, to an inaccurate description of all of the chemical equilibria and reactions.

Complexity

Thermodynamic systems may be constructed to be simple or complex but natural systems are almost always complex. The difficulty with system complexity will be illustrated by the examples provided by Stumm and Morgan [5]. It may be possible to reduce the complexity of a natural system into simple thermodynamic systems for inspection. Figure 5 cites six examples of "simple" thermodynamic systems. System (a) is, perhaps, the simplest model of a system that may not have application in the real world. It is an aqueous solution that is not in contact with any gaseous or solid phases. Chemical reactions are occurring within the boundaries of this system and that energy no doubt is being exchanged across the boundaries. The total mass of this system is fixed, but any change in the constituents of this system must be accomplished by some "external" factor. In any event, chemical reactions and the changes in chemical composition of constituents may be studied in system (a). An aqueous solution in contact with a gas phase is seen in system (b). It is a closed system but the phases are open to each other so

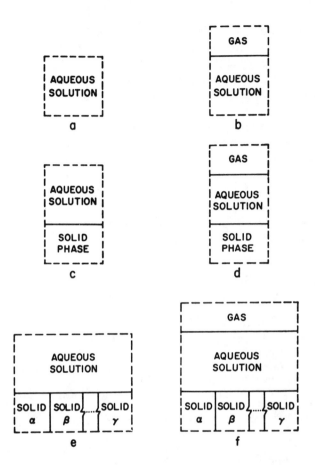

Figure 5. Conceptual thermodynamic systems (reproduced from Stumm and Morgan [5] courtesy of John Wiley & Sons, Inc.).

that matter and energy may be exchanged. Such a system may be exemplified by the exchange of carbon dioxide or oxygen and water molecules across the air–water interface. In natural systems, the reaeration of undersaturated waters with atmospheric oxygen is an appropriate example. System (c) represents a closed situation in which a solid phase and an aqueous solution are in contact. Energy and matter are permitted to be exchanged between the two phases. Such systems are often encountered in the real world. The exchange of phosphorus between bottom muds and eutrophic lake waters may be an example. This implies, however, that only one solid phase is present in the bottom muds, and such systems must not contain such volatile components as carbonates or sulfides. System (d) brings together the three phases of nature: gas, liquid and solid. The implication here, however, is that each phase contains only a single component. This is rarely the situation in the real world, but the simulation of such systems are valuable in the comprehension of natural conditions. The CO_2-H_2O-$CaCO_3$ system is often cited as an example. System (e) begins to approach real world situations where several solid phases are in contact with the aqueous phase. Since no gaseous phase is present, nonvolatile components must exist in the aqueous phase. This type of system is frequently encountered in nature, especially where clay minerals influence the water quality characteristics of surface and ground waters. System (f) is an attempt to model the real world and removes all of the limitations of systems (a) through (e). It is an excellent example of the complexity of several solid phases in contact with the aqueous and gaseous phases. Difficulties lie with the identification and separation of each of the solid phases, as cited above.

REFERENCES

1. Waser, J. *Basic Chemical Thermodynamics* (Menlo Park, CA: W. A. Benjamin, Inc., 1966).
2. Everdell, M. H. *Introduction to Chemical Thermodynamics* (New York: W. W. Norton & Co., 1965).
3. Klotz, I. *Introduction to Chemical Thermodynamics* (Menlo Park, CA: W. A. Benjamin, Inc., 1964).
4. Strong, L. E., and W. J. Stratton. *Chemical Energy* (New York: Van Nostrand Reinhold Company, 1965).
5. Stumm, W., and J. J. Morgan. *Aquatic Chemistry* (New York: John Wiley & Sons, Inc., 1970).
6. Robie, R. A., and D. R. Waldbaum. *Geol. Survey Bull. 1259*, U.S. Government Printing Office, Washington, DC (1968).
7. Robie, R. A., et al. *Geol. Survey Bull. 1452*, U.S. Government Printing Office, Washington, DC (1978).
8. Latimer, W. M. *Oxidation Potentials* (Englewood Cliffs, NJ: Prentice-Hall, Inc., 1952).
9. Krauskopf, K. *Introduction to Geochemistry* (New York: McGraw-Hill Book Co., 1967).
10. Rossini, F. D., et al. "Selected Values of Chemical Thermodynamic Properties," National Bureau of Standard Circ. 500, (1952). Partial revisions in National Bureau of Standards Technical Note 270-3, U.S. Department of Commerce, Washington, DC (1968).
11. Gibbs, J. W. *Trans. Conn. Acad.* III: 108 (October 1875–May 1876); III: 343 (May 1877–July 1878).
12. Findlay, A. *The Phase Rule and Its Applications.* 9th ed. A. N. Campbell and N. O. Smith, Eds. (New York: Dover Publications, Inc., 1951).
13. Hem, J. D. "Calculation and Use of Ion Activity," Geological Survey Water Supply Paper 1535-C, Washington, DC (1961).

CHAPTER 3

ACIDITY, ALKALINITY AND pH

ACID-BASE EQUILIBRIA

Brönsted Acids and Bases

Acidity, alkalinity and pH value are among the most common properties used to characterize water and wastewater. At the same time, these analyses are frequently misunderstood and used erroneously. In the Brönsted concept, an aqueous acid is a proton donor and an aqueous base is a proton acceptor. Furthermore, a proton must be transferred from one Brönsted base to another:

$$HA \; + \; \overline{B} \; = \; HB \; + \; \overline{A} \tag{1}$$

$$\text{Acid}_1 \quad \text{Base}_2 \quad \text{Acid}_2 \quad \text{Base}_1$$

Thus, acid–base reactions become proton transfers from one Brönsted base to another. If one places a proton donor into water, the proton transfer reaction is

$$HA + H_2O = H_3^+O + \overline{A} \tag{2}$$

In this case, the water molecule acts as the proton acceptor and competes with the base $\overline{A}$, for the proton. These proton transfers to the water molecule are called protolysis or protolytic reactions. In this case, the Brönsted acid on the right-hand side of Equation 2 is the hydronium ion, H_3^+O.

93

Protolysis Constant

There have been attempts to label acids or bases as "weak" or "strong." In the Brönsted system it is the tendency to donate or to accept a proton that denotes the strength of an acid or base. "Weak" acids have a weak proton-donating tendency, whereas a strong base would be one that has a strong tendency to accept a proton. Perhaps a better way to order the strengths of acids and bases is by the change in free energy involved in the proton transfer reaction. From Equation 2 one may obtain a protolysis constant:

$$K_a = \frac{[H_3^+O][\overline{A}]}{[HA][H_2O]} \tag{3}$$

In this case, K_a is the equilibrium constant K_{eq} from which the standard free energy change, ΔG^0, may be calculated from Equation 47 (Chapter 2), $\Delta G^0 = -1.364 \log K_a$. If the acid is hypochlorous acid (HOCl), the standard free energy change is $-1.364 \times -7.6 = +10.37$ kcal/mol. On the other hand, if the acid is acetic acid (HAc), the standard free energy change is $-1.364 \times -4.7 = +6.41$ kcal/mol. Thus, one may say that acetic acid is a "stronger" acid than hypochlorous acid because less energy is required to transfer 1 mole of protons from $A\bar{c}$ to H_2O than from OCl^- to H_2O. Conversely, one can say that OCl^- is a "stronger" base than $A\bar{c}$. (It might be noted here that since the ΔG^0 values are + for HOCl and HAc, the left to right reaction is not spontaneous but does occur to some extent.)

Table I gives the pK_a and pK_b values of some acids and bases frequently encountered in water and wastewater chemistry. The Brönsted acid–base equilibrium of water should be noted here:

$$H_2O + H_2O = H_3^+O + OH^- \tag{4}$$

This equation is the so-called autoprotolysis of water. The K_{eq} or K_w is given by

$$K_w = \frac{[H_3^+O][OH^-]}{[H_2O]^2} \tag{5}$$

Because the concentration of water is essentially constant in dilute aqueous solution, the $[H_2O]^2$ term in Equation 5 is given the value of 1. Thus, the K_w of water is $= [H_3^+O][OH^-]$. Likewise, we consider the hydrated proton equilibrium to be

Table I. Acidity and Basicity Constants (25°C) of Interest to Water
and Wastewater Chemistry [1]

Acids	pK_a	Bases	pK_b
HCl	~-3	Cl^-	17
H_2SO_4	~-3	HSO_4^-	17
HNO_3	-1	NO_3^-	15
H_3O^+	0	H_2O	14
H_3PO_4	2.1	$H_2PO_4^-$	11.9
CH_3COOH	4.7	CH_3COO^-	9.3
H_2CO_3	6.3	HCO_3^-	7.7
H_2S	7.1	HS^-	6.9
$H_2PO_4^-$	7.2	HPO_4^{2-}	6.8
HOCl	7.6	OCl^-	6.4
NH_4^+	9.3	NH_3	4.7
H_4SiO_4	9.5	$H_3SiO_4^-$	4.5
HCO_3^-	10.3	CO_3^{2-}	3.7
$H_3SiO_4^-$	12.6	$H_2SiO_4^{2-}$	1.4
H_2O	14	OH^-	0

$$H_2O + proton = H_3O^+ \tag{6}$$

The K_{eq} for this equation is placed by convention to be 1.0 so that $\Delta G^0 = 0$. Consequently, Equation 5 becomes

$$K_w = [H^+][OH^-] \tag{7}$$

At 25°C, $K_w = 1.008 \times 10^{-14}$. For K_w values at other temperatures see *Aquatic Chemistry* [1], p. 76. Furthermore, it can be shown that

$$K_w = K_a \cdot K_b \tag{8}$$

from which K_a or K_b may be calculated provided that one or the other value is given.

Distribution Diagrams—pH as the Master Variable

Cl_2-H_2O System

In accordance with Equation 2, a proton is transferred from the conjugate base, $\bar{A}$, to water. The extent of this transfer is regulated by the K_a expression, as noted in Equation 3. Two observations may be made from

Equation 2: (1) the conjugate base, $\bar{A}$, is distributed between HA and $\bar{A}$; and (2) the H$^+$ ion (symbol H$^+$ used for convenience) concentration affects the extent of this distribution. If one were to prepare a $10^{-4}M$ (5.25 mg/l) aqueous solution of hypochlorous acid (HOCl), a mass balance on OCl$^-$ would be

$$C_t = [HOCl] + [OCl^-] \tag{9}$$

where C_t is the total concentration, mol/l, of HOCl and OCl$^-$. The extent of the distribution of OCl$^-$ between HOCl and OCl$^-$ is regulated by

$$K_a = \frac{[H^+][OCl^-]}{[HOCl]} = 10^{-7.6} \ (25°C) \tag{10}$$

Consequently, the fraction present of HOCl at a given [H$^+$] may by computed from the following:

$$^\alpha HOCl = [HOCl]/C_t \tag{11}$$

Likewise, the fraction present of OCl$^-$ would be

$$^\alpha OCl^- = [OCl^-]/C_t \tag{12}$$

It must be noted, at a given [H$^+$], that

$$^\alpha HOCl + ^\alpha OCl^- = 1.0 \tag{13}$$

By combination of Equations 9 and 10, one obtains

$$^\alpha HOCl = \frac{[H^+]}{K_a + [H^+]} \tag{14}$$

and

$$^\alpha OCl^- = \frac{K_a}{K_a + [H^+]} \tag{15}$$

Thus, these fractions may be calculated at any $[H^+]$. For example, at a pH value of 7.0,

$$\alpha_{HOCl} = \frac{[H^+]}{K_a + [H^+]} = \frac{10^{-7.0}}{2.51 \times 10^{-8} + 10^{-7.0}} = \frac{10^{-7.0}}{1.251 \times 10^{-7}} = 0.800$$

$$\alpha_{OCl^-} = 1.000 - \alpha_{HOCl} = 1.000 - 0.800 = 0.200$$

Therefore, 80% of the solution is HOCl and 20% is OCl^-. If C_t is $10^{-4}M$, then

$$[HOCl] = \alpha_{HOCl} \, C_t = 0.8 \times 10^{-4}M$$

and

$$[OCl^-] = \alpha_{OCl^-} \, C_t = 0.2 \times 10^{-4}M$$

Sometimes it is convenient to show α_{HOCl} and α_{OCl^-} as a function of pH (the master variable) in a distribution diagram that is illustrated in Figure 1.

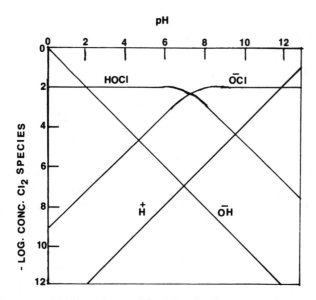

Figure 1.   Concentration distribution diagram for hypochlorous acid.

An excellent and very practical application of the α-value concept may be made with HOCl, OCl$^-$ and the disinfection of water. HOCl and OCl$^-$ differ in their ability to kill microorganisms. Morris [2] has provided an example, which is modified here. If an additive efficiency of disinfection is assumed for HOCl and OCl$^-$, one may produce a theoretical curve for the total amount of chlorine required to give a percentage kill of $E.$ $coli$, for example, for a specified contact time at various pH values. Let R = the total concentration of chlorine required for the stipulated kill, A = the concentration of HOCl necessary to produce this kill (this must be evaluated experimentally at a pH value where $^{\alpha}$HOCl $\cong 1.0$), and B = an efficiency ratio of [HOCl]:[OCl$^-$]. (This [OCl$^-$] would be the concentration necessary to produce the same percentage kill that must be evaluated experimentally at a pH value where $^{\alpha}$OCl$^- \cong 1.0$.) Then,

$$R = [HOCl] + [OCl^-] = C_t\,(^{\alpha}HOCl + {}^{\alpha}OCl^-) \qquad (16)$$

$$A = [HOCl] + B[OCl^-] = C_t\,(^{\alpha}HOCl + B\,^{\alpha}OCl^-) \qquad (17)$$

$$C_t = A/(^{\alpha}HOCl + B\,^{\alpha}OCl^-) \qquad (18)$$

Combining Equations 16 and 17 leads to

$$R = \frac{A}{(^{\alpha}HOCl + B\,^{\alpha}OCl^-)} \qquad (19)$$

For a kill of 99% of $E.$ $coli$ in 30 minutes at 2–5°C at a pH of 8.0, [HOCl] = 0.005 mg/l, [OCl$^-$] = 0.042 mg/l, B = 0.12. At pH 8.0, $^{\alpha}$HOCl = 0.286 and $^{\alpha}$OCl$^-$ = 0.714. Whereupon,

$$R = \frac{0.005}{(0.286 + 0.2 \times 0.714)} = \frac{0.005}{0.372} = 0.013 \text{ mg/l}$$

Thus, a total concentration of 0.013 mg/l Cl$_2$ is required for the stipulated kill. One may extend this to a plot of total chlorine vs pH, as seen in Figure 2. This curve is applicable, of course, only to the conditions cited thereon. One may construct a curve for other temperatures, contact times and other organisms.

CO_2-H_2O System

Carbonic acid (H_2CO_3) is a weak, diprotic acid of some importance in water chemistry. As seen below, total acidity and total alkalinity of aqueous systems are defined around this acid, H_2CO_3, which may be formed from

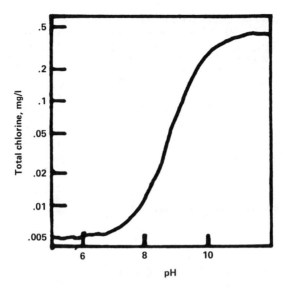

Figure 2.   Concentration of free available chlorine required for 99% kill of *E. coli* in 30 minutes at 2–5°C (reproduced from Morris [2] courtesy of John Wiley & Sons, Inc.).

$$CO_{2(g)} = CO_{2(aq)} \tag{20}$$

and

$$CO_{2(aq)} + H_2O = H_2CO_3 \tag{21}$$

In turn, H_2CO_3 enters into two Brönsted acid–base equilibria:

$$H_2CO_3 = H^+ + HCO_3^- \ ; \quad K_1 = \frac{[H^+][HCO_3^-]}{[H_2CO_3]} \tag{22}$$

$$HCO_3^- = H^+ + CO_3^{2-} \ ; \quad K_2 = \frac{[H^+][CO_3^{2-}]}{[HCO_3^-]} \tag{23}$$

The equilibrium constants for Equations 22 and 23 are given in Table II for several temperatures [3,4]. A mass balance on carbonate species is

$$C_t = [H_2CO_3] + [HCO_3^-] + [CO_3^{2-}] \tag{24}$$

Table II. Acidity Constants for H_2CO_3 and HCO_3^- at Several Temperatures [3,4]

Temperature (°C)	Equation 22 −log K_1	Equation 23 −log K_2
0	6.579	10.625
5	6.517	10.557
10	6.464	10.490
15	6.419	10.430
20	6.381	10.377
25	6.352	10.329
30	6.327	10.290
35	6.309	10.250
40	6.298	10.220
50	6.285	10.172

whereupon the distribution of species as influenced by the $[H^+]$ would be

$$\alpha_{H_2C} = \frac{[H_2CO_3]}{C_t} \tag{25}$$

$$\alpha_{HC^-} = \frac{[HCO_3^-]}{C_t} \tag{26}$$

$$\alpha_{C^{2-}} = \frac{[CO_3^{2-}]}{C_t} \tag{27}$$

These alpha values may be computed from various pH values by combining Equations 22, 23, 24, 25, 26 and 27. The resultant equations are as follows:

$$\alpha_{H_2C} = \frac{[H^+]^2}{[H^+]^2 + K_1[H^+] + K_1 K_2} \tag{28}$$

$$\alpha_{HC^-} = \frac{K_1[H^+]}{[H^+]^2 + K_1[H^+] + K_1 K_2} \tag{29}$$

$$\alpha_{C^{2-}} = \frac{K_1 K_2}{[H^+]^2 + K_1[H^+] + K_1 K_2} \tag{30}$$

Table IA of the Appendix gives the α-values for H_2CO_3, HCO_3^- and CO_3^{2-} that were computed from Equations 28, 29 and 30 using the 25°C K_1 and K_2 values from Table II. Note that $\alpha_{H_2C} + \alpha_{HC^-} + \alpha_{C^{2-}} = 1.0$.

Some useful qualitative interpretations based on the carbonate alpha values may be obtained through the use of a distribution diagram. In this case, the log α-value is plotted against pH as the master variable. To construct this plot it is convenient to divide the pH abscissa into three regions: 0.00–6.35 (I), 6.35–10.33 (II) and 10.33–14.00 (III). The boundaries between regions I and II, and II and III are established by the pK_1 and pK_2 values of H_2CO_3. In region I, the approximation that $[H^+]^2 \gg K_1[H^+] \gg K_1K_2$ in the denominators of Equations 28, 29 and 30 is made whereupon

$$\alpha_{H_2C} \simeq \frac{[H^+]^2}{[H^+]^2} \simeq 1.0 \tag{31}$$

$$\alpha_{HC^-} \simeq \frac{K_1}{[H^+]} \tag{32}$$

$$\alpha_{C^{2-}} \simeq \frac{K_1K_2}{[H^+]^2} \tag{33}$$

In region II, the approximation that $K_1[H^+] \gg [H^+]^2 \gg K_1K_2$ is made whereupon

$$\alpha_{H_2C} \simeq \frac{[H^+]}{K_1} \tag{34}$$

$$\alpha_{HC^-} \simeq \frac{K_1[H^+]}{K_1[H^+]} \simeq 1.0 \tag{35}$$

$$\alpha_{C^{2-}} \simeq \frac{K_2}{[H^+]} \tag{36}$$

In region III, the approximation that $K_1K_2 \gg K_1[H^+] \gg [H^+]^2$ is made whereupon

$$\alpha_{H_2C} \simeq \frac{[H^+]^2}{K_1K_2} \tag{37}$$

$$\alpha_{HC^-} \simeq \frac{[H^+]}{K_2} \tag{38}$$

$$\alpha_{C^{2-}} \simeq \frac{K_1K_2}{K_1K_2} \simeq 1.0 \tag{39}$$

When Equations 28 through 39 are placed in logarithmic form, the plots as seen in Figure 3 result. It should be noted that the above approximations are not valid as the equations approach the pH values equal to pK_1 and pK_2. In this case, Equations 28, 29 and 30 must be solved in full with the result that a curvature exists as the respective lines pass through the pH = pK_1, pK_2

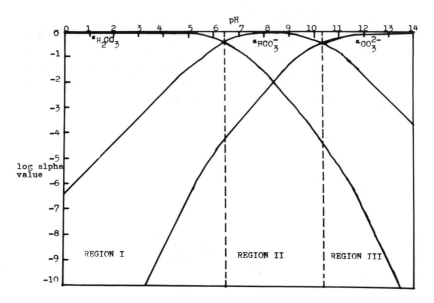

Figure 3.   Alpha value distribution diagram for carbonic acid.

boundaries. It should be obvious that H_2CO_3 is the predominant species in region I, HCO_3^- in region II and CO_3^{2-} in region III. Furthermore, the respective curves of H_2CO_3, HCO_3^- and CO_3^{2-} approach $\alpha = 1.0$ asymptotically in regions I, II and III. More comment and application of the alpha values for the carbonate system are made in the section under total acidity and alkalinity of natural waters.

H_3PO_4 System

Phosphoric acid (H_3PO_4) is a triprotic acid of some importance to water and wastewater chemistry because of the possible role of phosphorus in accelerating eutrophic processes. The appropriate acid–base equilibria are as follows:

$$H_3PO_4 = H^+ + H_2PO_4^- \quad ; \quad K_1 = \frac{[H^+][H_2PO_4^-]}{[H_3PO_4]} \tag{40}$$

$$H_2PO_4^- = H^+ + HPO_4^{2-} \quad ; \quad K_2 = \frac{[H^+][HPO_4^{2-}]}{[H_2PO_4^-]} \tag{41}$$

$$HPO_4^{2-} = H^+ + PO_4^{3-} \quad ; \quad K_3 = \frac{[H^+][PO_4^{3-}]}{[HPO_4^{2-}]} \tag{42}$$

Whereupon the mass balance on phosphate is

$$C_t = [H_3PO_4] + [H_2PO_4^-] + [HPO_4^{2-}] + [PO_4^{3-}] \tag{43}$$

Again, the alpha values for the various phosphate species are

$$\alpha_{H_3P} = \frac{[H_3PO_4]}{C_t} \tag{44}$$

$$\alpha_{H_2P^-} = \frac{[H_2PO_4^-]}{C_t} \tag{45}$$

$$\alpha_{HP^{2-}} = \frac{[HPO_4^{2-}]}{C_t} \tag{46}$$

$$\alpha_{P^{3-}} = \frac{[PO_4^{3-}]}{C_t} \tag{47}$$

Whereupon a combination of Equations 40 through 47 yields

$$\alpha_{H_3P} = \frac{[H^+]^3}{[H^+]^3 + K_1[H^+]^2 + K_1K_2[H^+] + K_1K_2K_3} \tag{48}$$

$$\alpha_{H_2P^-} = \frac{K_1[H^+]^2}{[H^+]^3 + K_1[H^+]^2 + K_1K_2[H^+] + K_1K_2K_3} \tag{49}$$

$$\alpha_{HP^{2-}} = \frac{K_1K_2[H^+]}{[H^+]^3 + K_1[H^+]^2 + K_1K_2[H^+] + K_1K_2K_3} \tag{50}$$

$$\alpha_{P^{3-}} = \frac{K_1K_2K_3}{[H^+]^3 + K_1[H^+]^2 + K_1K_2[H^+] + K_1K_2K_3} \tag{51}$$

Note that $\alpha_{H_3P} + \alpha_{H_2P^-} + \alpha_{HP^{2-}} + \alpha_{P^{3-}} = 1.0$ at any one pH value. The appropriate solution of Equations 48 through 51 will yield the $-\log \alpha$ vs pH plots, as seen in Figure 4. In this case of a triprotic acid, four regions—I, II, III and IV—may be devised by the boundaries of where the pH value is equal to pK_1, pK_2 and pK_3 of phosphoric acid. It follows that H_3PO_4 predominates in region I, $H_2PO_4^-$ in region II, HPO_4^{2-} in region III and PO_4^{3-} in region IV. This distribution of phosphate species with respect to pH becomes important when PO_4^{3-}, for example, enters into precipitation reactions. Table IIA

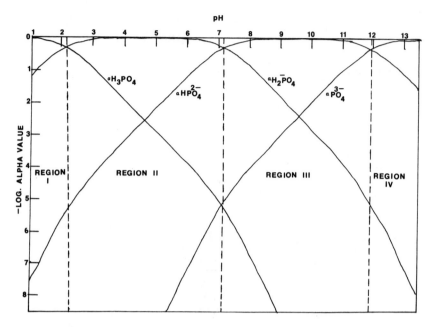

Figure 4.   Alpha value distribution diagram for phosphoric acid.

(Appendix) gives the alpha values for the four species in the phosphoric acid system at 25°C and 0.0 M ionic strength.

Proton Energy Level Diagrams

The strengths of acids and bases may now be ordered according to the standard free energy changes involved in the proton transfer. This order sometimes is illustrated in a proton energy level diagram shown in Figure 5. The scale on the left gives the energy required for the transfer of one mole of protons from a Brönsted acid to H_2O. As one moves from top to bottom, more energy is required and the acids become "weaker." Conversely, the right-hand scale gives the energy required for the transfer of one mole of protons from water to a Brönsted base. As one moves upward on this scale, the bases become "weaker." For a Brönsted acid to be "neutralized" by a Brönsted base, one must use a base whose proton transfer level lies below that of the acid. Another feature of Figure 5 is that pH may be a measure of the proton free energy level. For example, when $[HA] = [\bar{A}]$, the pH is given by the pK of HA. Thus, this pH value gives the average proton free energy level, which depends, of course, on the acid–base pair. Another

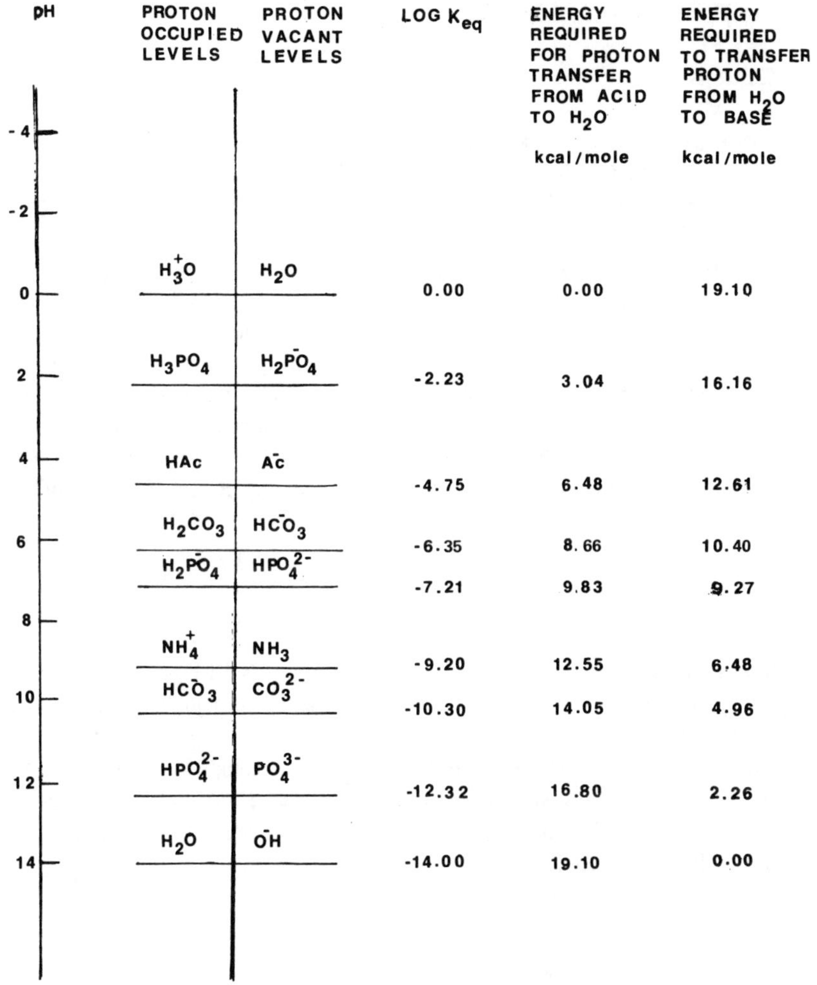

pH	PROTON OCCUPIED LEVELS	PROTON VACANT LEVELS	LOG K_{eq}	ENERGY REQUIRED FOR PROTON TRANSFER FROM ACID TO H_2O (kcal/mole)	ENERGY REQUIRED TO TRANSFER PROTON FROM H_2O TO BASE (kcal/mole)
0	H_3O^+	H_2O	0.00	0.00	19.10
2	H_3PO_4	$H_2PO_4^-$	-2.23	3.04	16.16
4	HAc	Ac^-	-4.75	6.48	12.61
6	H_2CO_3	HCO_3^-	-6.35	8.66	10.40
	$H_2PO_4^-$	HPO_4^{2-}	-7.21	9.83	9.27
8	NH_4^+	NH_3	-9.20	12.55	6.48
10	HCO_3^-	CO_3^{2-}	-10.30	14.05	4.96
12	HPO_4^{2-}	PO_4^{3-}	-12.32	16.80	2.26
14	H_2O	OH^-	-14.00	19.10	0.00

Figure 5. Proton energy level diagram.

interpretation of Figure 5 is that the energy levels on the left are occupied with protons. If, for example, an aqueous system contains several proton acceptors, e.g., CO_3^{2-}, OH^-, OCl^-, HCO_3^-, etc., the lower or unoccupied energy levels will be filled first on the addition of protons. The order of occupation will be OH^-, CO_3^{2-}, OCl^- and HCO_3^-. By the same reasoning, depopulation of proton levels will occur in the same order from high to low proton free energy levels.

ACIDITY AND ALKALINITY

Definitions

Classically, acidity and alkalinity in water and wastewater are determined by titration with standardized solutions of NaOH and H_2SO_4. The endpoints of these titrations usually are detected by the colorimetric indicators methyl orange and phenolphthalein. There are, however, some conceptual and operational errors in these titrations.

Operationally, acidity is defined as the capacity to neutralize the equivalent sum of all bases and alkalinity is the capacity to neutralize the equivalent sum of all acids. In the Brönsted system, acidity is the sum of the concentrations of the proton donors, whereas alkalinity is the sum of the concentrations of proton acceptors. From Equation 2, the proton donors in the monoprotic HA-H_2O system are HA and H_3^+O, whereas the proton acceptors are $\bar{A}$ and H_2O. The addition of a strong Brönsted base, such as NaOH, to Equation 2 yields

$$HA + NaOH = NaA + H_2O \tag{52}$$

If the quantity of NaOH is equivalent to the quantity of HA, then a measure of the proton-donating capacity is obtained, which is said to be the acidity of the system.

Another approach to acidity and alkalinity is through the concept of the proton condition of a system containing proton acceptors and proton donors. We have seen from Equation 9, for example, that the mass balance for a weak acid may be expressed as

$$C_t = [HA] + [\bar{A}] \tag{53}$$

In aqueous solutions where cations and anions are present, a charge balance or a condition of electroneutrality must exist. For the weak acid, HA, in water, the charged species are H^+, $\bar{A}$ and OH^-. Consequently, the charge balance is

$$[H^+] = [\bar{A}] + [OH^-] \tag{54}$$

A combination of Equations 53 and 54 yields

$$[H^+] + [HA] = C_t + [OH^-] \tag{55}$$

We now have an equation where the proton donors are summed on the left-hand side and the proton acceptors are summed on the right-hand side. Equation 55 may be rearranged as follows:

$$C_t = [H^+] + [HA] - [OH^-] \qquad (56)$$

If the sum of $[H^+]$ and $[HA] > [OH^-]$, C_t now becomes a measure of the proton-donating capacity of the system, hence acidity. If $[OH^-] > [H^+] + [HA]$, C_t becomes a measure of the proton-accepting capacity of the system, hence alkalinity. In the case of acidity, C_t would represent the quantity of a base, whereas in the case of alkalinity, C_t would represent the quantity of an acid.

The proton conditions at the beginning and end of the titration of HA are important also to the concepts of acidity and alkalinity. Equation 54 represents the proton condition at the beginning of the titration where the proton donors are balanced by the proton acceptors. Under these conditions, the acid-neutralizing capacity, hence alkalinity, of the system is zero. At the end of the titration of HA, the proton condition is obtained by substitution of the equivalent quantity of $[Na^+]$ (from NaOH) equal to C_t into Equation 56 which, when combined with Equation 53, yields

$$[HA] + [H^+] = [OH^-] \qquad (57)$$

In this case, the proton donors are balanced by the proton acceptor and the base-neutralizing capacity, hence acidity, of the system is zero. It follows that the quantity of $[Na^+]$ added to reach the conditions expressed by Equation 57 is a measure of the acidity.

Acidity and Alkalinity in the CO_2–H_2O System

In the measurement of acidity and alkalinity in natural waters, the proton donors and acceptors in the H_2CO_3 system must be considered. Equations 22 and 23 indicate that the proton donors are H_2CO_3 and HCO_3^- (H_3^+O understood), and the proton acceptors are HCO_3^- and CO_3^{2-} (OH^- understood). Since H_2CO_3 is a diprotic acid, three proton conditions may be cited: (1) at the beginning of the titration of H_2CO_3, (2) at the midpoint, and (3) at the end. These three points may be illustrated conveniently by a distribution diagram (similar to Figure 3) in which carbonate species concentration is plotted versus pH in Figure 6. In this diagram, the mass balance from Equation 24 on carbonate species is assumed to be $10^{-3}M$. At the beginning of a titration of H_2CO_3, the charge balance would be

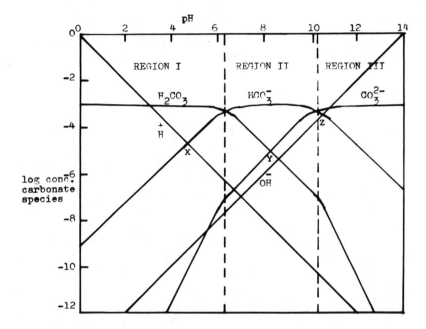

Figure 6.   Concentration distribution diagram for carbonic acid.

$$[H^+] = [HCO_3^-] + 2[CO_3^{2-}] + [OH^-] \tag{58}$$

The proton condition is obtained by combination of Equations 58 and 24, which yields

$$[H_2CO_3] + [H^+] = C_t + [CO_3^{2-}] + [OH^-] \tag{59}$$

Titration is begun now with a C_t solution of NaOH. At the mid-point one obtains a $10^{-3}M$ solution of NaHCO$_3$ for which the charge balance is

$$[Na^+] + [H^+] = [HCO_3^-] + 2[CO_3^{2-}] + [OH^-] \tag{60}$$

Since $C_t = [Na^+]$, the proton condition becomes

$$[H_2CO_3] + [H^+] = [CO_3^{2-}] + [OH^-] \tag{61}$$

Continuation of the titration to the end with NaOH yields a $10^{-3}M$ solution of Na_2CO_3 for which the charge balance is the same as Equation 60. In this case, however, $2C_t = [Na^+]$ which leads to the proton condition

$$2[H_2CO_3] + [HCO_3^-] + [H^+] = [OH^-] \tag{62}$$

Conceptual definitions for three types of alkalinity may be cited for the carbonate system. Caustic alkalinity may be defined from Equation 62 and point Z in Figure 6 as

$$[OH^--Alk] = [OH^-] - [H^+] - [HCO_3^-] - 2[H_2CO_3] \tag{63}$$

Carbonate alkalinity may be defined from Equation 61 and point Y in Figure 6 as

$$[CO_3^{2-}-Alk] = [CO_3^{2-}] + [OH^-] - [H^+] - [H_2CO_3] \tag{64}$$

Alkalinity (sometimes called total alkalinity) may be defined from Equation 58 and point X in Figure 6 as

$$[Alk] = [HCO_3^-] + 2[CO_3^{2-}] + [OH^-] - [H^+] \tag{65}$$

Likewise, conceptual definitions for three types of acidity may be cited for the carbonate system. Mineral acidity may be defined from Equation 58 and point X in Figure 6 as

$$[Min-Acy] = [H^+] - [HCO_3^-] - 2[CO_3^{2-}] - [OH^-] \tag{66}$$

Carbon dioxide acidity may be defined from Equation 61 and point Y in Figure 6 as

$$[CO_2-Acy] = [H_2CO_3] + [H^+] - [CO_3^{2-}] - [OH^-] \tag{67}$$

Acidity (sometimes called total acidity) may be defined from Equation 62 and point Z in Figure 6 as

$$[Acy] = 2[H_2CO_3] + [HCO_3^-] + [H^+] - [OH^-] \tag{68}$$

In aqueous carbonate systems, alkalinity becomes the equivalent sum of the proton acceptors, hence, the acid-neutralizing capacity (ANC), whereas acidity becomes the equivalent sum of the proton donors, hence, the base-neutralizing capacity (BNC). Alkalinity is a measure of the amount per liter of a strong acid required to reach a pH equal to that of a C_t molar solution of H_2CO_3 at endpoint X in Figure 6. Likewise, acidity is a measure of the quantity per liter of a strong base required to reach a pH equal to a C_t molar solution of Na_2CO_3 at endpoint Z in Figure 6. It should be noted that mineral acidity exists only to the left of endpoint X in a carbonate system. That is, if a mineral acid were added to the $10^{-3}M$ C_t solution of H_2CO_3 in Figure 6, the resultant pH values would lie between 0 and 4.7. Likewise, caustic alkalinity exists only to the right of endpoint Z in the carbonate system that would be 10.7 in Figure 6.

Calculation of Acidity and Alkalinity

The conceptual definitions of acidity and alkalinity expressed in Equations 63, 64, 65, 66, 67 and 68 may be combined with Equations 25, 26 and 27 to yield the following:

$$[OH^- - Alk] = [OH^-] - [H^+] - C_t(^\alpha HC^- - 2^\alpha H_2C) \tag{69}$$

$$[CO_3^{2-} - Alk] = [OH^-] - [H^+] + C_t(^\alpha C^{2-} - {}^\alpha H_2C) \tag{70}$$

$$[Alk] = [OH^-] - [H^+] + C_t(^\alpha HC^- + 2^\alpha C^2) \tag{71}$$

$$[Min-Acy] = [H^+] - [OH^-] - C_t(^\alpha HC^- - 2^\alpha C^{2-}) \tag{72}$$

$$[CO_2-Acy] = [H^+] - [OH^-] + C_t(^\alpha H_2C - {}^\alpha C^{2-}) \tag{73}$$

$$[Acy] = [H^+] - [OH^-] + C_t(2^\alpha H_2C + {}^\alpha HC^-) \tag{74}$$

These equations apply to aqueous systems closed to the atmosphere, and H_2CO_3 is considered to be a nonvolatile acid. There may be occasions, however, when a system open to the atmosphere must be considered. In this case, the equilibria expressed in Equations 20 and 21 are included in the acidity and alkalinity equations. The equilibrium expression for the combined Equations 20 and 21 is

$$K_{eq} = \frac{[H_2CO_3]}{[CO_{2(g)}]} \tag{75}$$

Dalton's law of partial pressure is used for the following $CO_{2(g)}$ term:

$$[CO_{2(g)}] = P_{CO_2}/RT \tag{76}$$

where $[CO_{2(g)}]$ = the concentration of CO_2 in the gas phase
P_{CO_2} = the partial pressure of CO_2
R = the ideal gas law constant
T = the absolute temperature.

Combination of Equations 75 and 76 yields

$$[H_2CO_3] = K_H P_{CO_2} \tag{77}$$

where $K_H = K_{eq}/RT$. The dimensions of K_H are mol/l/atm, for which values at several temperatures appear in Table III. In the alpha value system, Equation 77 may be rewritten as follows:

$$C_t{}^\alpha H_2 C = K_H P_{CO_2} \tag{78}$$

The various forms of acidity and alkalinity may be calculated from Equations 69 through 74 from the water quality analyses of pH, bicarbonate and temperature. For example, a spring water near Jefferson City, Tennessee contains 213 mg/l HCO_3^-, has a pH value of 7.4 and a temperature of 58°F [5]. (It is standard practice for the USGS to report $[HCO_3^-]$ rather than alkalinity.) Table IV shows the calculations of the mass balance on the carbonate species, the concentrations of H_2CO_3, CO_3^{2-}, alkalinity, acidity and CO_2-acidity. Also, the partial pressure of CO_2 required to maintain the $[H_2CO_3]$ is computed. This is a typical example of how one may obtain the carbonate chemistry, acidity and alkalinity for a natural water from only three water quality analyses. It should be noted that this water does not contain any mineral acidity, and caustic and carbonate alkalinities. This may

Table III. Henry's Law Constants for $CO_{2(g)}$ Solubility in Water [1]

Temperature (°C)	K_H mol/l/atm $\times 10^2$
0	7.69
5	6.40
10	5.36
15	4.55
20	3.92
25	3.45
30	3.02
40	2.29
50	1.91

Table IV. Calculations of Acidity and Alkalinity in the Carbonate System

1. Water Quality Analyses of a spring water: pH = 7.4, T = 58°F, $[HCO_3^-]$ = 213 mg/l.

2. At pH 7.4, $^\alpha H_2C$ = 0.0947, $^\alpha HC^-$ = 0.9044 and $^\alpha C^{2-}$ = 0.0008 (15°C values).

3. From Equation 26:

$$C_t = \frac{[HCO_3^-]}{^\alpha HC^-} = \frac{3.49 \times 10^{-3}M}{0.9044} = 3.86 \times 10^{-3}M$$

4. From Equation 25:

$$[H_2CO_3] = C_t \, ^\alpha H_2C = 3.86 \times 10^{-3} \times 0.0947 = 3.65 \times 10^{-4}M$$

5. From Equation 27:

$$[CO_3^{2-}] = C_t \, ^\alpha C^{2-} = 3.86 \times 10^{-3} \times 0.0008 = 3 \times 10^{-6}M$$

6. From Equation 71 ($[H^+]$ and $[OH^-]$ neglected):

$$[Alk] = C_t \, (^\alpha HC^- + 2^\alpha C^{2-}) = 3.86 \times 10^{-3} (0.9044 + 2 \times 0.0008)$$
$$[Alk] = 3.50 \times 10^{-3} \text{ eq/l}$$

7. From Equation 73 ($[H^+]$ and $[OH^-]$ neglected):

$$[CO_2\text{-Acy}] = C_t \, (^\alpha H_2C - ^\alpha C^{2-}) = 3.86 \times 10^{-3} (0.0947 - 0.0008)$$
$$[CO_2\text{-Acy}] = 3.62 \times 10^{-4} \text{ eq/l}$$

8. From Equation 74 ($[H^+]$ and $[OH^-]$ neglected):

$$[Acy] = C_t \, (2^\alpha H_2C + ^\alpha HC^-) = 3.86 \times 10^{-3} (2 \times 0.0947 + 0.9044)$$
$$[Acy] = 4.22 \times 10^{-3} \text{ eq/l}$$

9. From Equation 78:

$$P_{CO_2} = \frac{[H_2C]}{K_H} = \frac{3.65 \times 10^{-4}}{4.55 \times 10^{-2}} = 8.03 \times 10^{-3} \text{ atm}$$

be deduced from the three water quality analyses and the carbonate species distribution diagram. Also, the approximation may be made for this water that the $[HCO_3^-] \cong [Alk]$.

Often it is necessary to compute the alkalinities in the carbonate system for waters that have been softened by the so-called "lime-soda process." Table V gives a typical analysis of a softened water [6]. Sodium alkalinity (a term not commonly used now) is an operational definition for the arithmetic difference between total alkalinity and total hardness. To provide an example of how to compute the various forms of alkalinity in this water, it is assumed that the arithmetic difference of total alkalinity and caustic alkalinity yields the carbonate concentration. This assumption was necessary to obtain a C_t value for the carbonate species. The results of the calculations for the three types of alkalinity are seen in Table V. It is obvious that this approach is somewhat less tedious than the nomographic methods of Dye [7] as employed by *Standards Methods* [8]. See Figures 1A and 2A in the Appendix for two of Dye's nomographs.

Operational Errors in Analytical Determination

There are several operational errors in the analytical determination of acidity and alkalinity in the carbonate system of natural waters. In the alkalinity titration, the colorimetric indicator methyl orange is used frequently to detect endpoint X in Figure 6. The color change of methyl orange occurs over the pH range of 3.1 to 4.4 with the colors proceeding from red to pink to orange to yellow. The orange color is generally chosen as the endpoint. Figure 6 shows that the pH of a C_t solution of H_2CO_3 is concentration dependent. For example, a $10^{-1}M$ solution of H_2CO_3 would have a pH value of 3.67 (25°C), whereas a $10^{-5}M$ solution would have a pH of 5.67 (25°C). Consequently, there would be a tendency to overtitrate waters with low alkalinities. In the acidity titration, the colorimetric indicator, phenolphthalein, is used to detect endpoint Y in Figure 6. Therefore, the determination of H_2CO_3 acidity is operationally incorrect as the titration is carried to endpoint Z in the carbonate region. This error is, of course, automatically corrected by the equivalency of the NaOH titrant. Operationally, the endpoint in the H_2CO_3 acidity titration should be detected at a pH of 8.3 by phenolphthalein, whose color change occurs within the pH range 8.2 to 10.0 (colorless-red). A faint pink color usually designates the endpoint. The individual visual detection of this pink color is highly variable, and many analysts do not see any color change until pH values around 9.0 are reached. Thus, there is a tendency to overtitrate. Another operational error in the H_2CO_3 acidity determination is the loss of CO_2 to the atmosphere during sampling, transfer and titration. The exact loss of CO_2 is difficult to evaluate.

Table V. Calculations of Three Types of Alkalinity for a Lime–Soda Softened Water

1. Typical Analyses [6]

   Total hardness   = 16 mg/l as $CaCO_3$
   Total alkalinity = 140 mg/l as $CaCO_3$
   Sodium alkalinity = 124 mg/l as $CaCO_3$
   Caustic alkalinity = 70 mg/l as $CaCO_3$
   pH   = 11.2

2. At pH 11.2, $^{\alpha}H_2C = 2 \times 10^{-6}$, $^{\alpha}HCO_3^- = 0.118621$, $^{\alpha}CO_3^{2-} = 0.881377$ (25°C values).

3. C_t was computed to be $7.0 \times 10^{-4}M$

   $[H_2CO_3]$ was computed to be $1.4 \times 10^{-9}M$

   $[HCO_3^-]$ was computed to be $0.83 \times 10^{-4}M$

   $[CO_3^{2-}]$ was computed to be $6.17 \times 10^{-4}M$

4. From pH 11.2, $[OH^-] = 1.6 \times 10^{-3}M$

5. From Equation 69:

   $[OH\text{-}Alk] = [OH^-] - [H^+]^a - C_t(^{\alpha}HC^- - 2^{\alpha}H_2C)$

   $[OH\text{-}Alk] = 1.6 \times 10^{-3} - 7.0 \times 10^{-4} (0.118621 - 2 \times 2 \times 10^{-6})$

   $[OH\text{-}Alk] = 1.52 \times 10^{-3}$ eq/l

6. From Equation 70:

   $[CO_3^{2-}\text{-}Alk] = [OH^-] - [H^+]^a + C_t(^{\alpha}C^{2-} - ^{\alpha}H_2C)$

   $[CO_3^{2-}\text{-}Alk] = 1.6 \times 10^{-3} + 7.0 \times 10^{-4} (0.881377 - 0.000002)$

   $[CO_3^{2-}\text{-}Alk] = 2.22 \times 10^{-3}$ eq/l

7. From Equation 71:

   $[Alk] = [OH^-] - [H^+]^a + C_t(^{\alpha}HC^- + 2^{\alpha}C^{2-})$

   $[Alk] = 1.6 \times 10^{-3} + 7.0 \times 10^{-4} (1.881375)$

   $[Alk] = 2.92 \times 10^{-3}$ eq/l

8. Check, from Equation 65:

   $[Alk] = [HCO_3^-] + 2[CO_3^{2-}] + [OH^-] - [H^+]^a$

   $[Alk] = 0.083 \times 10^{-3} + 2 \times 0.617 \times 10^{-3} + 1.6 \times 10^{-3}$

   $[Alk] = 2.92 \times 10^{-3}$ eq/l

[a]Neglected in computation.

Other errors lie in the determination of caustic and carbonate alkalinities in the carbonate system. For example, caustic alkalinity is carried to the phenolphthalein endpoint Y and corrections are made for the presence of carbonate alkalinity. It should be obvious from Figure 6 that caustic alkalinity should be titrated to endpoint Z, whereupon carbonate alkalinity then would lie between endpoints Y and Z. Such a determination of caustic alkalinity could be easily accomplished since the indicator, alizarin yellow R operates in the pH range 10-12. A frequent error occurs where it is stated that one-half of the carbonate alkalinity is titrated at the phenolphthalein endpoint. Figure 6 shows that all of the carbonate alkalinity is titrated at endpoint Y, whereupon the bicarbonate alkalinity is titrated to endpoint X.

Graphic Approaches for Carbonate Equilibria

An extremely useful graphic approach has been provided by Deffeyes [9] for the solution of problems in carbonate equilibria. In this system, total carbonate (C_t) and alkalinity are employed as abscissa and ordinate with the pH value, P_{CO_2}, HCO_3^- and CO_3^{2-} presented as contours. For pH contours, Equation 65 is rearranged and divided through by the mass balance on the carbonates or C_t:

$$\frac{[Alk] + [H_3^+O] - [OH^-]}{C_t} = \frac{2[C^{2-}] + [HC^-]}{[H_2C] + [HC^-] + [C^{2-}]} \tag{79}$$

Substitution of K_1 and K_2 for carbonic acid into Equation 79 and neglecting $[H_3^+O]$ and $[OH^-]$ yields

$$[Alk] = \left(\frac{K_1[H_3^+O] + 2K_1K_2}{[H_3^+O]^2 + K_1[H_3^+O] + K_1K_2}\right)C_t \tag{80}$$

Alkalinity concentration (meq/l) may now be plotted against C_t (mM/l), with the terms in the brackets as the slope of the straight line. Equation 80 is solved at various concentrations of H_3^+O and plotted as pH contours in Figure 7. Interpretation is as follows: the addition of a mineral acid decreases the alkalinity without a change in C_t. Thus, the point moves downward in accordance with the number of equivalents of acid added. Additions of a strong base move the point upward in accordance with the number of equivalents added. Addition of $CO_{2(g)}$ increases C_t, of course, but the alkalinity does not change because the point moves to the right. To do this, however, the pH value must decrease. Addition of bicarbonate compounds adds equally to alkalinity and C_t along a line with a 45° slope. If carbonate is added, then the alkalinity is increased by two equivalents for every mole of C_t. In this case, the point moves upward along a 2:1 slope. Figure 7 is extremely helpful in demonstrating the various types of acidity and alkalinity in natural carbonate waters. It is also useful to determine the loss of $CaCO_3$ through precipitation and $CO_{2(g)}$ through evolution in natural waters where photosynthesis is active. Other useful diagrams are seen in Figures 8, 9 and 10,

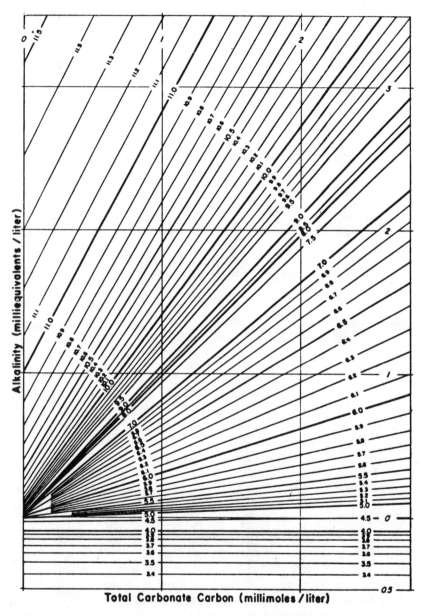

Figure 7. pH contours for the $CO_{2(g)}$-H_2O system (reproduced from Deffeyes [9] courtesy of the American Society of Limnology and Oceanography).

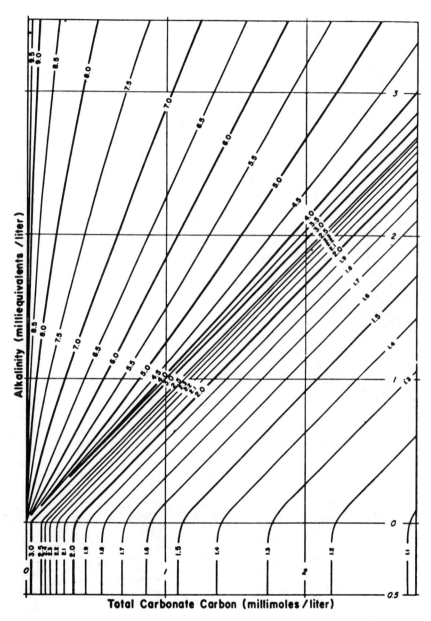

Figure 8.   Contours are the negative log of the $CO_{2(g)}$ pressure (in atm) (reproduced from Deffeyes [9] courtesy of the American Society of Limnology and Oceanography).

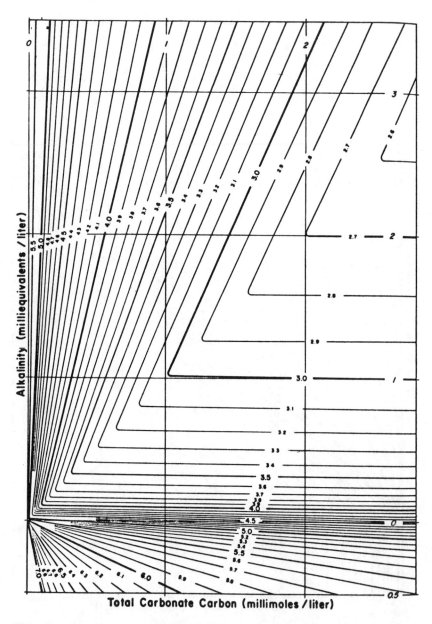

Figure 9.   Contours are the negative log of the HCO_3^- concentration in mol/l (reproduced from Deffeyes [9] courtesy of the American Society of Limnology and
Oceanography).

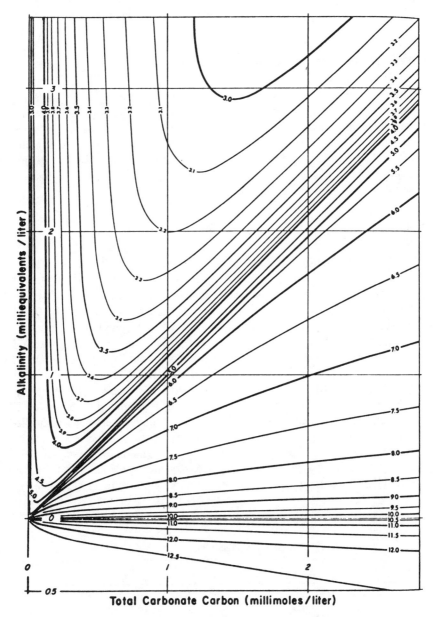

Figure 10.   Contours are the negative log of the CO_3^{2-} concentration in mol/l (reproduced from Deffeyes [9] courtesy of the American Society of Limnology and Oceanography).

where $CO_{2(g)}$, HCO_3^- and CO_3^{2-} contours, respectively, are given for the alkalinity versus C_t plot.

HYDRONIUM ION CONCENTRATION

Equation 4 above describes the transfer of a proton from one molecule of water to another that results in a hydrated hydrogen ion (i.e., hydronium ion) and a hydroxyl ion. The equilibrium constant for this reaction is given by Equation 7. By convention, the water of hydration on the hydronium ion is usually omitted and the term is written as H^+. In any event, an equal number of H^+ and OH^- ions are formed in Equation 4. A charge balance would be as follows:

$$[H^+] = [OH^-] \tag{81}$$

Substituting Equation 81 into Equation 7, we have $[H^+]^2 = 1.00 \times 10^{-14}$, whereupon $[H^+] = [OH^-] = 1.00 \times 10^{-7}$. If the hydrogen ion and hydroxyl ion concentrations are expressed as logarithmic functions, then: $-\log [H^+] = 7.00$ and $-\log [OH^-] = 7.00$. Sørensen [10] defined pH as the $-\log [H^+]$ (concentration, mol/l). In turn, pOH = $-\log [OH^-]$, whereupon pH + pOH = 14.00. These definitions led to the widespread use of the "pH scale" (0.00–14.00) to denote acidic and basic water solutions. If the $[H^+] > [OH^-]$, the solution is said to be acid, whereas if the $[OH^-] > [H^+]$, the solution is basic. By definition then, pure water is neutral because the $[H^+] = [OH^-]$. Qualitative criteria for acidic and basic solutions are summarized in Table VI.

In solutions containing foreign ions or in natural waters containing dissolved constituents, the active mass of the hydrogen ion does not equal its concentration. (See Chapter 2, p. 85, for a discussion of ionic strength, activity coefficients and activity.) Consequently, pH is redefined as $-\log (H^+) = -\log [H^+] \gamma_{H^+}$, where the ( ) now denote activity. This is more

Table VI. Criteria for Acid and Base Aqueous Solutions [11]

Acid	Base
$[H^+]^a > [OH^-]$ (by definition)	$[OH^-] > [H^+]$ (by definition)
$[H^+] > 10^{-7}$	$[OH^-] > 10^{-7}$
$[OH] < 10^{-7}$	$[H^+] < 10^{-7}$
pH < 7	pH > 7
pOH > 7	pOH < 7

^{a}Concentration in mol/l, 25°C.

of an operational than conceptual definition. Electrometric pH meters, for example, measure and record potential differences due to differences in (H^+) and not $[H^+]$. For a more detailed discussion on pH and all of its mysteries, the reader is referred to the excellent text of Bates [12].

BUFFER SYSTEMS IN NATURAL WATERS

Definition of a Buffer

Classically, buffers have been defined for aqueous solutions as follows: "Buffers are substances which, by their presence in solution increase the amount of acid or alkali that must be added to cause unit change in pH." This definition was proposed by Van Slyke [13], whose classic paper introduced the mathematical concepts for aqueous buffer systems. Implicit in this definition is the ability of a solution to resist the change in pH value through the addition of a base or an acid. Operationally, buffer systems have been constructed from the so-called "weak" acids and bases and their salts and are designed to resist the change in pH value in a particular system. This application of buffer systems tends to divert attention from the capacity (i.e., amount) portion of the definition. This capacity concept is developed more fully below.

For an aqueous system to resist a change in pH value due to the addition of a base, for example, there must be a continuous source of protons or hydrogen ions. This source is usually a "weak" acid, which dissociates to produce more hydrogen ions in the event that the original ions are consumed. In turn, this implies that the pH value of the system remains constant, or nearly so.

Classically, the pH value of an aqueous buffer system is calculated from the Henderson equation [14]:

$$pH = pK_a - \log \frac{[HA]}{[A^-]} \qquad (82)$$

where pK_a is the $-\log$ of the acidity constant and $[HA]$ and $[A^-]$ are the concentrations of the undissociated weak acid and its conjugate base, respectively. Equation 82 shows that the pH value will remain constant as long as the ratio of $[HA]/[A^-]$ remains constant or nearly so. For this ratio to remain constant, it is obvious that the concentrations of the two species should be equal. Consequently, the buffer effect is the greatest where the pH value is equal to the pK_a value. This may be seen in Figure 2, for example, for the monoprotic acid, HOCl. Equation 82 also suggests that the pH value of the buffer system is dependent on the pK_a value and is independent of the [HA] and [A$^-$]. However, the pH range over which a buffer is effective is somewhat narrow, as seen in Figure 2.

Neutralizing Capacity

As cited above, one of the definitions of an aqueous buffer includes a capacity factor. In this case, the acidic and basic neutralizing capacities of an aqueous buffer must be considered. In other words, capacity designates the amount or quantity of an acid or a base to cause a unit change in the pH value. In turn, this capacity factor is dependent on the respective concentrations of proton donors and proton acceptors in the system. For natural waters, the carbonate system is the most significant and is the model considered here. Equations 65 and 68 above, are the conceptual definitions of ANC (alkalinity) and BNC (acidity), respectively, for the carbonate system. Special note should be made of the relation between definitions of ANC and BNC and the titration endpoints X, Y and Z cited in Figure 6.

Buffer Value

Van Slyke [13] expressed the relation between the increment of strong base, C_b, added to a buffer solution and the resultant increment change in pH value as

$$\beta = \frac{dC_b}{dpH} \qquad (83)$$

β was offered as the buffer value (this term, instead of buffer intensity [1] or buffer index [11], is used in this text in deference to Van Slyke), which has a value of 1 when 1 liter of solution requires 1 gram equivalent of strong base to cause a unit change in pH value. Conversely, an increment of a strong acid can cause an incremental change in pH value:

$$\beta = -\frac{dC_a}{dpH} \qquad (84)$$

In Equations 83 and 84, β is always a positive value because the pH value is increased (i.e., +) when a base is added and is decreased (i.e., −) when a negative quantity of a base (i.e., dC_a) is added. If one solution has two times the buffer value of another, twice as much base or acid is needed to change the pH value by a given amount.

The significance of the β ratio may be simply illustrated by a hypothetical titration curve in Figure 11. For purposes of presentation, infinitesimal increments, ΔC_b and ΔpH, are used instead of the differential values. To increase the pH value of Solution 1 from 3 to 4, 0.1 gram equivalent per liter of NaOH is needed; therefore, $\Delta C_b = 0.1$. Substituting these values into Equation 83:

$$\beta = \frac{\Delta C_b}{\Delta pH} = \frac{0.1}{1.0} = 0.1 \text{ g eq/l}$$

To increase the pH value of Solution 2 from 3 to 4, 0.2 gram equivalent of NaOH is required whereupon

$$\beta = \frac{\Delta C_b}{\Delta pH} = \frac{0.2}{1.0} = 0.2 \text{ g eq/l}$$

Thus, Solution 2 has twice the buffer value and capacity of Solution 1.
    Another important concept of the buffer value may be observed in

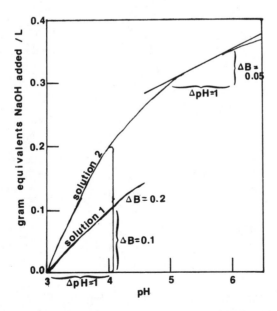

Figure 11.   Hypothetical titration curves for calculation of buffer values (reproduced from Van Slyke [13] courtesy of *J. Biol. Chem.*).

Figure 11 from the curve for Solution 2. As the pH value is increased from 4 to 5 and from 5 to 6, the curve becomes less steep. Consequently, the C_b value becomes less with the concomitant decrease in the β value. Thus, the buffer values change as the slope of the curve changes. Mathematically, this may be expressed as the slope of a straight line drawn tangentially to the curve. This provides a more accurate β value than the estimations from the incremental approach. Most texts [1,11] treat the buffer value at any point in a titration curve as inversely proportional to the slope because the plot is pH value on the ordinate scale and C_b is on the abscissa.

The buffer value of a diprotic acid (H_2CO_3, for example) may be derived from differentiation of the charge balance equation of its protolysis curve. For the carbonate system this is Equation 65, where the [Alk] term now replaces the [Na^+] term. The relationships between C_b, $-C_a$, [Alk], and $-$[Acy] may be seen below:

$$C_b = -C_a = [Alk] = -[Acy] = [HCO_3^-] + 2[CO_3^{2-}] + [OH^-] - [H^+] \qquad (85)$$

The relationships between β, dC_b, $-dC_a$, dpH, d[Alk] and $-$d[Acy] may be seen below:

$$\beta = \frac{dC_b}{dpH} = \frac{-dC_a}{dpH} = \frac{dAlk}{dpH} = \frac{-dAcy}{dpH} \qquad (86)$$

Equation 85 is rearranged so that it may be differentiated with respect to [H^+]:

$$[Alk] = C_t \alpha HC^- \left(1 + \frac{2K_2}{[H^+]}\right) + \frac{K_w}{[H^+]} - [H^+] \qquad (87)$$

Differentiating Equation 87 with C_t held constant and combining with Equation 86 yields the desired expression for the buffer value (in this case, with the appropriate α values in the carbonate system) [1]:

$$\beta = 2.303 \left(C_t[\alpha HC^-(\alpha H_2C + \alpha C^{2-}) + 4\alpha C^{2-}\alpha H_2C] + [H^+] + \frac{K_w}{[H^+]}\right) \qquad (88)$$

The reader is referred to the literature [1,11,13] for a more detailed treatment of the mathematics of the derivation of Equation 88.

Three additional buffer value expressions may be cited for aqueous carbonate solutions. Equation 88 is an expression where C_t is held constant in a closed system (to the atmosphere) and increments of strong base or acid are added. An expression may be derived for a carbonate system open to the atmosphere where $CO_{2(g)}$ is exchanged between the water and gaseous phases. Such a system in nature would be a lake where the photosynthetic and respiratory activities of algae would influence the pH and buffer values of the water through the exchange of $CO_{2(g)}$. In this case, the alkalinity is considered as a conservative quantity and $CO_{2(g)}$ is added or removed from the system. The equation for this type of system is [1]:

$$\beta' = \frac{-d[Acy]}{dpH} = -4.606 \left(\frac{\alpha_{HC^-}\,\alpha_{H_2C} + \alpha_{HC^-}\,\alpha_{C^{2-}} + 4\alpha_{H_2C}\,\alpha_{C^{2-}}}{(\alpha_{HC^-} + 2\alpha_{C^{2-}})^2} \right)[Alk] \qquad (89)$$

In systems where Na_2CO_3 or $CaCO_3$ may be treated as the addition of a base, [Acy] is held constant whereupon the expression is [1]

$$\beta'' = \frac{d[Alk]}{dpH} = 4.606 \left(\frac{\alpha_{HC^-}\,\alpha_{H_2C} + \alpha_{HC^-}\,\alpha_{C^{2-}} + 4\alpha_{H_2C}\,\alpha_{C^{2-}}}{(\alpha_{HC^-} + 2\alpha_{H_2C})^2} \right)[Acy] \qquad (90)$$

In systems open to the atmosphere where the partial pressure of $CO_{2(g)}$ remains constant, the buffer value for the addition of a strong acid or base may be computed from the following [1]:

$$\beta''' = \frac{d[Alk]}{dpH} = 2.303 \left\{ K_H P_{CO_2} \left(\frac{\alpha_{HC^-}(\alpha_{H_2C} + \alpha_{C^{2-}}) + 4\alpha_{C^{2-}}\,\alpha_{H_2C} + (\alpha_{HC^-} + 2\alpha_{C^{2-}})^2}{\alpha_{H_2C}} \right) \right\}$$
$$+ [H^+] + K_W/[H^+] \qquad (91)$$

This expression shows the buffering effects of $CO_{2(g)}$ on waters where the pH value exceeds 7.5.

Equation 88 indicates that the buffer value is a function of C_t and the $[H^+]$. Solution of Equation 88, holding C_t constant at $10^{-3}M$, with respect to $[H^+]$ yields, in graphic form (Figure 12), the values for the carbonate system [15]. It is noted that two minima are observed in the curve near the pH values of endpoints X and Y in the titration curve of the carbonate system. These two minima could be obtained also by the second derivative of Equation 87, which should be positive. A maximum β value is observed at the pH value of 6.3, which is very close to the pK_1 value for H_2CO_3 (or any other weak acid for that matter) may be found in the pH region of its pK values. A second maximum should appear in the curve at pH = pK_2, and a second minimum should appear at endpoint Z, but these are obscured because the $K_W/[H^+]$ ($[OH^-]$) term in Equation 87 becomes very large in relation to

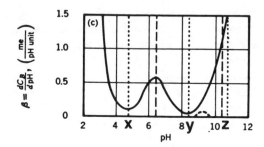

Figure 12.   The calculated buffer capacity for a $10^{-3}M$ carbonate system (reproduced from Weber and Stumm [15] courtesy of the American Water Works Association).

the other two terms. It should be emphasized that a minimum and not a maximum β value appears at a pH value of 8.3, as widely misconceived in the field of environmental engineering.

Buffer values were computed for some natural waters selected at random from Hem [5] and White et al. [16] to illustrate the practical application of Equations 88 and 89. The effects of pH value and C_t for the carbonate species on the buffer values are demonstrated in Table VII.

For example, waters 4 and 5 have the same pH value of 7.0, but water 5 with its higher C_t value has a higher β value. Water 6 has a C_t value nearly that of water 5, but it has a lower β value due to the higher pH value of 8.2. β values were computed from Equation 89 to demonstrate the effect of the addition or removal of $CO_{2(g)}$ on these waters.

An application of the β value concept may be illustrated by the softening of water 5, for example, by the adjustment of the pH value to 11.0 through

Table VII. Buffer Values of Some Natural Waters

Source of Water	pH	HCO$_3^-$ (mg/l)	C_t ($M \times 10^3$)	β^a (eq/dpH $\times 10^3$)	β'^b (eq/dpH $\times 10^3$)	Reference
1. Limestone	7.0	146	2.93	1.0	−2.5	5
2. Sandstone	7.5	320	5.62	0.8	−1.7	5
3. Surface	6.7	90	2.13	1.05	−0.3	5
4. Limestone	7.0	74	1.49	0.5	−1.25	16
5. Limestone	7.0	277	5.56	1.92	−4.7	16
6. Limestone	8.2	291	4.87	0.2	−0.5	16

[a]From Equation 88: C_t constant, addition of acid or base.
[b]From Equation 89: [Alk], addition or removal of CO_2. β' is recorded as a − number so that the addition of $CO_{2(g)}$ results in a + quantity.

the addition of NaOH. From Equation 86: $dC_b = \beta dpH = 1.92 \times 10^{-3}$ eq $\times$ $4.0 = 7.68 \times 10^{-3}$ eq NaOH would have to be added.

Another application may be demonstrated through the recarbonation of the softened water cited previously in Table V. From Equation 89, the β' value was computed to be -0.116 eq/dpH. $CO_{2(g)}$ may be added to lower the pH value and to stabilize the water against postprecipitation of $CaCO_{3(s)}$. If, for example, it is desired to lower the pH value to 8.0, then the amount of $CO_{2(g)}$ would be (from Equation 86): $-dC_a = \beta dpH = -0.116 \times 4.2 = 0.49$ eq.

For additional information about buffer systems in natural waters, the reader is referred to the excellent papers of Weber and Stumm [15] and Kleyn [17].

CONTENTS IN NATURAL WATERS

Fresh Waters

Natural waters, of course, have variable contents of acidity, alkalinity and hydronium ions. The major influence on these contents would be the geology with which the water contacts (pollution from wastewaters excepted). Table VIII provides some typical acidities, alkalinities and pH values of

Table VIII. Examples of the Acidity and Alkalinity Contents and pH Values of Natural Waters [16]

Source of Water	Geology	pH	Source of Acidity SO_4^{2-} (mg/l)	Source of Alkalinity HCO_3^- (mg/l)	CO_3^{2-} (mg/l)	SiO_2 [a] (mg/l)
Spring	Silicic volcanics	6.6	–	21	0	–
Spring	Rhyolite	7.9	–	131	0	–
Well	Granite	6.6	–	93	0	–
Well	Basalt	7.7	–	156	0	–
Well	Sandstone	6.2	–	18	0	–
Well	Sandstone	9.2	–	217[b]	33	11
Well	Shale	7.9	–	126	0	–
Spring	Limestone	7.7	–	168	0	–
Well	Dolomite	8.2	–	241	0	18
Spring	Alluvium	7.8	–	120	0	–
Spring	Volcanics	1.8	5710	0	0	–
Spring	Volcanics	1.9	1570	0	0	–

[a]USGS standard practice for reporting dissolved silica content.
[b]Na^+ content = 144 mg/l.

selected natural waters, mostly subsurface, from several geological sources [16]. In most instances, alkalinity may be assigned to the bicarbonate anion, with lesser amounts due to carbonate and silica. Unfortunately, the U.S. Geological Survey does not include acidity determinations in chemical water quality reports. Undoubtedly, dissolved carbon dioxide would provide the acidity for the bicarbonate waters. Occasionally, alkaline subsurface waters are encountered where sodium bicarbonate is the predominant constituent. Natural waters, especially volcanic and thermal waters, may acquire unusual chemical characteristics. Two examples are given in Table VIII, where the pH values are quite acid (1.8 and 1.9) that originate from naturally occurring sulfuric acid. Chapter 6 of this volume describes the chemical reactions that yield the acidic and alkaline constituents in natural waters. Additional chemical water quality data of rivers and lakes may be obtained from another U.S. Geological Survey report [18].

In addition to bicarbonate and carbonate, borates, phosphates and silicates may contribute to naturally occurring alkalinity:

$$[Alk] = [HCO_3^-] + 2[CO_3^{2-}] + [H_4BO_4^-] + [H_2PO_4^-] + 2[HPO_4^{2-}]$$
$$+ 3[PO_4^{3-}] + [H_3SiO_4^-] + [OH^-] - [H^+] \tag{92}$$

These three substances are, of course, minor constituents, but there are some natural waters where their contribution to total alkalinity may be significant. In sea water, for example, bicarbonate and carbonate alkalinity comprises about 95% of the total, with borate alkalinity constituting the remaining 5%. For a fresh water, Park and co-workers [19] determined the "alkalinity budget" of the Columbia River where bicarbonate represented, on the average, 94% of the total, with silicate, borates and phosphates constituting 1% of the total. Their data are given in Table IX. "Surplus" alkalinity was considered to be the total alkalinity less the bicarbonate, silicate, borate and phosphate alkalinities and constituted about 5% of the total. No satisfactory explanation was offered for the occurrence and source of this "surplus" alkalinity.

Measurements of Alkalinity, Acidity and pH Value

Total Alkalinity and pH Value

As discussed above, there are several conceptual and operational errors in the analytical determination of acidity and alkalinity in natural fresh waters. In addition, differences have been observed between field and laboratory determinations of the pH value and alkalinity that are ascribed to the carbonate system. Examples of these differences are provided by Roberson

Table IX. Alkalinity Components of the Columbia River [19]

River	Alkalinity (meq/l)				Carbonate Alkalinity (%) (Total) Alkalinity
	Total	Carbonate	Silicate	Surplus[a]	
Columbia	1.36	1.26	0.00	0.10	93
(main stream)	1.56	1.42	0.00	0.14	94
	1.56	1.47	0.00	0.09	94
	1.32	1.19	0.00	0.13	90
	1.18	1.10	0.00	0.08	93
	1.35	1.27	0.00	0.08	94
Willamette	0.55	0.52	0.00	0.03	95
Yakima	3.06	3.01	0.01	0.04	98
Snake	2.41	2.29	0.01	0.11	95
	2.45	2.37	0.01	0.07	97
	3.45	3.35	0.02	0.08	97
Clearwater	0.52	0.47	0.00	0.05	90
Spokane	0.34	0.30	0.00	0.04	88
Pend Oreille	1.60	1.54	0.00	0.06	96
	Avg 1.51			Avg 0.08	Avg 94%

[a]Surplus alkalinity = alkalinity − bicarbonate alkalinity − silicate alkalinity − borate alkalinity − phosphate alkalinity.

et al. [20], who examined the ground and surface waters of Sierra Nevada, California. A portion of their data is given in Table X, where the time interval between field and laboratory determinations ranged from 5 to 120 days. No consistent correlation was found between the time of storage and changes in the values of pH and alkalinity. There was some indication, however, that laboratory determinations made within one week showed somewhat smaller changes than samples stored longer periods of time. Also, there was no consistency in the changes toward always higher or lower values in the laboratory results. No doubt these changes were due to the exchange of $CO_{2(g)}$ between the atmosphere and the water. There seemed to be a tendency for waters with a "low" total carbonate content to gain CO_2 and to show a decrease in pH value, whereas waters with a "high" total carbonate content tended to lose $CO_{2(g)}$ to the atmosphere and to show an increase in pH value. The major significance of these data lies in their unreliability for use in geochemical equilibrium calculations (see Chapter 6).

Several suggestions have been offered for "accurate" determination of acidity and alkalinity in natural fresh waters, where the carbonate system predominates. One example is a procedure for the field measurement of pH value and alkalinity of natural waters by Barnes [21]. The basis of this

Table X. Comparison of Field and Laboratory Determinations of Water Samples
Collected in Sierra Nevada, California [20]

Source	pH		Alkalinity as HCO_3^- (mg/l)		Time Interval (days)	Temperature of Sample at Source (°F)
	Laboratory	Field	Laboratory	Field		
G[a]	6.8	6.8	740	738	90	86
G	7.2	7.1	67	71	16	51
G	7.4	7.2	105	107	16	49
SW	7.7	8.2	273	280	120	48
G	7.1	7.3	108	103	120	38
SW	7.3	7.4	47	49	6	46
SW	7.3	7.4	47	49	6	41
G	5.8	5.8	11	14	6	43
SW	6.3	6.8	5	8	20	48
SW	6.2	6.4	6	8	20	48
SW	7.2	7.2	21	23	20	42
G	6.6	6.8	101	103	7	48
G	6.7	6.7	72	75	7	60
G	6.5	6.6	77	67	6	47
G	7.4	7.4	84	95	6	53
G	5.6	5.8	14	14	6	51
G	6.8	6.9	34	32	6	62
G	7.0	6.9	97	100	6	51
G	5.4	5.8	12	14	5	52
G	5.6	5.8	17	20	5	44
G	7.5	6.7	36	40	90	50
G	7.5	7.9	88	90	90	50
G	6.6	6.6	91	92	90	45
G	6.6	6.6	73	76	90	45
SW	7.6	7.9	80	91	90	62
G	6.6	6.8	33	35	75	54
G	8.0	8.2	190	216	75	56
G	7.1	7.6	109	93	75	56
G	7.5	8.0	143	168	60	60
G	6.2	6.5	31	34	60	44
G	7.7	7.6	59	64	60	50
G	5.8	6.2	19	21	60	44
G	7.4	7.1	68	70	60	50
G	6.3	6.7	40	41	60	41
G	6.3	6.5	49	52	60	46
G	7.4	7.5	63	63	60	48
G	7.4	7.7	103	108	60	48
G	6.5	7.3	13	15	60	44
G	6.8	7.2	37	38	60	44
G	7.8	7.4	108	111	60	83
G	7.6	7.4	88	81	60	44
G	5.8	6.2	18	18	60	48
G	7.4	7.5	146	150	60	65
G	5.9	6.1	42	45	60	50
G	5.8	6.0	28	30	60	54
G	6.8	7.1	119	113	60	70

[a]G = ground water; SW = surface water.

procedure lies in an accurate and precise determination of the pH value via electrometric techniques. Also, alkalinity titrations may be made in terms of a "true" pH value. The total alkalinity in the carbonate system is, perhaps, the most important variable that influences the titration endpoint pH value, which may vary from 5.38 ($0°C$, 5 mg/l HCO_3^-) to 4.32 ($35°C$, 300 mg/l HCO_3^-). With proper precautions, Barnes claims that the pH value may be determined to ±0.02 units and the alkalinity value to ±0.06 mg/l HCO_3^- for natural fresh waters.

The major point of Barnes' procedure lies in calibrating the electrometric instrument at a pH value of 8.00 at $25°C$ (with National Bureau of Standards buffer solutions, of course) and then measuring the pH value of a pH = 4.00 buffer. The instrument was considered to be linear when the pH value of the latter buffer was in the 4.00 to 4.03 range. For field determinations of pH and alkalinity, these steps were found necessary by Barnes:

1. Measure water temperature with the thermometer in shade.
2. Set temperature compensator on meter to water temperature.
3. Wash electrodes in a stream of buffer whose nominal pH is near that of the water.
4. Immerse electrodes 1 inch in the buffer.
5. Balance and set meter at the nearest integral value of the buffer pH.
6. Rinse electrodes in water.
7. Collect water sample in beaker.
8. Measure the pH of the sample.
9. Pipet 50 ml of sample into dry beaker.
10. Titrate with standard acid ($0.01639N$ H_2SO_4) and record pH with each increment of acid in the range from pH 5 to pH 4.
11. Rinse electrodes in water.
12. Wash electrodes in a stream of pH_4 buffer.
13. Measure pH of pH_4 buffer.
14. Replace electrodes and reagent bottles in water.

After the titration is completed, the "true" pH value of the endpoint between pH 5 and 4 is computed from an arithmetic equation, which averages the "observed" pH values of the water and the "observed" and "true" pH values of the two buffer solutions. This equation (not presented here) is mathematically incorrect because the arithmetic average of pH values is meaningless. Rather, an average of molar concentrations of the hydronium ion would be correct. Nonetheless, Barnes [21] does offer some excellent suggestions for field measurement of the pH value.

Stumm and Morgan [1] discuss the difficulties of alkalinity and acidity titrations. As indicated above, the operational endpoint of an acidic titration of alkalinity is not generally in precise accord with the calculated pH values of the equivalence point because $CO_{2(g)}$ is lost during the analysis. Again, the quantity of $CO_{2(g)}$ lost throughout the titration depends on sample handling, amount of agitation and the degree of oversaturation achieved at the lower pH values. "It has been the authors' experience, however, that relative CO_2 losses are negligible if smooth stirring (magnetic stirrer, for example) is maintained throughout the course of the titration." No data were given by Stumm and Morgan to support this statement. Nonetheless, the pH values for the endpoint at room temperature are in accord with this equation:

$$[H^+] = (C_tK_1 + K_W)^{0.5} \tag{93}$$

which may be derived from Equation 58.

Stumm and Morgan discuss further the precision of the pH value at the endpoint relative to the equivalence point for waters of "low" alkalinities (defined by *Standard Methods* [8] as less than 20 mg/l CaCO₃). These authors have confidence in the procedure developed by Gran [22] to determine the equivalence points with precision and with a small relative error. This is based on the principle that added increments of mineral acid linearly increase [H⁺]. Larson and Henley [23] and Thomas and Lynch [24] have used a modification of Gran's technique for fresh waters, whereas Dyrssen et al. [25,26] have used it for sea waters.

The Gran plot essentially takes Equations 53, 54 and 56 to produce two straight line equations, one before the equivalence point and one after the equivalence point. The intersection of these two linear plots yields the volume of titrant, usually H₂SO₄ for alkalinity, required to reach the equivalence point. Figure 13 [26], an example from sea water, shows a Gran plot for location of the equivalence point in the titration of carbonate alkalinity in sea water. A full mathematical treatment of the Gran technique is given in

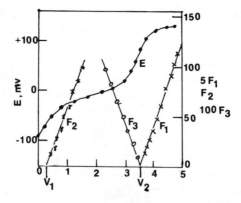

Figure 13.   EMF titration curve for 154 g of seawater. Characterization of equivalence points by Gran method. F₁, F₂ and F₃ are Gran functions for finding the equivalence points V₁ and V₂ (reproduced from Dyrssen and Sillen [26] courtesy of *Tellus*).

the literature [1,22]. This technique appears at first glance to be straight-forward; however, considerable laboratory experience is required to yield data giving linear plots before and after the equivalence point.

Loewenthal and Marais [27] concur with the often reported condition that the total alkalinity endpoint is affected minimally by temperature and ionic strength, but is affected by C_t. These authors agree with Stumm and Morgan that titration to a predetermined endpoint would give incorrect results, but that a titration to the corresponding inflection point in the titration curve would not cause an error in the alkalinity measurement since $CO_{2(g)}$ loss does not directly affect the results. Again, there is the contention that "smooth" stirring with a magnetic stirrer would cause negligible CO_2 losses. Since there is a "low" value for the minimum buffer capacity associated with the H_2CO_3 endpoint (X on Figure 6), as seen in Figure 12, this gives a clearly defined inflection point in the titration curve and a corresponding low titration error. It is not clear from these authors, however, that this latter statement applies to natural fresh waters of "low" alkalinities.

Loewenthal and Marais [27] prefer to determine the H_2CO_3 equivalence point by direct titration between preselected pH values for field applications. First, the pH value of the water is adjusted to 6.0 with a standard strong acid. Second, the alkalinity required to titrate the sample from this pH value to 4.5 is measured. Third, the H_2CO_3 equivalence point is obtained directly by reference to the measured alkalinity from pH 6.0 to 4.5 to the graph in Figure 14, which is developed on these theoretical considerations [27]:

    (a) the pH of the H_2CO_3 equivalence point is dependent principally on the total carbonic species concentration and is virtually independent of ionic strength and temperature.
    (b) by measuring the alkalinity between any two fixed pH values, C_t can be closely calculated.
    (c) for any C_t value the H_2CO_3 equivalence point pH can be calculated from basic theory.

That pH_e is virtually independent of ionic strength and temperature may be seen in Figure 14 [27], and the dependence on C_t may be seen from Figure 6 (endpoint X). That C_t can be calculated from basic theory is developed from Equation 88 or from Equation 93, which can be approximated very closely by

$$pH_e = \frac{1}{2} pC_t + \frac{1}{2} pK_1 \tag{94}$$

After the pH_e value is determined from Figure 14, then C_t is calculated from Equation 94. This latter calculation is repeated for a number of C_t values until a theoretical plot of alkalinity from pH 6.0 to 4.5 against the H_2CO_3

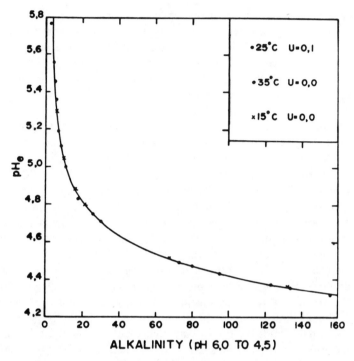

Figure 14.   Variation of $H_2CO_3^*$ equivalence point, pH_e, with alkalinity between pH 6.0 and 4.5 (reproduced from Loewenthal and Marais [27] courtesy of Ann Arbor Science Publishers, Inc.).

equivalence point pH is obtained (Figure 14). Subsequently, "accurate alkalinity measurements" are obtained in this manner:

1.   Adjust the pH of the sample to pH 6.0.
2.   Titrate to pH 4.5 and record the alkalinity measured, and read a pH_e value from Figure 14.
3.   Titrate a new sample of the water to the endpoint pH value determined in 2 to obtain total alkalinity.

This technique of Loewenthal and Marais [27] appears to be useful for waters between alkalinities of 20 and 100 on Figure 14. Above and below these numbers the curve is asymptotic, which limits its application, especially for low alkalinity waters.

Kemp [28] uses the predetermined pH endpoint approach in determining total alkalinity of natural waters. However, Kemp has more confidence in conductance measurements than in electrometric titrations. Empirically, there is a significant correlation between conductivity and C_t, which led to the regression equation:

$$C_t = 7.63 \times 10^{-6}C \tag{95}$$

where C = the conductivity in μmhos/cm. Combining Equations 94 and 95 gives

$$pH_e = 5.74 - \frac{1}{2}\log C \tag{96}$$

This allows the endpoint pH value of the total alkalinity to be estimated from the conductivity of water. Apparently, Kemp had to develop this relationship by an independent determination of total alkalinity by titration.

These methods of determining total alkalinity in the carbonate system cited above and the calculation of three forms of alkalinity given in Table V are somewhat at variance with *Standard Methods* [8]. The variance lies largely in the endpoints selected for the total and phenolphthalein alkalinities. *Standard Methods* persists in using an endpoint pH value of 8.3 for the latter alkalinity. This is proper when a potentiometric titration is employed but highly inaccurate when the colorimetric indicator is used (discussed above in the section entitled Operational Errors in Analytical Determination, p. 113).

Total Acidity

Total acidity in the carbonate system is defined by Equation 68:

$$[Acy] = 2[H_2CO_3] + [HCO_3^-] + [H^+] - [OH^-] \tag{68}$$

The most significant term in this equation is the first one or $2[H_2CO_3]$, the "free" CO_2 acidity. There is no direct manner to measure rigorously by alkalimetric titration the individual value of $[H_2CO_3]$. For most waters without mineral acidity (Equation 66) and with pH values below 7.6, CO_2-acidity is almost identical to $[H_2CO_3]$. Stumm and Morgan [1] claim that "in most natural waters the determination of CO_2-acidity is only slightly less precise than an alkalinity measurement. For example, CO_2-acidity can be determined in a water containing 10^{-3} eq/l of alkalinity and 5×10^{-4} eq/l of CO_2-acidity with a relative error of approximately 4% provided the endpoint can be defined with 0.3 pH unit of the true equivalence point pH. However, the error increases quite rapidly as the ratio of CO_2-acidity to alkalinity decreases." Again, the source of these statements is not documented but is believed to be reasonably accurate.

In waters containing "high" hardness and alkalinity, the solubility product constant of $CaCO_{3(s)}$ is frequently exceeded prior to the phenolphthalein endpoint. If $CaCO_{3(s)}$ precipitates before the endpoint is reached, positive errors occur because this reaction occurs:

$$Ca^{2+} + HCO_3^- = CaCO_{3(s)} + H^+ \tag{97}$$

Another source of error in CO_2 determinations is its loss to the atmosphere during sampling and transfer. Stumm and Morgan [1] suggest, without documentation, that it is most expedient to determine pH and water temperature in the field. Since alkalinity is not changed by the loss of CO_2, this quantity can be determined after samples have been taken to the laboratory. Consequently, the free CO_2 can be calculated from the measured pH values and alkalinity with the appropriate constants (see Table IV). This calculated value is valid, of course, at the temperature measured in the field and corrected for activity. Kemp [29] also discusses this problem but does not offer a viable alternative solution to the determination of free CO_2 in natural waters.

APPENDIX

Table IA. Alpha Values for the $CO_{2(aq)}$-H_2O System at 25°C and 0.0M Ionic Strength

pH	αH_2CO_3	αHCO_3^-	αCO_3^{2-}	pH	αH_2CO_3	αHCO_3^-	αCO_3^{2-}
0.00	1.000000	0.000000	0.000000	0.75	0.999997	0.000003	0.000000
0.05	1.000000	0.000001	0.000000	0.80	0.999997	0.000003	0.000000
0.10	0.999999	0.000001	0.000000	0.85	0.999997	0.000003	0.000000
0.15	0.999999	0.000001	0.000000	0.90	0.999996	0.000004	0.000000
0.20	0.999999	0.000001	0.000000	0.95	0.999996	0.000004	0.000000
0.25	0.999999	0.000001	0.000000	1.00	0.999996	0.000004	0.000000
0.30	0.999999	0.000001	0.000000	1.05	0.999995	0.000005	0.000000
0.35	0.999999	0.000001	0.000000	1.10	0.999994	0.000006	0.000000
0.40	0.999999	0.000001	0.000000	1.15	0.999994	0.000006	0.000000
0.45	0.999999	0.000001	0.000000	1.20	0.999993	0.000007	0.000000
0.50	0.999999	0.000001	0.000000	1.25	0.999992	0.000008	0.000000
0.55	0.999998	0.000002	0.000000	1.30	0.999991	0.000009	0.000000
0.60	0.999998	0.000002	0.000000	1.35	0.999990	0.000010	0.000000
0.65	0.999998	0.000002	0.000000	1.40	0.999989	0.000011	0.000000
0.70	0.999998	0.000002	0.000000	1.45	0.999987	0.000013	0.000000

Table IA, continued

pH	αH_2CO_3	αHCO_3^-	αCO_3^{2-}	pH	αH_2CO_3	αHCO_3^-	αCO_3^{2-}
1.50	0.999986	0.000014	0.000000	3.75	0.997500	0.002500	0.000000
1.55	0.999984	0.000016	0.000000	3.80	0.997196	0.002804	0.000000
1.60	0.999982	0.000018	0.000000	3.85	0.996855	0.003145	0.000000
1.65	0.999980	0.000020	0.000000	3.90	0.996473	0.003527	0.000000
1.70	0.999978	0.000022	0.000000	3.95	0.996044	0.003956	0.000000
1.75	0.999975	0.000025	0.000000	4.00	0.995563	0.004437	0.000000
1.80	0.999972	0.000028	0.000000	4.05	0.995025	0.004975	0.000000
1.85	0.999968	0.000032	0.000000	4.10	0.994421	0.005579	0.000000
1.90	0.999965	0.000035	0.000000	4.15	0.993744	0.006256	0.000000
1.95	0.999960	0.000040	0.000000	4.20	0.992986	0.007014	0.000000
2.00	0.999955	0.000045	0.000000	4.25	0.992137	0.007863	0.000000
2.05	0.999950	0.000050	0.000000	4.30	0.991186	0.008814	0.000000
2.10	0.999944	0.000056	0.000000	4.35	0.990122	0.009878	0.000000
2.15	0.999937	0.000063	0.000000	4.40	0.988930	0.011070	0.000000
2.20	0.999929	0.000071	0.000000	4.45	0.987595	0.012404	0.000000
2.25	0.999921	0.000079	0.000000	4.50	0.986103	0.013897	0.000000
2.30	0.999911	0.000089	0.000000	4.55	0.984434	0.015566	0.000000
2.35	0.999900	0.000100	0.000000	4.60	0.982567	0.017433	0.000000
2.40	0.999888	0.000112	0.000000	4.65	0.980482	0.019518	0.000000
2.45	0.999874	0.000126	0.000000	4.70	0.978152	0.021848	0.000000
2.50	0.999859	0.000141	0.000000	4.75	0.975552	0.024448	0.000000
2.55	0.999842	0.000158	0.000000	4.80	0.972650	0.027350	0.000000
2.60	0.999823	0.000177	0.000000	4.85	0.969415	0.030585	0.000000
2.65	0.999801	0.000199	0.000000	4.90	0.965810	0.034189	0.000000
2.70	0.999777	0.000223	0.000000	4.95	0.961798	0.038202	0.000000
2.75	0.999749	0.000251	0.000000	5.00	0.957336	0.042664	0.000000
2.80	0.999719	0.000281	0.000000	5.05	0.952378	0.047622	0.000000
2.85	0.999685	0.000315	0.000000	5.10	0.946875	0.053124	0.000000
2.90	0.999646	0.000354	0.000000	5.15	0.940777	0.059222	0.000000
2.95	0.999603	0.000397	0.000000	5.20	0.934028	0.065972	0.000000
3.00	0.999555	0.000445	0.000000	5.25	0.926569	0.073431	0.000001
3.05	0.999500	0.000500	0.000000	5.30	0.918340	0.081659	0.000001
3.10	0.999439	0.000561	0.000000	5.35	0.909280	0.090719	0.000001
3.15	0.999371	0.000629	0.000000	5.40	0.899325	0.100674	0.000001
3.20	0.999294	0.000706	0.000000	5.45	0.888411	0.111587	0.000001
3.25	0.999208	0.000792	0.000000	5.50	0.876477	0.123521	0.000002
3.30	0.999112	0.000888	0.000000	5.55	0.863463	0.136535	0.000002
3.35	0.999003	0.000997	0.000000	5.60	0.849313	0.150684	0.000003
3.40	0.998882	0.001118	0.000000	5.65	0.833979	0.166018	0.000003
3.45	0.998746	0.001254	0.000000	5.70	0.817419	0.182576	0.000004
3.50	0.998593	0.001407	0.000000	5.75	0.799605	0.200390	0.000005
3.55	0.998421	0.001579	0.000000	5.80	0.780519	0.219474	0.000006
3.60	0.998229	0.001771	0.000000	5.85	0.760161	0.239831	0.000008
3.65	0.998013	0.001987	0.000000	5.90	0.738547	0.261444	0.000010
3.70	0.997771	0.002229	0.000000	5.95	0.715713	0.284275	0.000012

Table IA, continued

pH	αH_2CO_3	αHCO_3^-	αCO_3^{2-}	pH	αH_2CO_3	αHCO_3^-	αCO_3^{2-}
6.00	0.691717	0.308268	0.000014	8.25	0.012359	0.979475	0.008166
6.05	0.666640	0.333343	0.000018	8.30	0.011019	0.979816	0.009165
6.10	0.640582	0.359397	0.000021	8.35	0.009822	0.979894	0.010284
6.15	0.613667	0.386307	0.000026	8.40	0.008752	0.979711	0.011537
6.20	0.586039	0.413930	0.000031	8.45	0.007797	0.979265	0.012939
6.25	0.557859	0.442104	0.000037	8.50	0.006944	0.978549	0.014507
6.30	0.529301	0.470655	0.000044	8.55	0.006182	0.977557	0.016261
6.35	0.500549	0.499398	0.000052	8.60	0.005503	0.976276	0.018221
6.40	0.471794	0.528144	0.000062	8.65	0.004896	0.974693	0.020411
6.45	0.443224	0.556703	0.000074	8.70	0.004355	0.972788	0.022857
6.50	0.415024	0.584889	0.000087	8.75	0.003873	0.970541	0.025587
6.55	0.387370	0.612528	0.000102	8.80	0.003442	0.967926	0.028631
6.60	0.360423	0.639458	0.000119	8.85	0.003058	0.964917	0.032025
6.65	0.334326	0.665534	0.000139	8.90	0.002716	0.961479	0.035805
6.70	0.309205	0.690632	0.000162	8.95	0.002411	0.957579	0.040011
6.75	0.285163	0.714649	0.000188	9.00	0.002139	0.953175	0.044686
6.80	0.262279	0.737503	0.000218	9.05	0.001896	0.948225	0.049878
6.85	0.240613	0.759135	0.000252	9.10	0.001680	0.942683	0.055637
6.90	0.220202	0.779508	0.000290	9.15	0.001488	0.936496	0.062016
6.95	0.201063	0.798604	0.000334	9.20	0.001316	0.929612	0.069072
7.00	0.183195	0.816422	0.000383	9.25	0.001163	0.921973	0.076863
7.05	0.166584	0.832978	0.000438	9.30	0.001027	0.913521	0.085451
7.10	0.151199	0.848300	0.000501	9.35	0.000906	0.904195	0.094899
7.15	0.137001	0.862428	0.000571	9.40	0.000799	0.893932	0.105270
7.20	0.123940	0.875410	0.000650	9.45	0.000703	0.882670	0.116627
7.25	0.111962	0.887299	0.000740	9.50	0.000618	0.870351	0.129031
7.30	0.101007	0.898153	0.000840	9.55	0.000542	0.856917	0.142541
7.35	0.091013	0.908034	0.000953	9.60	0.000475	0.842317	0.157208
7.40	0.081916	0.917004	0.001080	9.65	0.000415	0.826505	0.173080
7.45	0.073655	0.925123	0.001222	9.70	0.000362	0.809447	0.190190
7.50	0.066165	0.932453	0.001382	9.75	0.000316	0.791119	0.208565
7.55	0.059387	0.939051	0.001562	9.80	0.000274	0.771512	0.228214
7.60	0.053262	0.944974	0.001764	9.85	0.000238	0.750632	0.249130
7.65	0.047736	0.950274	0.001990	9.90	0.000206	0.728505	0.271289
7.70	0.042757	0.955000	0.002244	9.95	0.000178	0.705178	0.294645
7.75	0.038274	0.959197	0.002529	10.00	0.000153	0.680718	0.319130
7.80	0.034244	0.962908	0.002848	10.05	0.000131	0.655215	0.344654
7.85	0.030623	0.966170	0.003207	10.10	0.000122	0.628781	0.371107
7.90	0.027374	0.969018	0.003609	10.15	0.000096	0.601549	0.398356
7.95	0.024459	0.971482	0.004059	10.20	0.000081	0.573671	0.426248
8.00	0.021846	0.973589	0.004564	10.25	0.000069	0.545314	0.454618
8.05	0.019506	0.975363	0.005131	10.30	0.000058	0.516658	0.483284
8.10	0.017411	0.976824	0.005765	10.35	0.000049	0.487890	0.512061
8.15	0.015536	0.977988	0.006476	10.40	0.000041	0.459201	0.540758
8.20	0.013859	0.978868	0.007273	10.45	0.000034	0.430779	0.569187

Table IA, continued

pH	αH_2CO_3	αHCO_3^-	αCO_3^{2-}	pH	αH_2CO_3	αHCO_3^-	αCO_3^{2-}
10.50	0.000029	0.402805	0.597166	12.25	0.000000	0.011853	0.986147
10.55	0.000024	0.375449	0.624527	12.30	0.000000	0.010577	0.989423
10.60	0.000020	0.348865	0.651115	12.35	0.000000	0.009438	0.990562
10.65	0.000016	0.323189	0.676795	12.40	0.000000	0.008420	0.991580
10.70	0.000013	0.298536	0.701450	12.45	0.000000	0.007511	0.992489
10.75	0.000011	0.274999	0.724990	12.50	0.000000	0.006700	0.993300
10.80	0.000009	0.242650	0.747341	12.55	0.000000	0.005976	0.994024
10.85	0.000007	0.231536	0.768456	12.60	0.000000	0.005329	0.994671
10.90	0.000006	0.211687	0.788307	12.65	0.000000	0.004753	0.885247
10.95	0.000005	0.193112	0.806883	12.70	0.000000	0.004238	0.995762
11.00	0.000004	0.175804	0.824192	12.75	0.000000	0.003779	0.996221
11.05	0.000003	0.159739	0.840257	12.80	0.000000	0.003369	0.996631
11.10	0.000003	0.144885	0.855112	12.85	0.000000	0.003004	0.996996
11.15	0.000002	0.131196	0.868802	12.90	0.000000	0.002678	0.997322
11.20	0.000002	0.118621	0.881377	12.95	0.000000	0.002388	0.997612
11.25	0.000001	0.107103	0.892896	13.00	0.000000	0.002129	0.997871
11.30	0.000001	0.096580	0.903419	13.05	0.000000	0.001897	0.998103
11.35	0.000001	0.086991	0.913008	13.10	0.000000	0.001691	0.998309
11.40	0.000001	0.078271	0.921728	13.15	0.000000	0.001508	0.998492
11.45	0.000001	0.070358	0.929641	13.20	0.000000	0.001344	0.998656
11.50	0.000000	0.063190	0.936809	13.25	0.000000	0.001198	0.998802
11.55	0.000000	0.056708	0.943291	13.30	0.000000	0.001068	0.998932
11.60	0.000000	0.050855	0.949145	13.35	0.000000	0.000952	0.999048
11.65	0.000000	0.045577	0.954423	13.40	0.000000	0.000848	0.999152
11.70	0.000000	0.040822	0.959177	13.45	0.000000	0.000756	0.999244
11.75	0.000000	0.036545	0.963455	13.50	0.000000	0.000674	0.999326
11.80	0.000000	0.032701	0.967299	13.55	0.000000	0.000601	0.999399
11.85	0.000000	0.029249	0.970751	13.60	0.000000	0.000536	0.999464
11.90	0.000000	0.026151	0.973849	13.65	0.000000	0.000477	0.999523
11.95	0.000000	0.023374	0.976626	13.70	0.000000	0.000425	0.999575
12.00	0.000000	0.020885	0.979115	13.75	0.000000	0.000379	0.999621
12.05	0.000000	0.018656	0.981344	13.80	0.000000	0.000338	0.999662
12.10	0.000000	0.016661	0.983339	13.85	0.000000	0.000301	0.999699
12.15	0.000000	0.014876	0.985124	13.90	0.000000	0.000268	0.999732
12.20	0.000000	0.013280	0.986720	13.95	0.000000	0.000239	0.999761
				14.00	0.000000	0.000213	0.999787

Table IIA. Alpha Values of Orthophosphate System from pH 0.00 to 14.00 at 25°C, Ionic Strength, 0.0M

pH Value	Alpha 3 H_3PO_4	Alpha 2 $H_2PO_4^-$	Alpha 1 HPO_4^{2-}	Alpha 0 PO_4^{3-}
0.00	0.9925	0.0075	0.0000	0.0000
0.01	0.9923	0.0077	0.0000	0.0000
0.02	0.9921	0.0079	0.0000	0.0000
0.03	0.9919	0.0081	0.0000	0.0000
0.04	0.9918	0.0082	0.0000	0.0000
0.05	0.9916	0.0084	0.0000	0.0000
0.06	0.9914	0.0086	0.0000	0.0000
0.07	0.9912	0.0088	0.0000	0.0000
0.08	0.9910	0.0090	0.0000	0.0000
0.09	0.9908	0.0092	0.0000	0.0000
0.10	0.9905	0.0095	0.0000	0.0000
0.11	0.9903	0.0097	0.0000	0.0000
0.12	0.9901	0.0099	0.0000	0.0000
0.13	0.9899	0.0101	0.0000	0.0000
0.14	0.9896	0.0104	0.0000	0.0000
0.15	0.9894	0.0106	0.0000	0.0000
0.16	0.9892	0.0108	0.0000	0.0000
0.17	0.9889	0.0111	0.0000	0.0000
0.18	0.9886	0.0114	0.0000	0.0000
0.19	0.9884	0.0116	0.0000	0.0000
0.20	0.9881	0.0119	0.0000	0.0000
0.21	0.9878	0.0122	0.0000	0.0000
0.22	0.9876	0.0124	0.0000	0.0000
0.23	0.9873	0.0127	0.0000	0.0000
0.24	0.9870	0.0130	0.0000	0.0000
0.25	0.9867	0.0133	0.0000	0.0000

pH Value	Alpha 3	Alpha 2	Alpha 1	Alpha 0
0.56	0.9732	0.0268	0.0000	0.0000
0.57	0.9726	0.0274	0.0000	0.0000
0.58	0.9720	0.0280	0.0000	0.0000
0.59	0.9713	0.0287	0.0000	0.0000
0.60	0.9707	0.0293	0.0000	0.0000
0.61	0.9700	0.0300	0.0000	0.0000
0.62	0.9693	0.0307	0.0000	0.0000
0.63	0.9687	0.0313	0.0000	0.0000
0.64	0.9679	0.0321	0.0000	0.0000
0.65	0.9672	0.0328	0.0000	0.0000
0.66	0.9665	0.0335	0.0000	0.0000
0.67	0.9657	0.0343	0.0000	0.0000
0.68	0.9650	0.0350	0.0000	0.0000
0.69	0.9642	0.0358	0.0000	0.0000
0.70	0.9634	0.0366	0.0000	0.0000
0.71	0.9626	0.0374	0.0000	0.0000
0.72	0.9617	0.0383	0.0000	0.0000
0.73	0.9609	0.0391	0.0000	0.0000
0.74	0.9600	0.0400	0.0000	0.0000
0.75	0.9591	0.0409	0.0000	0.0000
0.76	0.9582	0.0418	0.0000	0.0000
0.77	0.9572	0.0428	0.0000	0.0000
0.78	0.9563	0.0437	0.0000	0.0000
0.79	0.9553	0.0447	0.0000	0.0000
0.80	0.9543	0.0457	0.0000	0.0000

pH				
0.26	0.9864	0.0136	0.0000	0.0000
0.27	0.9861	0.0139	0.0000	0.0000
0.28	0.9858	0.0142	0.0000	0.0000
0.29	0.9854	0.0146	0.0000	0.0000
0.30	0.9851	0.0149	0.0000	0.0000
0.31	0.9847	0.0153	0.0000	0.0000
0.32	0.9844	0.0156	0.0000	0.0000
0.33	0.9840	0.0160	0.0000	0.0000
0.34	0.9837	0.0163	0.0000	0.0000
0.35	0.9833	0.0167	0.0000	0.0000
0.36	0.9829	0.0171	0.0000	0.0000
0.37	0.9825	0.0175	0.0000	0.0000
0.38	0.9821	0.0179	0.0000	0.0000
0.39	0.9817	0.0183	0.0000	0.0000
0.40	0.9813	0.0187	0.0000	0.0000
0.41	0.9809	0.0191	0.0000	0.0000
0.42	0.9804	0.0196	0.0000	0.0000
0.43	0.9800	0.0200	0.0000	0.0000
0.44	0.9795	0.0205	0.0000	0.0000
0.45	0.9791	0.0209	0.0000	0.0000
0.46	0.9786	0.0214	0.0000	0.0000
0.47	0.9781	0.0219	0.0000	0.0000
0.48	0.9776	0.0224	0.0000	0.0000
0.49	0.9771	0.0229	0.0000	0.0000
0.50	0.9766	0.0234	0.0000	0.0000
0.51	0.9760	0.0240	0.0000	0.0000
0.52	0.9755	0.0245	0.0000	0.0000
0.53	0.9749	0.0251	0.0000	0.0000
0.54	0.9744	0.0256	0.0000	0.0000
0.55	0.9738	0.0262	0.0000	0.0000

pH				
0.81	0.9533	0.0467	0.0000	0.0000
0.82	0.9523	0.0477	0.0000	0.0000
0.83	0.9512	0.0488	0.0000	0.0000
0.84	0.9501	0.0499	0.0000	0.0000
0.85	0.9490	0.0510	0.0000	0.0000
0.86	0.9479	0.0521	0.0000	0.0000
0.87	0.9468	0.0532	0.0000	0.0000
0.88	0.9456	0.0544	0.0000	0.0000
0.89	0.9444	0.0556	0.0000	0.0000
0.90	0.9432	0.0568	0.0000	0.0000
0.91	0.9419	0.0581	0.0000	0.0000
0.92	0.9406	0.0594	0.0000	0.0000
0.93	0.9394	0.0606	0.0000	0.0000
0.94	0.9380	0.0620	0.0000	0.0000
0.95	0.9367	0.0633	0.0000	0.0000
0.96	0.9353	0.0647	0.0000	0.0000
0.97	0.9339	0.0661	0.0000	0.0000
0.98	0.9325	0.0676	0.0000	0.0000
0.99	0.9310	0.0690	0.0000	0.0000
1.00	0.9295	0.0705	0.0000	0.0000
1.01	0.9280	0.0720	0.0000	0.0000
1.02	0.9264	0.0736	0.0000	0.0000
1.03	0.9248	0.0752	0.0000	0.0000
1.04	0.9232	0.0768	0.0000	0.0000
1.05	0.9216	0.0784	0.0000	0.0000
1.06	0.9199	0.0801	0.0000	0.0000
1.07	0.9182	0.0818	0.0000	0.0000
1.08	0.9164	0.0836	0.0000	0.0000
1.09	0.9146	0.0854	0.0000	0.0000
1.10	0.9128	0.0872	0.0000	0.0000

Table IIA, continued

pH Value	Alpha 3	Alpha 2	Alpha 1	Alpha 0
1.11	0.9110	0.0890	0.0000	0.0000
1.12	0.9091	0.0909	0.0000	0.0000
1.13	0.9072	0.0928	0.0000	0.0000
1.14	0.9052	0.0948	0.0000	0.0000
1.15	0.9032	0.0968	0.0000	0.0000
1.16	0.9012	0.0988	0.0000	0.0000
1.17	0.8991	0.1009	0.0000	0.0000
1.18	0.8970	0.1030	0.0000	0.0000
1.19	0.8949	0.1051	0.0000	0.0000
1.20	0.8927	0.1073	0.0000	0.0000
1.21	0.8905	0.1095	0.0000	0.0000
1.22	0.8882	0.1118	0.0000	0.0000
1.23	0.8859	0.1141	0.0000	0.0000
1.24	0.8835	0.1165	0.0000	0.0000
1.25	0.8811	0.1189	0.0000	0.0000
1.26	0.8787	0.1213	0.0000	0.0000
1.27	0.8762	0.1238	0.0000	0.0000
1.28	0.8737	0.1263	0.0000	0.0000
1.29	0.8711	0.1289	0.0000	0.0000
1.30	0.8685	0.1315	0.0000	0.0000
1.31	0.8659	0.1341	0.0000	0.0000
1.32	0.8632	0.1368	0.0000	0.0000
1.33	0.8605	0.1395	0.0000	0.0000
1.34	0.8577	0.1423	0.0000	0.0000
1.35	0.8548	0.1452	0.0000	0.0000

pH Value	Alpha 3	Alpha 2	Alpha 1	Alpha 0
1.66	0.7425	0.2575	0.0000	0.0000
1.67	0.7381	0.2619	0.0000	0.0000
1.68	0.7336	0.2664	0.0000	0.0000
1.69	0.7291	0.2709	0.0000	0.0000
1.70	0.7245	0.2755	0.0000	0.0000
1.71	0.7199	0.2801	0.0000	0.0000
1.72	0.7153	0.2847	0.0000	0.0000
1.73	0.7105	0.2895	0.0000	0.0000
1.74	0.7058	0.2942	0.0000	0.0000
1.75	0.7010	0.2990	0.0000	0.0000
1.76	0.6961	0.3039	0.0000	0.0000
1.77	0.6912	0.3088	0.0000	0.0000
1.78	0.6863	0.3137	0.0000	0.0000
1.79	0.6813	0.3187	0.0000	0.0000
1.80	0.6763	0.3237	0.0000	0.0000
1.81	0.6712	0.3288	0.0000	0.0000
1.82	0.6661	0.3339	0.0000	0.0000
1.83	0.6610	0.3390	0.0000	0.0000
1.84	0.6558	0.3442	0.0000	0.0000
1.85	0.6506	0.3494	0.0000	0.0000
1.86	0.6454	0.3546	0.0000	0.0000
1.87	0.6401	0.3599	0.0000	0.0000
1.88	0.6347	0.3653	0.0000	0.0000
1.89	0.6294	0.3706	0.0000	0.0000
1.90	0.6240	0.3760	0.0000	0.0000

				pH		
0.0000	0.0000	0.3814	0.6186	1.91	0.0000	0.0000
0.0000	0.0000	0.3869	0.6131	1.92	0.0000	0.0000
0.0000	0.0000	0.3923	0.6077	1.93	0.0000	0.0000
0.0000	0.0000	0.3978	0.6022	1.94	0.0000	0.0000
0.0000	0.0000	0.4034	0.5966	1.95	0.0000	0.0000
0.0000	0.0000	0.4089	0.5911	1.96	0.0000	0.0000
0.0000	0.0000	0.4145	0.5855	1.97	0.0000	0.0000
0.0000	0.0000	0.4201	0.5799	1.98	0.0000	0.0000
0.0000	0.0000	0.4257	0.5743	1.99	0.0000	0.0000
0.0000	0.0000	0.4314	0.5686	2.00	0.0000	0.0000
0.0000	0.0000	0.4370	0.5630	2.01	0.0000	0.0000
0.0000	0.0000	0.4427	0.5573	2.02	0.0000	0.0000
0.0000	0.0000	0.4484	0.5516	2.03	0.0000	0.0000
0.0000	0.0000	0.4541	0.5459	2.04	0.0000	0.0000
0.0000	0.0000	0.4598	0.5402	2.05	0.0000	0.0000
0.0000	0.0000	0.4655	0.5345	2.06	0.0000	0.0000
0.0000	0.0000	0.4712	0.5287	2.07	0.0000	0.0000
0.0000	0.0000	0.4770	0.5230	2.08	0.0000	0.0000
0.0000	0.0000	0.4827	0.5173	2.09	0.0000	0.0000
0.0000	0.0000	0.4885	0.5115	2.10	0.0000	0.0000
0.0000	0.0000	0.4942	0.5058	2.11	0.0000	0.0000
0.0000	0.0000	0.5000	0.5000	2.12	0.0000	0.0000
0.0000	0.0000	0.5058	0.4942	2.13	0.0000	0.0000
0.0000	0.0000	0.5115	0.4885	2.14	0.0000	0.0000
0.0000	0.0000	0.5173	0.4827	2.15	0.0000	0.0000
0.0000	0.0000	0.5230	0.4770	2.16	0.0000	0.0000
0.0000	0.0000	0.5287	0.4712	2.17	0.0000	0.0000
0.0000	0.0000	0.5345	0.4655	2.18	0.0000	0.0000
0.0000	0.0000	0.5402	0.4598	2.19	0.0000	0.0000
0.0000	0.0000	0.5459	0.4541	2.20	0.0000	0.0000

pH						
1.36	0.8519	0.1481	0.0000	0.0000	0.0000	0.0000
1.37	0.8490	0.1510	0.0000	0.0000	0.0000	0.0000
1.38	0.8460	0.1540	0.0000	0.0000	0.0000	0.0000
1.39	0.8430	0.1570	0.0000	0.0000	0.0000	0.0000
1.40	0.8400	0.1600	0.0000	0.0000	0.0000	0.0000
1.41	0.8368	0.1632	0.0000	0.0000	0.0000	0.0000
1.42	0.8337	0.1663	0.0000	0.0000	0.0000	0.0000
1.43	0.8304	0.1696	0.0000	0.0000	0.0000	0.0000
1.44	0.8272	0.1728	0.0000	0.0000	0.0000	0.0000
1.45	0.8239	0.1761	0.0000	0.0000	0.0000	0.0000
1.46	0.8205	0.1795	0.0000	0.0000	0.0000	0.0000
1.47	0.8171	0.1829	0.0000	0.0000	0.0000	0.0000
1.48	0.8136	0.1864	0.0000	0.0000	0.0000	0.0000
1.49	0.8101	0.1899	0.0000	0.0000	0.0000	0.0000
1.50	0.8065	0.1935	0.0000	0.0000	0.0000	0.0000
1.51	0.8029	0.1971	0.0000	0.0000	0.0000	0.0000
1.52	0.7992	0.2008	0.0000	0.0000	0.0000	0.0000
1.53	0.7955	0.2045	0.0000	0.0000	0.0000	0.0000
1.54	0.7917	0.2083	0.0000	0.0000	0.0000	0.0000
1.55	0.7879	0.2121	0.0000	0.0000	0.0000	0.0000
1.56	0.7841	0.2159	0.0000	0.0000	0.0000	0.0000
1.57	0.7801	0.2199	0.0000	0.0000	0.0000	0.0000
1.58	0.7762	0.2238	0.0000	0.0000	0.0000	0.0000
1.59	0.7721	0.2279	0.0000	0.0000	0.0000	0.0000
1.60	0.7681	0.2319	0.0000	0.0000	0.0000	0.0000
1.61	0.7639	0.2361	0.0000	0.0000	0.0000	0.0000
1.62	0.7597	0.2403	0.0000	0.0000	0.0000	0.0000
1.63	0.7555	0.2445	0.0000	0.0000	0.0000	0.0000
1.64	0.7512	0.2488	0.0000	0.0000	0.0000	0.0000
1.65	0.7469	0.2531	0.0000	0.0000	0.0000	0.0000

Table IIA, continued

pH Value	Alpha 3	Alpha 2	Alpha 1	Alpha 0	pH Value	Alpha 3	Alpha 2	Alpha 1	Alpha 0
2.21	0.4484	0.5516	0.0000	0.0000	2.76	0.1864	0.8136	0.0000	0.0000
2.22	0.4427	0.5573	0.0000	0.0000	2.77	0.1829	0.8171	0.0000	0.0000
2.23	0.4370	0.5630	0.0000	0.0000	2.78	0.1795	0.8205	0.0000	0.0000
2.24	0.4314	0.5686	0.0000	0.0000	2.79	0.1761	0.8238	0.0000	0.0000
2.25	0.4257	0.5743	0.0000	0.0000	2.80	0.1728	0.8272	0.0000	0.0000
2.26	0.4201	0.5799	0.0000	0.0000	2.81	0.1695	0.8304	0.0000	0.0000
2.27	0.4145	0.5855	0.0000	0.0000	2.82	0.1663	0.8336	0.0000	0.0000
2.28	0.4089	0.5911	0.0000	0.0000	2.83	0.1632	0.8368	0.0000	0.0000
2.29	0.4034	0.5966	0.0000	0.0000	2.84	0.1600	0.8399	0.0000	0.0000
2.30	0.3978	0.6022	0.0000	0.0000	2.85	0.1570	0.8430	0.0000	0.0000
2.31	0.3923	0.6077	0.0000	0.0000	2.86	0.1539	0.8460	0.0000	0.0000
2.32	0.3869	0.6131	0.0000	0.0000	2.87	0.1510	0.8490	0.0000	0.0000
2.33	0.3814	0.6186	0.0000	0.0000	2.88	0.1480	0.8519	0.0000	0.0000
2.34	0.3760	0.6240	0.0000	0.0000	2.89	0.1452	0.8548	0.0000	0.0000
2.35	0.3706	0.6294	0.0000	0.0000	2.90	0.1423	0.8576	0.0000	0.0000
2.36	0.3653	0.6347	0.0000	0.0000	2.91	0.1395	0.8604	0.0000	0.0000
2.37	0.3599	0.6401	0.0000	0.0000	2.92	0.1368	0.8632	0.0000	0.0000
2.38	0.3546	0.6453	0.0000	0.0000	2.93	0.1341	0.8658	0.0000	0.0000
2.39	0.3494	0.6506	0.0000	0.0000	2.94	0.1315	0.8685	0.0000	0.0000
2.40	0.3442	0.6558	0.0000	0.0000	2.95	0.1288	0.8711	0.0000	0.0000
2.41	0.3390	0.6610	0.0000	0.0000	2.96	0.1263	0.8737	0.0001	0.0000
2.42	0.3339	0.6661	0.0000	0.0000	2.97	0.1238	0.8762	0.0001	0.0000
2.43	0.3288	0.6712	0.0000	0.0000	2.98	0.1213	0.8787	0.0001	0.0000
2.44	0.3237	0.6763	0.0000	0.0000	2.99	0.1189	0.8811	0.0001	0.0000
2.45	0.3187	0.6813	0.0000	0.0000	3.00	0.1165	0.8835	0.0001	0.0000

				pH		
0.0000	0.0001	0.8858	0.1141	3.01	0.0000	0.0000
0.0000	0.0001	0.8881	0.1118	3.02	0.0000	0.0000
0.0000	0.0001	0.8904	0.1095	3.03	0.0000	0.0000
0.0000	0.0001	0.8926	0.1073	3.04	0.0000	0.0000
0.0000	0.0001	0.8948	0.1051	3.05	0.0000	0.0000
0.0000	0.0001	0.8970	0.1030	3.06	0.0000	0.0000
0.0000	0.0001	0.8991	0.1009	3.07	0.0000	0.0000
0.0000	0.0001	0.9011	0.0988	3.08	0.0000	0.0000
0.0000	0.0001	0.9032	0.0968	3.09	0.0000	0.0000
0.0000	0.0001	0.9051	0.0948	3.10	0.0000	0.0000
0.0000	0.0001	0.9071	0.0928	3.11	0.0000	0.0000
0.0000	0.0001	0.9090	0.0909	3.12	0.0000	0.0000
0.0000	0.0001	0.9109	0.0890	3.13	0.0000	0.0000
0.0000	0.0001	0.9128	0.0872	3.14	0.0000	0.0000
0.0000	0.0001	0.9146	0.0854	3.15	0.0000	0.0000
0.0000	0.0001	0.9163	0.0836	3.16	0.0000	0.0000
0.0000	0.0001	0.9181	0.0818	3.17	0.0000	0.0000
0.0000	0.0001	0.9198	0.0801	3.18	0.0000	0.0000
0.0000	0.0001	0.9215	0.0784	3.19	0.0000	0.0000
0.0000	0.0001	0.9231	0.0768	3.20	0.0000	0.0000
0.0000	0.0001	0.9247	0.0752	3.21	0.0000	0.0000
0.0000	0.0001	0.9263	0.0736	3.22	0.0000	0.0000
0.0000	0.0001	0.9279	0.0720	3.23	0.0000	0.0000
0.0000	0.0001	0.9294	0.0705	3.24	0.0000	0.0000
0.0000	0.0001	0.9309	0.0690	3.25	0.0000	0.0000
0.0000	0.0001	0.9323	0.0675	3.26	0.0000	0.0000
0.0000	0.0001	0.9338	0.0661	3.27	0.0000	0.0000
0.0000	0.0001	0.9352	0.0647	3.28	0.0000	0.0000
0.0000	0.0001	0.9366	0.0633	3.29	0.0000	0.0000
0.0000	0.0001	0.9379	0.0620	3.30	0.0000	0.0000

				pH
0.0000	0.0000	0.6863	0.3137	2.46
0.0000	0.0000	0.6912	0.3088	2.47
0.0000	0.0000	0.6961	0.3039	2.48
0.0000	0.0000	0.7010	0.2990	2.49
0.0000	0.0000	0.7058	0.2942	2.50
0.0000	0.0000	0.7105	0.2895	2.51
0.0000	0.0000	0.7152	0.2847	2.52
0.0000	0.0000	0.7199	0.2801	2.53
0.0000	0.0000	0.7245	0.2755	2.54
0.0000	0.0000	0.7291	0.2709	2.55
0.0000	0.0000	0.7336	0.2664	2.56
0.0000	0.0000	0.7381	0.2619	2.57
0.0000	0.0000	0.7425	0.2575	2.58
0.0000	0.0000	0.7469	0.2531	2.59
0.0000	0.0000	0.7512	0.2488	2.60
0.0000	0.0000	0.7555	0.2445	2.61
0.0000	0.0000	0.7597	0.2402	2.62
0.0000	0.0000	0.7639	0.2361	2.63
0.0000	0.0000	0.7680	0.2319	2.64
0.0000	0.0000	0.7721	0.2279	2.65
0.0000	0.0000	0.7761	0.2238	2.66
0.0000	0.0000	0.7801	0.2199	2.67
0.0000	0.0000	0.7840	0.2159	2.68
0.0000	0.0000	0.7879	0.2121	2.69
0.0000	0.0000	0.7917	0.2082	2.70
0.0000	0.0000	0.7955	0.2045	2.71
0.0000	0.0000	0.7992	0.2008	2.72
0.0000	0.0000	0.8029	0.1971	2.73
0.0000	0.0000	0.8065	0.1935	2.74
0.0000	0.0000	0.8101	0.1899	2.75

Table IIA, continued

pH Value	Alpha 0	Alpha 1	Alpha 2	Alpha 3
3.31	0.0000	0.0001	0.9392	0.0606
3.32	0.0000	0.0001	0.9405	0.0593
3.33	0.0000	0.0001	0.9418	0.0581
3.34	0.0000	0.0001	0.9430	0.0568
3.35	0.0000	0.0001	0.9443	0.0556
3.36	0.0000	0.0001	0.9455	0.0544
3.37	0.0000	0.0001	0.9466	0.0532
3.38	0.0000	0.0001	0.9478	0.0521
3.39	0.0000	0.0001	0.9489	0.0510
3.40	0.0000	0.0002	0.9500	0.0499
3.41	0.0000	0.0002	0.9511	0.0488
3.42	0.0000	0.0002	0.9521	0.0477
3.43	0.0000	0.0002	0.9532	0.0467
3.44	0.0000	0.0002	0.9542	0.0457
3.45	0.0000	0.0002	0.9552	0.0447
3.46	0.0000	0.0002	0.9561	0.0437
3.47	0.0000	0.0002	0.9571	0.0428
3.48	0.0000	0.0002	0.9580	0.0418
3.49	0.0000	0.0002	0.9589	0.0409
3.50	0.0000	0.0002	0.9598	0.0400
3.51	0.0000	0.0002	0.9607	0.0391
3.52	0.0000	0.0002	0.9615	0.0383
3.53	0.0000	0.0002	0.9624	0.0374
3.54	0.0000	0.0002	0.9632	0.0366
3.55	0.0000	0.0002	0.9640	0.0358

pH Value	Alpha 0	Alpha 1	Alpha 2	Alpha 3
3.86	0.0000	0.0004	0.9817	0.0179
3.87	0.0000	0.0005	0.9821	0.0175
3.88	0.0000	0.0005	0.9825	0.0171
3.89	0.0000	0.0005	0.9828	0.0167
3.90	0.0000	0.0005	0.9832	0.0163
3.91	0.0000	0.0005	0.9835	0.0160
3.92	0.0000	0.0005	0.9839	0.0156
3.93	0.0000	0.0005	0.9842	0.0152
3.94	0.0000	0.0005	0.9846	0.0149
3.95	0.0000	0.0006	0.9849	0.0146
3.96	0.0000	0.0006	0.9852	0.0142
3.97	0.0000	0.0006	0.9855	0.0139
3.98	0.0000	0.0006	0.9858	0.0136
3.99	0.0000	0.0006	0.9861	0.0133
4.00	0.0000	0.0006	0.9864	0.0130
4.01	0.0000	0.0006	0.9867	0.0127
4.02	0.0000	0.0007	0.9869	0.0124
4.03	0.0000	0.0007	0.9872	0.0121
4.04	0.0000	0.0007	0.9874	0.0119
4.05	0.0000	0.0007	0.9877	0.0116
4.06	0.0000	0.0007	0.9879	0.0113
4.07	0.0000	0.0007	0.9882	0.0111
4.08	0.0000	0.0007	0.9884	0.0108
4.09	0.0000	0.0008	0.9886	0.0106
4.10	0.0000	0.0008	0.9889	0.0104

				pH
0.0000	0.0008	0.9891	0.0101	4.11
0.0000	0.0008	0.9893	0.0099	4.12
0.0000	0.0008	0.9895	0.0097	4.13
0.0000	0.0009	0.9897	0.0095	4.14
0.0000	0.0009	0.9899	0.0092	4.15
0.0000	0.0009	0.9901	0.0090	4.16
0.0000	0.0009	0.9903	0.0088	4.17
0.0000	0.0009	0.9904	0.0086	4.18
0.0000	0.0010	0.9906	0.0084	4.19
0.0000	0.0010	0.9908	0.0082	4.20
0.0000	0.0010	0.9909	0.0081	4.21
0.0000	0.0010	0.9911	0.0079	4.22
0.0000	0.0011	0.9912	0.0077	4.23
0.0000	0.0011	0.9914	0.0075	4.24
0.0000	0.0011	0.9915	0.0074	4.25
0.0000	0.0011	0.9917	0.0072	4.26
0.0000	0.0012	0.9918	0.0070	4.27
0.0000	0.0012	0.9919	0.0069	4.28
0.0000	0.0012	0.9921	0.0067	4.29
0.0000	0.0012	0.9922	0.0066	4.30
0.0000	0.0013	0.9923	0.0064	4.31
0.0000	0.0013	0.9924	0.0063	4.32
0.0000	0.0013	0.9925	0.0061	4.33
0.0000	0.0014	0.9926	0.0060	4.34
0.0000	0.0014	0.9928	0.0058	4.35
0.0000	0.0014	0.9929	0.0057	4.36
0.0000	0.0015	0.9929	0.0056	4.37
0.0000	0.0015	0.9930	0.0055	4.38
0.0000	0.0015	0.9931	0.0053	4.39
0.0000	0.0016	0.9932	0.0052	4.40

				pH
0.0000	0.0002	0.9648	0.0350	3.56
0.0000	0.0002	0.9655	0.0343	3.57
0.0000	0.0002	0.9663	0.0335	3.58
0.0000	0.0002	0.9670	0.0328	3.59
0.0000	0.0002	0.9677	0.0320	3.60
0.0000	0.0002	0.9684	0.0313	3.61
0.0000	0.0003	0.9691	0.0306	3.62
0.0000	0.0003	0.9698	0.0300	3.63
0.0000	0.0003	0.9704	0.0293	3.64
0.0000	0.0003	0.9711	0.0287	3.65
0.0000	0.0003	0.9717	0.0280	3.66
0.0000	0.0003	0.9723	0.0274	3.67
0.0000	0.0003	0.9729	0.0268	3.68
0.0000	0.0003	0.9735	0.0262	3.69
0.0000	0.0003	0.9741	0.0256	3.70
0.0000	0.0003	0.9746	0.0251	3.71
0.0000	0.0003	0.9752	0.0245	3.72
0.0000	0.0003	0.9757	0.0240	3.73
0.0000	0.0003	0.9762	0.0234	3.74
0.0000	0.0003	0.9768	0.0229	3.75
0.0000	0.0004	0.9773	0.0224	3.76
0.0000	0.0004	0.9777	0.0219	3.77
0.0000	0.0004	0.9782	0.0214	3.78
0.0000	0.0004	0.9787	0.0209	3.79
0.0000	0.0004	0.9792	0.0205	3.80
0.0000	0.0004	0.9796	0.0200	3.81
0.0000	0.0004	0.9800	0.0196	3.82
0.0000	0.0004	0.9805	0.0191	3.83
0.0000	0.0004	0.9809	0.0187	3.84
0.0000	0.0004	0.9813	0.0183	3.85

Table IIA, continued

pH Value	Alpha 3	Alpha 2	Alpha 1	Alpha 0
4.41	0.0051	0.9933	0.0016	0.0000
4.42	0.0050	0.9934	0.0016	0.0000
4.43	0.0049	0.9934	0.0017	0.0000
4.44	0.0048	0.9935	0.0017	0.0000
4.45	0.0046	0.9936	0.0018	0.0000
4.46	0.0045	0.9936	0.0018	0.0000
4.47	0.0044	0.9937	0.0019	0.0000
4.48	0.0043	0.9938	0.0019	0.0000
4.49	0.0042	0.9938	0.0019	0.0000
4.50	0.0041	0.9939	0.0020	0.0000
4.51	0.0040	0.9939	0.0020	0.0000
4.52	0.0040	0.9940	0.0021	0.0000
4.53	0.0039	0.9940	0.0021	0.0000
4.54	0.0038	0.9940	0.0022	0.0000
4.55	0.0037	0.9941	0.0022	0.0000
4.56	0.0036	0.9941	0.0023	0.0000
4.57	0.0035	0.9941	0.0023	0.0000
4.58	0.0034	0.9942	0.0024	0.0000
4.59	0.0034	0.9942	0.0024	0.0000
4.60	0.0033	0.9942	0.0025	0.0000
4.61	0.0032	0.9942	0.0026	0.0000
4.62	0.0031	0.9942	0.0026	0.0000
4.63	0.0031	0.9943	0.0027	0.0000
4.64	0.0030	0.9943	0.0027	0.0000
4.65	0.0029	0.9943	0.0028	0.0000
4.96	0.0014	0.9929	0.0057	0.0000
4.97	0.0014	0.9928	0.0058	0.0000
4.98	0.0014	0.9926	0.0060	0.0000
4.99	0.0013	0.9925	0.0061	0.0000
5.00	0.0013	0.9924	0.0063	0.0000
5.01	0.0013	0.9923	0.0064	0.0000
5.02	0.0012	0.9922	0.0066	0.0000
5.03	0.0012	0.9921	0.0067	0.0000
5.04	0.0012	0.9919	0.0069	0.0000
5.05	0.0012	0.9918	0.0070	0.0000
5.06	0.0011	0.9917	0.0072	0.0000
5.07	0.0011	0.9915	0.0074	0.0000
5.08	0.0011	0.9914	0.0075	0.0000
5.09	0.0011	0.9912	0.0077	0.0000
5.10	0.0010	0.9911	0.0079	0.0000
5.11	0.0010	0.9909	0.0081	0.0000
5.12	0.0010	0.9908	0.0082	0.0000
5.13	0.0010	0.9906	0.0084	0.0000
5.14	0.0009	0.9904	0.0086	0.0000
5.15	0.0009	0.9903	0.0088	0.0000
5.16	0.0009	0.9901	0.0090	0.0000
5.17	0.0009	0.9899	0.0092	0.0000
5.18	0.0009	0.9897	0.0095	0.0000
5.19	0.0008	0.9895	0.0097	0.0000
5.20	0.0008	0.9893	0.0099	0.0000

pH					pH				
4.66	0.0000	0.0029	0.9943	0.0029	5.21	0.0008	0.9891	0.0101	0.0000
4.67	0.0000	0.0029	0.9943	0.0028	5.22	0.0008	0.9889	0.0104	0.0000
4.68	0.0000	0.0030	0.9943	0.0027	5.23	0.0008	0.9886	0.0106	0.0000
4.69	0.0000	0.0031	0.9943	0.0027	5.24	0.0007	0.9884	0.0108	0.0000
4.70	0.0000	0.0031	0.9942	0.0026	5.25	0.0007	0.9882	0.0111	0.0000
4.71	0.0000	0.0032	0.9942	0.0026	5.26	0.0007	0.9879	0.0113	0.0000
4.72	0.0000	0.0033	0.9942	0.0025	5.27	0.0007	0.9877	0.0116	0.0000
4.73	0.0000	0.0034	0.9942	0.0024	5.28	0.0007	0.9874	0.0119	0.0000
4.74	0.0000	0.0034	0.9942	0.0024	5.29	0.0007	0.9872	0.0121	0.0000
4.75	0.0000	0.0035	0.9941	0.0023	5.30	0.0007	0.9869	0.0124	0.0000
4.76	0.0000	0.0036	0.9941	0.0023	5.31	0.0006	0.9867	0.0127	0.0000
4.77	0.0000	0.0037	0.9941	0.0022	5.32	0.0006	0.9864	0.0130	0.0000
4.78	0.0000	0.0038	0.9940	0.0022	5.33	0.0006	0.9861	0.0133	0.0000
4.79	0.0000	0.0039	0.9940	0.0021	5.34	0.0006	0.9858	0.0136	0.0000
4.80	0.0000	0.0040	0.9940	0.0021	5.35	0.0006	0.9855	0.0139	0.0000
4.81	0.0000	0.0040	0.9939	0.0020	5.36	0.0006	0.9852	0.0142	0.0000
4.82	0.0000	0.0041	0.9939	0.0020	5.37	0.0006	0.9849	0.0146	0.0000
4.83	0.0000	0.0042	0.9938	0.0019	5.38	0.0005	0.9846	0.0149	0.0000
4.84	0.0000	0.0043	0.9938	0.0019	5.39	0.0005	0.9842	0.0152	0.0000
4.85	0.0000	0.0044	0.9937	0.0019	5.40	0.0005	0.9839	0.0156	0.0000
4.86	0.0000	0.0045	0.9937	0.0018	5.41	0.0005	0.9835	0.0160	0.0000
4.87	0.0000	0.0046	0.9936	0.0018	5.42	0.0005	0.9832	0.0163	0.0000
4.88	0.0000	0.0048	0.9935	0.0017	5.43	0.0005	0.9828	0.0167	0.0000
4.89	0.0000	0.0049	0.9934	0.0017	5.44	0.0005	0.9825	0.0171	0.0000
4.90	0.0000	0.0050	0.9934	0.0016	5.45	0.0005	0.9821	0.0175	0.0000
4.91	0.0000	0.0051	0.9933	0.0016	5.46	0.0004	0.9817	0.0179	0.0000
4.92	0.0000	0.0052	0.9932	0.0016	5.47	0.0004	0.9813	0.0183	0.0000
4.93	0.0000	0.0053	0.9931	0.0015	5.48	0.0004	0.9809	0.0187	0.0000
4.94	0.0000	0.0055	0.9930	0.0015	5.49	0.0004	0.9805	0.0191	0.0000
4.95	0.0000	0.0056	0.9929	0.0015	5.50	0.0004	0.9800	0.0196	0.0000

Table IIA, continued

pH Value	Alpha 3	Alpha 2	Alpha 1	Alpha 0	pH Value	Alpha 0	Alpha 1	Alpha 2	Alpha 3
5.51	0.0004	0.9796	0.0200	0.0000	6.06	0.0000	0.0675	0.9324	0.0001
5.52	0.0004	0.9792	0.0205	0.0000	6.07	0.0000	0.0690	0.9309	0.0001
5.53	0.0004	0.9787	0.0209	0.0000	6.08	0.0000	0.0705	0.9294	0.0001
5.54	0.0004	0.9782	0.0214	0.0000	6.09	0.0000	0.0720	0.9279	0.0001
5.55	0.0004	0.9777	0.0219	0.0000	6.10	0.0000	0.0736	0.9263	0.0001
5.56	0.0004	0.9773	0.0224	0.0000	6.11	0.0000	0.0752	0.9247	0.0001
5.57	0.0003	0.9768	0.0229	0.0000	6.12	0.0000	0.0768	0.9231	0.0001
5.58	0.0003	0.9762	0.0234	0.0000	6.13	0.0000	0.0784	0.9215	0.0001
5.59	0.0003	0.9757	0.0240	0.0000	6.14	0.0000	0.0801	0.9198	0.0001
5.60	0.0003	0.9752	0.0245	0.0000	6.15	0.0000	0.0818	0.9181	0.0001
5.61	0.0003	0.9746	0.0251	0.0000	6.16	0.0000	0.0836	0.9163	0.0001
5.62	0.0003	0.9741	0.0256	0.0000	6.17	0.0000	0.0854	0.9146	0.0001
5.63	0.0003	0.9735	0.0262	0.0000	6.18	0.0000	0.0872	0.9128	0.0001
5.64	0.0003	0.9729	0.0268	0.0000	6.19	0.0000	0.0890	0.9109	0.0001
5.65	0.0003	0.9723	0.0274	0.0000	6.20	0.0000	0.0909	0.9090	0.0001
5.66	0.0003	0.9717	0.0280	0.0000	6.21	0.0000	0.0928	0.9071	0.0001
5.67	0.0003	0.9711	0.0287	0.0000	6.22	0.0000	0.0948	0.9051	0.0001
5.68	0.0003	0.9704	0.0293	0.0000	6.23	0.0000	0.0968	0.9032	0.0001
5.69	0.0003	0.9698	0.0300	0.0000	6.24	0.0000	0.0988	0.9011	0.0001
5.70	0.0003	0.9691	0.0306	0.0000	6.25	0.0000	0.1009	0.8991	0.0001
5.71	0.0002	0.9684	0.0313	0.0000	6.26	0.0000	0.1030	0.8970	0.0001
5.72	0.0002	0.9677	0.0320	0.0000	6.27	0.0000	0.1051	0.8948	0.0001
5.73	0.0002	0.9670	0.0328	0.0000	6.28	0.0000	0.1073	0.8926	0.0001
5.74	0.0002	0.9663	0.0335	0.0000	6.29	0.0000	0.1095	0.8904	0.0001
5.75	0.0002	0.9655	0.0343	0.0000	6.30	0.0000	0.1118	0.8881	0.0001

pH				
5.76	0.0000	0.0350	0.9648	0.0002
5.77	0.0000	0.0358	0.9640	0.0002
5.78	0.0000	0.0366	0.9632	0.0002
5.79	0.0000	0.0374	0.9624	0.0002
5.80	0.0000	0.0383	0.9615	0.0002
5.81	0.0000	0.0391	0.9607	0.0002
5.82	0.0000	0.0400	0.9598	0.0002
5.83	0.0000	0.0409	0.9589	0.0002
5.84	0.0000	0.0418	0.9580	0.0002
5.85	0.0000	0.0428	0.9571	0.0002
5.86	0.0000	0.0437	0.9561	0.0002
5.87	0.0000	0.0447	0.9552	0.0002
5.88	0.0000	0.0457	0.9542	0.0002
5.89	0.0000	0.0467	0.9532	0.0002
5.90	0.0000	0.0477	0.9521	0.0002
5.91	0.0000	0.0488	0.9511	0.0002
5.92	0.0000	0.0499	0.9500	0.0002
5.93	0.0000	0.0510	0.9489	0.0001
5.94	0.0000	0.0521	0.9478	0.0001
5.95	0.0000	0.0532	0.9466	0.0001
5.96	0.0000	0.0544	0.9455	0.0001
5.97	0.0000	0.0556	0.9443	0.0001
5.98	0.0000	0.0568	0.9430	0.0001
5.99	0.0000	0.0581	0.9418	0.0001
6.00	0.0000	0.0593	0.9405	0.0001
6.01	0.0000	0.0606	0.9392	0.0001
6.02	0.0000	0.0620	0.9379	0.0001
6.03	0.0000	0.0633	0.9366	0.0001
6.04	0.0000	0.0647	0.9352	0.0001
6.05	0.0000	0.0661	0.9338	0.0001
6.31	0.0000	0.1141	0.8858	0.0001
6.32	0.0000	0.1165	0.8835	0.0001
6.33	0.0000	0.1189	0.8811	0.0001
6.34	0.0000	0.1213	0.8787	0.0001
6.35	0.0000	0.1238	0.8762	0.0001
6.36	0.0000	0.1263	0.8737	0.0001
6.37	0.0000	0.1288	0.8711	0.0000
6.38	0.0000	0.1315	0.8685	0.0000
6.39	0.0000	0.1341	0.8658	0.0000
6.40	0.0000	0.1368	0.8632	0.0000
6.41	0.0000	0.1395	0.8604	0.0000
6.42	0.0000	0.1423	0.8576	0.0000
6.43	0.0000	0.1452	0.8548	0.0000
6.44	0.0000	0.1480	0.8519	0.0000
6.45	0.0000	0.1510	0.8490	0.0000
6.46	0.0000	0.1539	0.8460	0.0000
6.47	0.0000	0.1570	0.8430	0.0000
6.48	0.0000	0.1600	0.8399	0.0000
6.49	0.0000	0.1632	0.8368	0.0000
6.50	0.0000	0.1663	0.8336	0.0000
6.51	0.0000	0.1695	0.8304	0.0000
6.52	0.0000	0.1728	0.8272	0.0000
6.53	0.0000	0.1761	0.8238	0.0000
6.54	0.0000	0.1795	0.8205	0.0000
6.55	0.0000	0.1829	0.8171	0.0000
6.56	0.0000	0.1864	0.8136	0.0000
6.57	0.0000	0.1899	0.8101	0.0000
6.58	0.0000	0.1935	0.8065	0.0000
6.59	0.0000	0.1971	0.8029	0.0000
6.60	0.0000	0.2008	0.7992	0.0000

Table IIA, continued

pH Value	Alpha 3	Alpha 2	Alpha 1	Alpha 0
6.61	0.0000	0.7955	0.2045	0.0000
6.62	0.0000	0.7917	0.2082	0.0000
6.63	0.0000	0.7879	0.2121	0.0000
6.64	0.0000	0.7840	0.2159	0.0000
6.65	0.0000	0.7801	0.2199	0.0000
6.66	0.0000	0.7761	0.2238	0.0000
6.67	0.0000	0.7721	0.2279	0.0000
6.68	0.0000	0.7680	0.2319	0.0000
6.69	0.0000	0.7639	0.2361	0.0000
6.70	0.0000	0.7597	0.2402	0.0000
6.71	0.0000	0.7555	0.2445	0.0000
6.72	0.0000	0.7512	0.2488	0.0000
6.73	0.0000	0.7469	0.2531	0.0000
6.74	0.0000	0.7425	0.2575	0.0000
6.75	0.0000	0.7381	0.2619	0.0000
6.76	0.0000	0.7336	0.2664	0.0000
6.77	0.0000	0.7291	0.2709	0.0000
6.78	0.0000	0.7245	0.2755	0.0000
6.79	0.0000	0.7199	0.2801	0.0000
6.80	0.0000	0.7152	0.2847	0.0000
6.81	0.0000	0.7105	0.2895	0.0000
6.82	0.0000	0.7058	0.2942	0.0000
6.83	0.0000	0.7010	0.2990	0.0000
6.84	0.0000	0.6961	0.3039	0.0000
6.85	0.0000	0.6912	0.3088	0.0000

pH Value	Alpha 3	Alpha 2	Alpha 1	Alpha 0
7.16	0.0000	0.5230	0.4770	0.0000
7.17	0.0000	0.5173	0.4827	0.0000
7.18	0.0000	0.5115	0.4885	0.0000
7.19	0.0000	0.5058	0.4942	0.0000
7.20	0.0000	0.5000	0.5000	0.0000
7.21	0.0000	0.4942	0.5058	0.0000
7.22	0.0000	0.4885	0.5115	0.0000
7.23	0.0000	0.4827	0.5173	0.0000
7.24	0.0000	0.4770	0.5230	0.0000
7.25	0.0000	0.4712	0.5287	0.0000
7.26	0.0000	0.4655	0.5345	0.0000
7.27	0.0000	0.4598	0.5402	0.0000
7.28	0.0000	0.4541	0.5459	0.0000
7.29	0.0000	0.4484	0.5516	0.0000
7.30	0.0000	0.4427	0.5573	0.0000
7.31	0.0000	0.4370	0.5630	0.0000
7.32	0.0000	0.4314	0.5686	0.0000
7.33	0.0000	0.4257	0.5743	0.0000
7.34	0.0000	0.4201	0.5799	0.0000
7.35	0.0000	0.4145	0.5855	0.0000
7.36	0.0000	0.4089	0.5911	0.0000
7.37	0.0000	0.4034	0.5966	0.0000
7.38	0.0000	0.3978	0.6022	0.0000
7.39	0.0000	0.3923	0.6077	0.0000
7.40	0.0000	0.3869	0.6131	0.0000

				pH	
0.0000	0.6186	0.3814	0.0000	7.41	0.0000
0.0000	0.6240	0.3760	0.0000	7.42	0.0000
0.0000	0.6294	0.3706	0.0000	7.43	0.0000
0.0000	0.6347	0.3653	0.0000	7.44	0.0000
0.0000	0.6401	0.3599	0.0000	7.45	0.0000
0.0000	0.6453	0.3546	0.0000	7.46	0.0000
0.0000	0.6506	0.3494	0.0000	7.47	0.0000
0.0000	0.6558	0.3442	0.0000	7.48	0.0000
0.0000	0.6610	0.3390	0.0000	7.49	0.0000
0.0000	0.6661	0.3339	0.0000	7.50	0.0000
0.0000	0.6712	0.3288	0.0000	7.51	0.0000
0.0000	0.6763	0.3237	0.0000	7.52	0.0000
0.0000	0.6813	0.3187	0.0000	7.53	0.0000
0.0000	0.6863	0.3137	0.0000	7.54	0.0000
0.0000	0.6912	0.3088	0.0000	7.55	0.0000
0.0000	0.6961	0.3039	0.0000	7.56	0.0000
0.0000	0.7010	0.2990	0.0000	7.57	0.0000
0.0000	0.7058	0.2942	0.0000	7.58	0.0000
0.0000	0.7105	0.2895	0.0000	7.59	0.0000
0.0000	0.7152	0.2847	0.0000	7.60	0.0000
0.0000	0.7199	0.2801	0.0000	7.61	0.0000
0.0000	0.7245	0.2755	0.0000	7.62	0.0000
0.0000	0.7291	0.2709	0.0000	7.63	0.0000
0.0000	0.7336	0.2664	0.0000	7.64	0.0000
0.0000	0.7381	0.2619	0.0000	7.65	0.0000
0.0000	0.7425	0.2575	0.0000	7.66	0.0000
0.0000	0.7469	0.2531	0.0000	7.67	0.0000
0.0000	0.7512	0.2488	0.0000	7.68	0.0000
0.0000	0.7555	0.2445	0.0000	7.69	0.0000
0.0000	0.7597	0.2402	0.0000	7.70	0.0000

				pH	
0.0000	0.3137	0.6863	0.0000	6.86	0.0000
0.0000	0.3187	0.6813	0.0000	6.87	0.0000
0.0000	0.3237	0.6763	0.0000	6.88	0.0000
0.0000	0.3288	0.6712	0.0000	6.89	0.0000
0.0000	0.3339	0.6661	0.0000	6.90	0.0000
0.0000	0.3390	0.6610	0.0000	6.91	0.0000
0.0000	0.3442	0.6558	0.0000	6.92	0.0000
0.0000	0.3494	0.6506	0.0000	6.93	0.0000
0.0000	0.3546	0.6453	0.0000	6.94	0.0000
0.0000	0.3599	0.6401	0.0000	6.95	0.0000
0.0000	0.3653	0.6347	0.0000	6.96	0.0000
0.0000	0.3706	0.6294	0.0000	6.97	0.0000
0.0000	0.3760	0.6240	0.0000	6.98	0.0000
0.0000	0.3814	0.6186	0.0000	6.99	0.0000
0.0000	0.3869	0.6131	0.0000	7.00	0.0000
0.0000	0.3923	0.6077	0.0000	7.01	0.0000
0.0000	0.3978	0.6022	0.0000	7.02	0.0000
0.0000	0.4034	0.5966	0.0000	7.03	0.0000
0.0000	0.4089	0.5911	0.0000	7.04	0.0000
0.0000	0.4145	0.5855	0.0000	7.05	0.0000
0.0000	0.4201	0.5799	0.0000	7.06	0.0000
0.0000	0.4257	0.5743	0.0000	7.07	0.0000
0.0000	0.4314	0.5686	0.0000	7.08	0.0000
0.0000	0.4370	0.5630	0.0000	7.09	0.0000
0.0000	0.4427	0.5573	0.0000	7.10	0.0000
0.0000	0.4484	0.5516	0.0000	7.11	0.0000
0.0000	0.4541	0.5459	0.0000	7.12	0.0000
0.0000	0.4598	0.5402	0.0000	7.13	0.0000
0.0000	0.4655	0.5345	0.0000	7.14	0.0000
0.0000	0.4712	0.5287	0.0000	7.15	0.0000

Table IIA, continued

pH Value	Alpha 0	Alpha 1	Alpha 2	Alpha 3
8.26	0.0001	0.9198	0.0801	0.0000
8.27	0.0001	0.9215	0.0784	0.0000
8.28	0.0001	0.9231	0.0768	0.0000
8.29	0.0001	0.9248	0.0752	0.0000
8.30	0.0001	0.9263	0.0736	0.0000
8.31	0.0001	0.9279	0.0720	0.0000
8.32	0.0001	0.9294	0.0705	0.0000
8.33	0.0001	0.9309	0.0690	0.0000
8.34	0.0001	0.9324	0.0675	0.0000
8.35	0.0001	0.9338	0.0661	0.0000
8.36	0.0001	0.9352	0.0647	0.0000
8.37	0.0001	0.9366	0.0633	0.0000
8.38	0.0001	0.9379	0.0620	0.0000
8.39	0.0001	0.9393	0.0606	0.0000
8.40	0.0001	0.9405	0.0593	0.0000
8.41	0.0001	0.9418	0.0581	0.0000
8.42	0.0001	0.9431	0.0568	0.0000
8.43	0.0001	0.9443	0.0556	0.0000
8.44	0.0001	0.9455	0.0544	0.0000
8.45	0.0001	0.9466	0.0532	0.0000
8.46	0.0001	0.9478	0.0521	0.0000
8.47	0.0001	0.9489	0.0510	0.0000
8.48	0.0001	0.9500	0.0499	0.0000
8.49	0.0001	0.9511	0.0488	0.0000
8.50	0.0001	0.9521	0.0477	0.0000

pH Value	Alpha 3	Alpha 2	Alpha 1	Alpha 0
7.71	0.0000	0.2361	0.7639	0.0000
7.72	0.0000	0.2319	0.7680	0.0000
7.73	0.0000	0.2279	0.7721	0.0000
7.74	0.0000	0.2238	0.7761	0.0000
7.75	0.0000	0.2199	0.7801	0.0000
7.76	0.0000	0.2159	0.7840	0.0000
7.77	0.0000	0.2121	0.7879	0.0000
7.78	0.0000	0.2082	0.7917	0.0000
7.79	0.0000	0.2045	0.7955	0.0000
7.80	0.0000	0.2008	0.7992	0.0000
7.81	0.0000	0.1971	0.8029	0.0000
7.82	0.0000	0.1935	0.8065	0.0000
7.83	0.0000	0.1899	0.8101	0.0000
7.84	0.0000	0.1864	0.8136	0.0000
7.85	0.0000	0.1829	0.8171	0.0000
7.86	0.0000	0.1795	0.8205	0.0000
7.87	0.0000	0.1761	0.8238	0.0000
7.88	0.0000	0.1728	0.8272	0.0000
7.89	0.0000	0.1696	0.8304	0.0000
7.90	0.0000	0.1663	0.8336	0.0000
7.91	0.0000	0.1632	0.8368	0.0000.
7.92	0.0000	0.1600	0.8399	0.0000
7.93	0.0000	0.1570	0.8430	0.0000
7.94	0.0000	0.1539	0.8460	0.0000
7.95	0.0000	0.1510	0.8490	0.0000

				pH
0.0000	0.8519	0.1480	0.0000	7.96
0.0000	0.8548	0.1452	0.0000	7.97
0.0000	0.8576	0.1423	0.0000	7.98
0.0000	0.8604	0.1395	0.0000	7.99
0.0000	0.8632	0.1368	0.0000	8.00
0.0000	0.8659	0.1341	0.0000	8.01
0.0000	0.8685	0.1315	0.0000	8.02
0.0000	0.8711	0.1288	0.0000	8.03
0.0000	0.8737	0.1263	0.0000	8.04
0.0000	0.8762	0.1238	0.0000	8.05
0.0000	0.8787	0.1213	0.0000	8.06
0.0000	0.8811	0.1189	0.0000	8.07
0.0000	0.8835	0.1165	0.0000	8.08
0.0000	0.8858	0.1141	0.0000	8.09
0.0000	0.8881	0.1118	0.0000	8.10
0.0000	0.8904	0.1095	0.0000	8.11
0.0000	0.8926	0.1073	0.0000	8.12
0.0000	0.8948	0.1051	0.0000	8.13
0.0000	0.8970	0.1030	0.0000	8.14
0.0000	0.8991	0.1009	0.0000	8.15
0.0000	0.9011	0.0988	0.0000	8.16
0.0000	0.9032	0.0968	0.0000	8.17
0.0000	0.9052	0.0948	0.0000	8.18
0.0000	0.9071	0.0928	0.0000	8.19
0.0000	0.9090	0.0909	0.0000	8.20
0.0000	0.9109	0.0890	0.0000	8.21
0.0000	0.9128	0.0872	0.0000	8.22
0.0000	0.9146	0.0854	0.0000	8.23
0.0000	0.9164	0.0836	0.0000	8.24
0.0000	0.9181	0.0818	0.0000	8.25

				pH
0.0001	0.9532	0.0467	0.0000	8.51
0.0001	0.9542	0.0457	0.0000	8.52
0.0001	0.9552	0.0447	0.0000	8.53
0.0001	0.9561	0.0437	0.0000	8.54
0.0002	0.9571	0.0428	0.0000	8.55
0.0002	0.9580	0.0418	0.0000	8.56
0.0002	0.9589	0.0409	0.0000	8.57
0.0002	0.9598	0.0400	0.0000	8.58
0.0002	0.9607	0.0391	0.0000	8.59
0.0002	0.9615	0.0383	0.0000	8.60
0.0002	0.9624	0.0374	0.0000	8.61
0.0002	0.9632	0.0366	0.0000	8.62
0.0002	0.9640	0.0358	0.0000	8.63
0.0002	0.9648	0.0350	0.0000	8.64
0.0002	0.9655	0.0343	0.0000	8.65
0.0002	0.9663	0.0335	0.0000	8.66
0.0002	0.9670	0.0328	0.0000	8.67
0.0002	0.9677	0.0320	0.0000	8.68
0.0002	0.9684	0.0313	0.0000	8.69
0.0002	0.9691	0.0306	0.0000	8.70
0.0002	0.9698	0.0300	0.0000	8.71
0.0002	0.9705	0.0293	0.0000	8.72
0.0002	0.9711	0.0287	0.0000	8.73
0.0002	0.9717	0.0280	0.0000	8.74
0.0002	0.9724	0.0274	0.0000	8.75
0.0003	0.9730	0.0268	0.0000	8.76
0.0003	0.9735	0.0262	0.0000	8.77
0.0003	0.9741	0.0256	0.0000	8.78
0.0003	0.9747	0.0251	0.0000	8.79
0.0003	0.9752	0.0245	0.0000	8.80

Table IIA, continued

pH Value	Alpha 0	Alpha 1	Alpha 2	Alpha 3
8.81	0.0003	0.9758	0.0240	0.0000
8.82	0.0003	0.9763	0.0234	0.0000
8.83	0.0003	0.9768	0.0229	0.0000
8.84	0.0003	0.9773	0.0224	0.0000
8.85	0.0003	0.9778	0.0219	0.0000
8.86	0.0003	0.9783	0.0214	0.0000
8.87	0.0003	0.9788	0.0209	0.0000
8.88	0.0003	0.9792	0.0205	0.0000
8.89	0.0003	0.9797	0.0200	0.0000
8.90	0.0003	0.9801	0.0196	0.0000
8.91	0.0004	0.9805	0.0191	0.0000
8.92	0.0004	0.9809	0.0187	0.0000
8.93	0.0004	0.9814	0.0183	0.0000
8.94	0.0004	0.9818	0.0179	0.0000
8.95	0.0004	0.9821	0.0175	0.0000
8.96	0.0004	0.9825	0.0171	0.0000
8.97	0.0004	0.9829	0.0167	0.0000
8.98	0.0004	0.9833	0.0163	0.0000
8.99	0.0004	0.9836	0.0160	0.0000
9.00	0.0004	0.9840	0.0156	0.0000
9.01	0.0004	0.9843	0.0152	0.0000
9.02	0.0005	0.9846	0.0149	0.0000
9.03	0.0005	0.9850	0.0146	0.0000
9.04	0.0005	0.9853	0.0142	0.0000
9.05	0.0005	0.9856	0.0139	0.0000

pH Value	Alpha 0	Alpha 1	Alpha 2	Alpha 3
9.36	0.0010	0.9921	0.0069	0.0000
9.37	0.0010	0.9923	0.0067	0.0000
9.38	0.0011	0.9924	0.0066	0.0000
9.39	0.0011	0.9925	0.0064	0.0000
9.40	0.0011	0.9926	0.0063	0.0000
9.41	0.0011	0.9927	0.0061	0.0000
9.42	0.0012	0.9929	0.0060	0.0000
9.43	0.0012	0.9930	0.0058	0.0000
9.44	0.0012	0.9931	0.0057	0.0000
9.45	0.0013	0.9932	0.0056	0.0000
9.46	0.0013	0.9933	0.0055	0.0000
9.47	0.0013	0.9934	0.0053	0.0000
9.48	0.0013	0.9934	0.0052	0.0000
9.49	0.0014	0.9935	0.0051	0.0000
9.50	0.0014	0.9936	0.0050	0.0000
9.51	0.0014	0.9937	0.0049	0.0000
9.52	0.0015	0.9938	0.0048	0.0000
9.53	0.0015	0.9938	0.0046	0.0000
9.54	0.0015	0.9939	0.0045	0.0000
9.55	0.0016	0.9940	0.0044	0.0000
9.56	0.0016	0.9940	0.0043	0.0000
9.57	0.0016	0.9941	0.0042	0.0000
9.58	0.0017	0.9942	0.0041	0.0000
9.59	0.0017	0.9942	0.0041	0.0000
9.60	0.0018	0.9943	0.0040	0.0000

pH				
9.06	0.0005	0.9859	0.0136	0.0000
9.07	0.0005	0.9862	0.0133	0.0000
9.08	0.0005	0.9865	0.0130	0.0000
9.09	0.0005	0.9867	0.0127	0.0000
9.10	0.0006	0.9870	0.0124	0.0000
9.11	0.0006	0.9873	0.0121	0.0000
9.12	0.0006	0.9875	0.0119	0.0000
9.13	0.0006	0.9878	0.0116	0.0000
9.14	0.0006	0.9880	0.0113	0.0000
9.15	0.0006	0.9883	0.0111	0.0000
9.16	0.0006	0.9885	0.0108	0.0000
9.17	0.0007	0.9888	0.0106	0.0000
9.18	0.0007	0.9890	0.0104	0.0000
9.19	0.0007	0.9892	0.0101	0.0000
9.20	0.0007	0.9894	0.0099	0.0000
9.21	0.0007	0.9896	0.0097	0.0000
9.22	0.0007	0.9898	0.0095	0.0000
9.23	0.0008	0.9900	0.0092	0.0000
9.24	0.0008	0.9902	0.0090	0.0000
9.25	0.0008	0.9904	0.0088	0.0000
9.26	0.0008	0.9906	0.0086	0.0000
9.27	0.0008	0.9907	0.0084	0.0000
9.28	0.0008	0.9909	0.0082	0.0000
9.29	0.0009	0.9911	0.0081	0.0000
9.30	0.0009	0.9912	0.0079	0.0000
9.31	0.0009	0.9914	0.0077	0.0000
9.32	0.0009	0.9916	0.0075	0.0000
9.33	0.0009	0.9917	0.0074	0.0000
9.34	0.0010	0.9918	0.0072	0.0000
9.35	0.0010	0.9920	0.0070	0.0000

pH				
9.61	0.0018	0.9943	0.0039	0.0000
9.62	0.0019	0.9944	0.0038	0.0000
9.63	0.0019	0.9944	0.0037	0.0000
9.64	0.0019	0.9945	0.0036	0.0000
9.65	0.0020	0.9945	0.0035	0.0000
9.66	0.0020	0.9945	0.0034	0.0000
9.67	0.0021	0.9946	0.0034	0.0000
9.68	0.0021	0.9946	0.0033	0.0000
9.69	0.0022	0.9946	0.0032	0.0000
9.70	0.0022	0.9946	0.0031	0.0000
9.71	0.0023	0.9946	0.0031	0.0000
9.72	0.0023	0.9947	0.0030	0.0000
9.73	0.0024	0.9947	0.0029	0.0000
9.74	0.0024	0.9947	0.0029	0.0000
9.75	0.0025	0.9947	0.0028	0.0000
9.76	0.0026	0.9947	0.0027	0.0000
9.77	0.0026	0.9947	0.0027	0.0000
9.78	0.0027	0.9947	0.0026	0.0000
9.79	0.0027	0.9947	0.0026	0.0000
9.80	0.0028	0.9947	0.0025	0.0000
9.81	0.0029	0.9947	0.0024	0.0000
9.82	0.0029	0.9947	0.0024	0.0000
9.83	0.0030	0.9947	0.0023	0.0000
9.84	0.0031	0.9946	0.0023	0.0000
9.85	0.0031	0.9946	0.0022	0.0000
9.86	0.0032	0.9946	0.0022	0.0000
9.87	0.0033	0.9946	0.0021	0.0000
9.88	0.0034	0.9946	0.0021	0.0000
9.89	0.0034	0.9945	0.0020	0.0000
9.90	0.0035	0.9945	0.0020	0.0000

Table IIA, continued

pH Value	Alpha 0	Alpha 1	Alpha 2	Alpha 3
9.91	0.0036	0.9945	0.0019	0.0000
9.92	0.0037	0.9944	0.0019	0.0000
9.93	0.0038	0.9944	0.0019	0.0000
9.94	0.0039	0.9943	0.0018	0.0000
9.95	0.0040	0.9943	0.0018	0.0000
9.96	0.0041	0.9942	0.0017	0.0000
9.97	0.0041	0.9942	0.0017	0.0000
9.98	0.0042	0.9941	0.0016	0.0000
9.99	0.0043	0.9940	0.0016	0.0000
10.00	0.0044	0.9940	0.0016	0.0000
10.01	0.0045	0.9939	0.0015	0.0000
10.02	0.0046	0.9938	0.0015	0.0000
10.03	0.0048	0.9938	0.0015	0.0000
10.04	0.0049	0.9937	0.0014	0.0000
10.05	0.0050	0.9936	0.0014	0.0000
10.06	0.0051	0.9935	0.0014	0.0000
10.07	0.0052	0.9934	0.0013	0.0000
10.08	0.0053	0.9934	0.0013	0.0000
10.09	0.0055	0.9933	0.0013	0.0000
10.10	0.0056	0.9932	0.0013	0.0000
10.11	0.0057	0.9931	0.0012	0.0000
10.12	0.0058	0.9930	0.0012	0.0000
10.13	0.0060	0.9929	0.0012	0.0000
10.14	0.0061	0.9927	0.0011	0.0000
10.15	0.0063	0.9926	0.0011	0.0000

pH Value	Alpha 0	Alpha 1	Alpha 2	Alpha 3
10.46	0.0127	0.9867	0.0005	0.0000
10.47	0.0130	0.9865	0.0005	0.0000
10.48	0.0133	0.9862	0.0005	0.0000
10.49	0.0136	0.9859	0.0005	0.0000
10.50	0.0139	0.9856	0.0005	0.0000
10.51	0.0142	0.9853	0.0005	0.0000
10.52	0.0146	0.9850	0.0005	0.0000
10.53	0.0149	0.9846	0.0005	0.0000
10.54	0.0152	0.9843	0.0004	0.0000
10.55	0.0156	0.9840	0.0004	0.0000
10.56	0.0160	0.9836	0.0004	0.0000
10.57	0.0163	0.9833	0.0004	0.0000
10.58	0.0167	0.9829	0.0004	0.0000
10.59	0.0171	0.9825	0.0004	0.0000
10.60	0.0175	0.9821	0.0004	0.0000
10.61	0.0179	0.9818	0.0004	0.0000
10.62	0.0183	0.9814	0.0004	0.0000
10.63	0.0187	0.9809	0.0004	0.0000
10.64	0.0191	0.9805	0.0004	0.0000
10.65	0.0196	0.9801	0.0003	0.0000
10.66	0.0200	0.9797	0.0003	0.0000
10.67	0.0205	0.9792	0.0003	0.0000
10.68	0.0209	0.9788	0.0003	0.0000
10.69	0.0214	0.9783	0.0003	0.0000
10.70	0.0219	0.9778	0.0003	0.0000

pH				
10.71	0.0000	0.0003	0.9773	0.0224
10.72	0.0000	0.0003	0.9768	0.0229
10.73	0.0000	0.0003	0.9763	0.0234
10.74	0.0000	0.0003	0.9758	0.0240
10.75	0.0000	0.0003	0.9752	0.0245
10.76	0.0000	0.0003	0.9747	0.0251
10.77	0.0000	0.0003	0.9741	0.0256
10.78	0.0000	0.0003	0.9735	0.0262
10.79	0.0000	0.0003	0.9730	0.0268
10.80	0.0000	0.0002	0.9724	0.0274
10.81	0.0000	0.0002	0.9717	0.0280
10.82	0.0000	0.0002	0.9711	0.0287
10.83	0.0000	0.0002	0.9705	0.0293
10.84	0.0000	0.0002	0.9698	0.0300
10.85	0.0000	0.0002	0.9691	0.0306
10.86	0.0000	0.0002	0.9685	0.0313
10.87	0.0000	0.0002	0.9677	0.0320
10.88	0.0000	0.0002	0.9670	0.0328
10.89	0.0000	0.0002	0.9663	0.0335
10.90	0.0000	0.0002	0.9656	0.0343
10.91	0.0000	0.0002	0.9648	0.0350
10.92	0.0000	0.0002	0.9640	0.0358
10.93	0.0000	0.0002	0.9632	0.0366
10.94	0.0000	0.0002	0.9624	0.0374
10.95	0.0000	0.0002	0.9616	0.0383
10.96	0.0000	0.0002	0.9607	0.0391
10.97	0.0000	0.0002	0.9598	0.0400
10.98	0.0000	0.0002	0.9589	0.0409
10.99	0.0000	0.0002	0.9580	0.0418
11.00	0.0000	0.0002	0.9571	0.0428

		pH				
0.0000	0.0000	10.16	0.0000	0.0011	0.9925	0.0064
0.0000	0.0000	10.17	0.0000	0.0011	0.9924	0.0066
0.0000	0.0000	10.18	0.0000	0.0010	0.9923	0.0067
0.0000	0.0000	10.19	0.0000	0.0010	0.9921	0.0069
0.0000	0.0000	10.20	0.0000	0.0010	0.9920	0.0070
0.0000	0.0000	10.21	0.0000	0.0010	0.9918	0.0072
0.0000	0.0000	10.22	0.0000	0.0009	0.9917	0.0074
0.0000	0.0000	10.23	0.0000	0.0009	0.9916	0.0075
0.0000	0.0000	10.24	0.0000	0.0009	0.9914	0.0077
0.0000	0.0000	10.25	0.0000	0.0009	0.9912	0.0079
0.0000	0.0000	10.26	0.0000	0.0009	0.9911	0.0081
0.0000	0.0000	10.27	0.0000	0.0008	0.9909	0.0082
0.0000	0.0000	10.28	0.0000	0.0008	0.9907	0.0084
0.0000	0.0000	10.29	0.0000	0.0008	0.9906	0.0086
0.0000	0.0000	10.30	0.0000	0.0008	0.9904	0.0088
0.0000	0.0000	10.31	0.0000	0.0008	0.9902	0.0090
0.0000	0.0000	10.32	0.0000	0.0008	0.9900	0.0092
0.0000	0.0000	10.33	0.0000	0.0007	0.9898	0.0095
0.0000	0.0000	10.34	0.0000	0.0007	0.9896	0.0097
0.0000	0.0000	10.35	0.0000	0.0007	0.9894	0.0099
0.0000	0.0000	10.36	0.0000	0.0007	0.9892	0.0101
0.0000	0.0000	10.37	0.0000	0.0007	0.9890	0.0104
0.0000	0.0000	10.38	0.0000	0.0007	0.9888	0.0106
0.0000	0.0000	10.39	0.0000	0.0006	0.9885	0.0108
0.0000	0.0000	10.40	0.0000	0.0006	0.9883	0.0111
0.0000	0.0000	10.41	0.0000	0.0006	0.9880	0.0113
0.0000	0.0000	10.42	0.0000	0.0006	0.9878	0.0116
0.0000	0.0000	10.43	0.0000	0.0006	0.9875	0.0119
0.0000	0.0000	10.44	0.0000	0.0006	0.9873	0.0121
0.0000	0.0000	10.45	0.0000	0.0006	0.9870	0.0124

Table IIA, continued

pH Value	Alpha 0	Alpha 1	Alpha 2	Alpha 3
11.01	0.0437	0.9561	0.0001	0.0000
11.02	0.0447	0.9552	0.0001	0.0000
11.03	0.0457	0.9542	0.0001	0.0000
11.04	0.0467	0.9532	0.0001	0.0000
11.05	0.0477	0.9521	0.0001	0.0000
11.06	0.0488	0.9511	0.0001	0.0000
11.07	0.0499	0.9500	0.0001	0.0000
11.08	0.0510	0.9489	0.0001	0.0000
11.09	0.0521	0.9478	0.0001	0.0000
11.10	0.0532	0.9466	0.0001	0.0000
11.11	0.0544	0.9455	0.0001	0.0000
11.12	0.0556	0.9443	0.0001	0.0000
11.13	0.0568	0.9431	0.0001	0.0000
11.14	0.0581	0.9418	0.0001	0.0000
11.15	0.0593	0.9406	0.0001	0.0000
11.16	0.0606	0.9393	0.0001	0.0000
11.17	0.0620	0.9379	0.0001	0.0000
11.18	0.0633	0.9366	0.0001	0.0000
11.19	0.0647	0.9352	0.0001	0.0000
11.20	0.0661	0.9338	0.0001	0.0000
11.21	0.0675	0.9324	0.0001	0.0000
11.22	0.0690	0.9309	0.0001	0.0000
11.23	0.0705	0.9294	0.0001	0.0000
11.24	0.0720	0.9279	0.0001	0.0000
11.25	0.0736	0.9263	0.0001	0.0000

pH Value	Alpha 0	Alpha 1	Alpha 2	Alpha 3
11.56	0.1395	0.8604	0.0000	0.0000
11.57	0.1423	0.8576	0.0000	0.0000
11.58	0.1452	0.8548	0.0000	0.0000
11.59	0.1480	0.8519	0.0000	0.0000
11.60	0.1510	0.8490	0.0000	0.0000
11.61	0.1539	0.8460	0.0000	0.0000
11.62	0.1570	0.8430	0.0000	0.0000
11.63	0.1600	0.8399	0.0000	0.0000
11.64	0.1632	0.8368	0.0000	0.0000
11.65	0.1663	0.8336	0.0000	0.0000
11.66	0.1695	0.8304	0.0000	0.0000
11.67	0.1728	0.8272	0.0000	0.0000
11.68	0.1761	0.8238	0.0000	0.0000
11.69	0.1795	0.8205	0.0000	0.0000
11.70	0.1829	0.8171	0.0000	0.0000
11.71	0.1864	0.8136	0.0000	0.0000
11.72	0.1899	0.8101	0.0000	0.0000
11.73	0.1935	0.8065	0.0000	0.0000
11.74	0.1971	0.8029	0.0000	0.0000
11.75	0.2008	0.7992	0.0000	0.0000
11.76	0.2045	0.7955	0.0000	0.0000
11.77	0.2082	0.7917	0.0000	0.0000
11.78	0.2121	0.7879	0.0000	0.0000
11.79	0.2159	0.7840	0.0000	0.0000
11.80	0.2199	0.7801	0.0000	0.0000

				pH		
0.2238	0.7761	0.0000	0.0000	11.81	0.0000	0.0000
0.2279	0.7721	0.0000	0.0000	11.82	0.0000	0.0000
0.2319	0.7680	0.0000	0.0000	11.83	0.0000	0.0000
0.2361	0.7639	0.0000	0.0000	11.84	0.0000	0.0000
0.2402	0.7597	0.0000	0.0000	11.85	0.0000	0.0000
0.2445	0.7555	0.0000	0.0000	11.86	0.0000	0.0000
0.2487	0.7512	0.0000	0.0000	11.87	0.0000	0.0000
0.2531	0.7469	0.0000	0.0000	11.88	0.0000	0.0000
0.2575	0.7425	0.0000	0.0000	11.89	0.0000	0.0000
0.2619	0.7381	0.0000	0.0000	11.90	0.0000	0.0000
0.2664	0.7336	0.0000	0.0000	11.91	0.0000	0.0000
0.2709	0.7291	0.0000	0.0000	11.92	0.0000	0.0000
0.2754	0.7245	0.0000	0.0000	11.93	0.0000	0.0000
0.2801	0.7199	0.0000	0.0000	11.94	0.0000	0.0000
0.2847	0.7153	0.0000	0.0000	11.95	0.0000	0.0000
0.2894	0.7105	0.0000	0.0000	11.96	0.0000	0.0000
0.2942	0.7058	0.0000	0.0000	11.97	0.0000	0.0000
0.2990	0.7010	0.0000	0.0000	11.98	0.0000	0.0000
0.3039	0.6961	0.0000	0.0000	11.99	0.0000	0.0000
0.3088	0.6912	0.0000	0.0000	12.00	0.0000	0.0000
0.3137	0.6863	0.0000	0.0000	12.01	0.0000	0.0000
0.3187	0.6813	0.0000	0.0000	12.02	0.0000	0.0000
0.3237	0.6763	0.0000	0.0000	12.03	0.0000	0.0000
0.3287	0.6712	0.0000	0.0000	12.04	0.0000	0.0000
0.3338	0.6661	0.0000	0.0000	12.05	0.0000	0.0000
0.3390	0.6610	0.0000	0.0000	12.06	0.0000	0.0000
0.3442	0.6558	0.0000	0.0000	12.07	0.0000	0.0000
0.3494	0.6506	0.0000	0.0000	12.08	0.0000	0.0000
0.3546	0.6454	0.0000	0.0000	12.09	0.0000	0.0000
0.3599	0.6401	0.0000	0.0000	12.10	0.0000	0.0000

				pH		
0.0752	0.9248	0.0001	0.0000	11.26	0.0000	0.0000
0.0768	0.9231	0.0001	0.0000	11.27	0.0000	0.0000
0.0784	0.9215	0.0001	0.0000	11.28	0.0000	0.0000
0.0801	0.9198	0.0001	0.0000	11.29	0.0000	0.0000
0.0818	0.9181	0.0001	0.0000	11.30	0.0000	0.0000
0.0836	0.9164	0.0001	0.0000	11.31	0.0000	0.0000
0.0853	0.9146	0.0001	0.0000	11.32	0.0000	0.0000
0.0872	0.9128	0.0001	0.0000	11.33	0.0000	0.0000
0.0890	0.9109	0.0001	0.0000	11.34	0.0000	0.0000
0.0909	0.9090	0.0001	0.0000	11.35	0.0000	0.0000
0.0928	0.9071	0.0001	0.0000	11.36	0.0000	0.0000
0.0948	0.9052	0.0001	0.0000	11.37	0.0000	0.0000
0.0968	0.9032	0.0001	0.0000	11.38	0.0000	0.0000
0.0988	0.9011	0.0001	0.0000	11.39	0.0000	0.0000
0.1009	0.8991	0.0001	0.0000	11.40	0.0000	0.0000
0.1030	0.8970	0.0001	0.0000	11.41	0.0000	0.0000
0.1051	0.8948	0.0001	0.0000	11.42	0.0000	0.0000
0.1073	0.8926	0.0001	0.0000	11.43	0.0000	0.0000
0.1095	0.8904	0.0001	0.0000	11.44	0.0000	0.0000
0.1118	0.8881	0.0000	0.0000	11.45	0.0000	0.0000
0.1141	0.8858	0.0000	0.0000	11.46	0.0000	0.0000
0.1165	0.8835	0.0000	0.0000	11.47	0.0000	0.0000
0.1189	0.8811	0.0000	0.0000	11.48	0.0000	0.0000
0.1213	0.8787	0.0000	0.0000	11.49	0.0000	0.0000
0.1238	0.8762	0.0000	0.0000	11.50	0.0000	0.0000
0.1263	0.8737	0.0000	0.0000	11.51	0.0000	0.0000
0.1288	0.8711	0.0000	0.0000	11.52	0.0000	0.0000
0.1314	0.8685	0.0000	0.0000	11.53	0.0000	0.0000
0.1341	0.8659	0.0000	0.0000	11.54	0.0000	0.0000
0.1368	0.8632	0.0000	0.0000	11.55	0.0000	0.0000

Table IIA, continued

pH Value	Alpha 3	Alpha 2	Alpha 1	Alpha 0	pH Value	Alpha 3	Alpha 2	Alpha 1	Alpha 0
12.11	0.0000	0.0000	0.6348	0.3652	12.66	0.0000	0.0000	0.3288	0.6712
12.12	0.0000	0.0000	0.6294	0.3706	12.67	0.0000	0.0000	0.3237	0.6763
12.13	0.0000	0.0000	0.6240	0.3760	12.68	0.0000	0.0000	0.3187	0.6813
12.14	0.0000	0.0000	0.6186	0.3814	12.69	0.0000	0.0000	0.3137	0.6863
12.15	0.0000	0.0000	0.6131	0.3869	12.70	0.0000	0.0000	0.3088	0.6912
12.16	0.0000	0.0000	0.6077	0.3923	12.71	0.0000	0.0000	0.3039	0.6961
12.17	0.0000	0.0000	0.6022	0.3978	12.72	0.0000	0.0000	0.2990	0.7010
12.18	0.0000	0.0000	0.5966	0.4034	12.73	0.0000	0.0000	0.2942	0.7058
12.19	0.0000	0.0000	0.5911	0.4089	12.74	0.0000	0.0000	0.2895	0.7105
12.20	0.0000	0.0000	0.5855	0.4145	12.75	0.0000	0.0000	0.2847	0.7152
12.21	0.0000	0.0000	0.5799	0.4201	12.76	0.0000	0.0000	0.2801	0.7199
12.22	0.0000	0.0000	0.5743	0.4257	12.77	0.0000	0.0000	0.2755	0.7245
12.23	0.0000	0.0000	0.5686	0.4313	12.78	0.0000	0.0000	0.2709	0.7291
12.24	0.0000	0.0000	0.5630	0.4370	12.79	0.0000	0.0000	0.2664	0.7336
12.25	0.0000	0.0000	0.5573	0.4427	12.80	0.0000	0.0000	0.2619	0.7381
12.26	0.0000	0.0000	0.5516	0.4484	12.81	0.0000	0.0000	0.2575	0.7425
12.27	0.0000	0.0000	0.5459	0.4541	12.82	0.0000	0.0000	0.2531	0.7469
12.28	0.0000	0.0000	0.5402	0.4598	12.83	0.0000	0.0000	0.2488	0.7512
12.29	0.0000	0.0000	0.5345	0.4655	12.84	0.0000	0.0000	0.2445	0.7555
12.30	0.0000	0.0000	0.5288	0.4712	12.85	0.0000	0.0000	0.2403	0.7597
12.31	0.0000	0.0000	0.5230	0.4770	12.86	0.0000	0.0000	0.2361	0.7639
12.32	0.0000	0.0000	0.5173	0.4827	12.87	0.0000	0.0000	0.2320	0.7680
12.33	0.0000	0.0000	0.5115	0.4885	12.88	0.0000	0.0000	0.2279	0.7721
12.34	0.0000	0.0000	0.5058	0.4942	12.89	0.0000	0.0000	0.2238	0.7762
12.35	0.0000	0.0000	0.5000	0.5000	12.90	0.0000	0.0000	0.2199	0.7801

pH				
12.91	0.7840	0.2160	0.0000	0.0000
12.92	0.7879	0.2121	0.0000	0.0000
12.93	0.7917	0.2083	0.0000	0.0000
12.94	0.7955	0.2045	0.0000	0.0000
12.95	0.7992	0.2008	0.0000	0.0000
12.96	0.8029	0.1971	0.0000	0.0000
12.97	0.8065	0.1935	0.0000	0.0000
12.98	0.8101	0.1899	0.0000	0.0000
12.99	0.8136	0.1864	0.0000	0.0000
13.00	0.8171	0.1829	0.0000	0.0000
13.01	0.8205	0.1795	0.0000	0.0000
13.02	0.8239	0.1761	0.0000	0.0000
13.03	0.8272	0.1728	0.0000	0.0000
13.04	0.8304	0.1696	0.0000	0.0000
13.05	0.8337	0.1663	0.0000	0.0000
13.06	0.8368	0.1632	0.0000	0.0000
13.07	0.8399	0.1601	0.0000	0.0000
13.08	0.8430	0.1570	0.0000	0.0000
13.09	0.8460	0.1540	0.0000	0.0000
13.10	0.8490	0.1510	0.0000	0.0000
13.11	0.8519	0.1481	0.0000	0.0000
13.12	0.8548	0.1452	0.0000	0.0000
13.13	0.8577	0.1423	0.0000	0.0000
13.14	0.8604	0.1396	0.0000	0.0000
13.15	0.8632	0.1368	0.0000	0.0000
13.16	0.8659	0.1341	0.0000	0.0000
13.17	0.8685	0.1315	0.0000	0.0000
13.18	0.8711	0.1289	0.0000	0.0000
13.19	0.8737	0.1263	0.0000	0.0000
13.20	0.8762	0.1238	0.0000	0.0000

pH				
12.36	0.5057	0.4943	0.0000	0.0000
12.37	0.5115	0.4885	0.0000	0.0000
12.38	0.5172	0.4827	0.0000	0.0000
12.39	0.5230	0.4770	0.0000	0.0000
12.40	0.5287	0.4713	0.0000	0.0000
12.41	0.5345	0.4655	0.0000	0.0000
12.42	0.5402	0.4598	0.0000	0.0000
12.43	0.5459	0.4541	0.0000	0.0000
12.44	0.5516	0.4484	0.0000	0.0000
12.45	0.5573	0.4427	0.0000	0.0000
12.46	0.5630	0.4370	0.0000	0.0000
12.47	0.5686	0.4314	0.0000	0.0000
12.48	0.5743	0.4257	0.0000	0.0000
12.49	0.5799	0.4201	0.0000	0.0000
12.50	0.5855	0.4145	0.0000	0.0000
12.51	0.5911	0.4089	0.0000	0.0000
12.52	0.5966	0.4034	0.0000	0.0000
12.53	0.6021	0.3978	0.0000	0.0000
12.54	0.6076	0.3923	0.0000	0.0000
12.55	0.6131	0.3869	0.0000	0.0000
12.56	0.6186	0.3814	0.0000	0.0000
12.57	0.6240	0.3760	0.0000	0.0000
12.58	0.6294	0.3706	0.0000	0.0000
12.59	0.6347	0.3653	0.0000	0.0000
12.60	0.6400	0.3599	0.0000	0.0000
12.61	0.6453	0.3547	0.0000	0.0000
12.62	0.6506	0.3494	0.0000	0.0000
12.63	0.6558	0.3442	0.0000	0.0000
12.64	0.6610	0.3390	0.0000	0.0000
12.65	0.6661	0.3339	0.0000	0.0000

Table IIA, continued

pH Value	Alpha 0	Alpha 1	Alpha 2	Alpha 3
13.21	0.8787	0.1213	0.0000	0.0000
13.22	0.8811	0.1189	0.0000	0.0000
13.23	0.8835	0.1165	0.0000	0.0000
13.24	0.8859	0.1141	0.0000	0.0000
13.25	0.8882	0.1118	0.0000	0.0000
13.26	0.8904	0.1096	0.0000	0.0000
13.27	0.8927	0.1073	0.0000	0.0000
13.28	0.8949	0.1051	0.0000	0.0000
13.29	0.8970	0.1030	0.0000	0.0000
13.30	0.8991	0.1009	0.0000	0.0000
13.31	0.9012	0.0988	0.0000	0.0000
13.32	0.9032	0.0968	0.0000	0.0000
13.33	0.9052	0.0948	0.0000	0.0000
13.34	0.9072	0.0928	0.0000	0.0000
13.35	0.9091	0.0909	0.0000	0.0000
13.36	0.9110	0.0890	0.0000	0.0000
13.37	0.9128	0.0872	0.0000	0.0000
13.38	0.9146	0.0854	0.0000	0.0000
13.39	0.9164	0.0836	0.0000	0.0000
13.40	0.9182	0.0818	0.0000	0.0000
13.41	0.9199	0.0801	0.0000	0.0000
13.42	0.9216	0.0784	0.0000	0.0000
13.43	0.9232	0.0768	0.0000	0.0000
13.44	0.9248	0.0752	0.0000	0.0000
13.45	0.9264	0.0736	0.0000	0.0000

pH Value	Alpha 0	Alpha 1	Alpha 2	Alpha 3
13.61	0.9479	0.0521	0.0000	0.0000
13.62	0.9490	0.0510	0.0000	0.0000
13.63	0.9501	0.0499	0.0000	0.0000
13.64	0.9512	0.0488	0.0000	0.0000
13.65	0.9523	0.0477	0.0000	0.0000
13.66	0.9533	0.0467	0.0000	0.0000
13.67	0.9543	0.0457	0.0000	0.0000
13.68	0.9553	0.0447	0.0000	0.0000
13.69	0.9563	0.0437	0.0000	0.0000
13.70	0.9572	0.0428	0.0000	0.0000
13.71	0.9582	0.0418	0.0000	0.0000
13.72	0.9591	0.0409	0.0000	0.0000
13.73	0.9600	0.0400	0.0000	0.0000
13.74	0.9609	0.0391	0.0000	0.0000
13.75	0.9617	0.0383	0.0000	0.0000
13.76	0.9626	0.0374	0.0000	0.0000
13.77	0.9634	0.0366	0.0000	0.0000
13.78	0.9642	0.0358	0.0000	0.0000
13.79	0.9650	0.0350	0.0000	0.0000
13.80	0.9657	0.0343	0.0000	0.0000
13.81	0.9665	0.0335	0.0000	0.0000
13.82	0.9672	0.0328	0.0000	0.0000
13.83	0.9679	0.0321	0.0000	0.0000
13.84	0.9687	0.0313	0.0000	0.0000
13.85	0.9693	0.0307	0.0000	0.0000

pH						
13.46	0.0000	0.0000	0.0720	0.9280	0.0000	0.0000
13.47	0.0000	0.0000	0.0705	0.9295	0.0000	0.0000
13.48	0.0000	0.0000	0.0690	0.9310	0.0000	0.0000
13.49	0.0000	0.0000	0.0676	0.9324	0.0000	0.0000
13.50	0.0000	0.0000	0.0661	0.9339	0.0000	0.0000
13.51	0.0000	0.0000	0.0647	0.9353	0.0000	0.0000
13.52	0.0000	0.0000	0.0633	0.9367	0.0000	0.0000
13.53	0.0000	0.0000	0.0620	0.9380	0.0000	0.0000
13.54	0.0000	0.0000	0.0607	0.9393	0.0000	0.0000
13.55	0.0000	0.0000	0.0594	0.9406	0.0000	0.0000
13.56	0.0000	0.0000	0.0581	0.9419	0.0000	0.0000
13.57	0.0000	0.0000	0.0568	0.9432	0.0000	0.0000
13.58	0.0000	0.0000	0.0556	0.9444	0.0000	0.0000
13.59	0.0000	0.0000	0.0544	0.9456	0.0000	0.0000
13.60	0.0000	0.0000	0.0532	0.9468	0.0000	0.0000

pH				
13.86	0.0000	0.0000	0.0300	0.9700
13.87	0.0000	0.0000	0.0293	0.9707
13.88	0.0000	0.0000	0.0287	0.9713
13.89	0.0000	0.0000	0.0280	0.9720
13.90	0.0000	0.0000	0.0274	0.9726
13.91	0.0000	0.0000	0.0268	0.9732
13.92	0.0000	0.0000	0.0262	0.9738
13.93	0.0000	0.0000	0.0256	0.9744
13.94	0.0000	0.0000	0.0251	0.9749
13.95	0.0000	0.0000	0.0245	0.9755
13.96	0.0000	0.0000	0.0240	0.9760
13.97	0.0000	0.0000	0.0234	0.9766
13.98	0.0000	0.0000	0.0229	0.9771
13.99	0.0000	0.0000	0.0224	0.9776
14.00	0.0000	0.0000	0.0219	0.9781

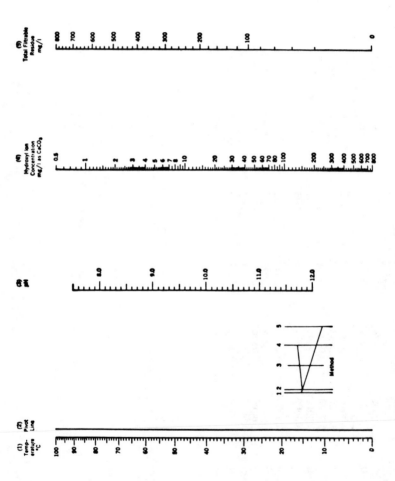

Figure 1A. Nomograph for evaluation of [OH⁻] (reproduced from *Standard Methods* [8] courtesy of the American Public Health Association).

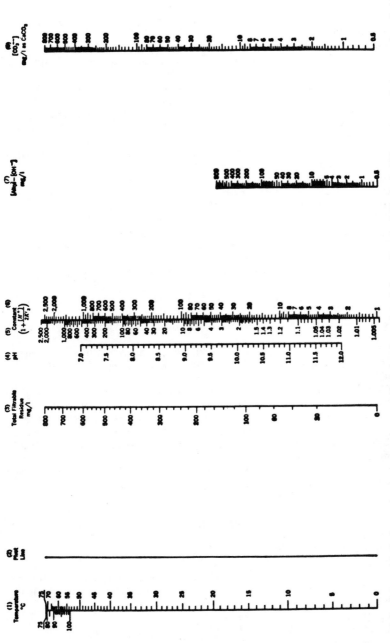

Figure 2A. Nomograph for evaluation of carbonate alkalinity (reproduced from *Standard Methods* [8] courtesy of the American Public Health Association).

REFERENCES

1. Stumm, W., and J. J. Morgan. *Aquatic Chemistry* (New York: John Wiley & Sons, Inc., 1970).
2. Morris, J. C. "Disinfection," in *Water Purification and Wastewater Treatment and Disposal*, Vol. 2, G. M. Fair, J. C. Geyer and D. A. Okun, Eds. (New York: John Wiley & Sons, Inc., 1968).
3. Harned, H. S., and R. Davies, Jr. *J. Am. Chem. Soc.* 65:2030 (1943).
4. Harned, H. S., and S. R. Scholes. *J. Am. Chem. Soc.* 63:1706 (1941).
5. Hem, J. D. "Study and Interpretation of the Chemical Characteristics of Natural Water," *Geol. Survey Water Supply Paper 1473*, 2nd ed., Washington, DC (1970).
6. Nordell, E. *Water Treatment for Industrial and Other Uses* (New York: Van Nostrand Reinhold Company, 1961).
7. Dye, J. F. *J. Am. Water Works Assoc.* 44:356 (1952).
8. *Standard Methods for the Examination of Water and Wastewater*, 14th ed. (New York: American Public Health Association, 1976).
9. Deffeyes, K. S. *Limnol. Oceanog.* 10:412 (1965).
10. Sørensen, S. P. L. *Biochem. Z.* 21:131 (1909); 21:201 (1909).
11. Butler, J. N. *Ionic Equilibrium* (Reading, MA: Addison-Wesley Publishing Co., Inc., 1964).
12. Bates, R. G. *Determination of pH* (New York: John Wiley & Sons, Inc., 1954).
13. Van Slyke, D. D. *J. Biol. Chem.* 52:525 (1922).
14. Henderson, L. J. *Am. J. Physiol.* 21:169 (1908).
15. Weber, W. J. Jr., and W. Stumm. *J. Am. Water Works Assoc.* 55(12):1553 (1963).
16. White, D. E., et al. "Data of Geochemistry, Chapter F, Chemical Composition of Subsurface Waters," *Geol. Survey Prof. Paper 440F*, Washington, DC (1963).
17. Kleyn, H. F. W. *Int. J. Air Water Poll.* 9:401 (1965).
18. Livingstone, D. A. "Data of Geochemistry, Chapter G, Chemical Composition of Rivers and Lakes," *Geol. Survey Prof.*, Washington, DC (1963).
19. Park, P. K., et al. *Limnol. Oceanog.* 14(4):559 (1969).
20. Roberson, C. E., et al. Geological Survey Professional Paper 475C, Washington, DC (1963), p. C212.
21. Barnes, I. "Field Measurement of Alkalinity and pH," *Geol. Survey Water Supply Paper 1535-H*, Washington, DC (1964).
22. Gran. G. *Analyst* 77:661 (1952).
23. Larson, T. E., and L. Henley. *Anal. Chem.* 27:851 (1965).
24. Thomas, J. F., and J. J. Lynch. *J. Am. Water Works Assoc.* 52:255 (1960).
25. Dyrssen, D. *Acta Chem. Scand.* 19:1265 (1965).
26. Dyrssen, D., and L. G. Sillen. *Tellus* 19:110 (1967).
27. Loewenthal, R. E., and G. v. R. Marais. *Carbonate Chemistry of Aquatic Systems—Theory and Applications* (Ann Arbor, MI: Ann Arbor Science Publishers, Inc., 1976).
28. Kemp, P. H. *Water Res.* 5:413 (1971).
29. Kemp, P. H. *Water Res.* 5:611 (1971).

CHAPTER 4

OXIDATION–REDUCTION CONDITIONS OF NATURAL WATERS AND ENVIRONS

INTRODUCTION

Many reactions in aquatic environments involve the transfer of electrons from one element to another or from one compound to another. The appropriate oxidation–reduction stability relationships are considered briefly here. However, it must be indicated that concentrations of reducible and oxidizable species may not be the same as those predicted from thermodynamic equilibrium models. There are several reasons for this discrepancy. Many oxidation–reduction (redox) reactions are slow. Further, there may be great differences in the redox environment throughout a specific body of water. For example, in a relatively deep lake, the surface waters in contact with oxygen of the atmosphere may be at oxidizing levels. The bottom waters, in contact with sediments, may be void of oxygen and may have reducing levels. In between these two strata of water there may be intermediate and localized oxidation–reduction levels where various biological systems are operating in nonequilibrium conditions. Also, biological systems may create a microenvironment whose redox levels are quite different than the surrounding macroenvironment. Briefly, the dynamics of aqueous systems must be understood thoroughly before a quantitative description of redox conditions can be developed. It is, therefore, the rates of approach to redox equilibria that are pertinent, rather than the description of the total equilibrium compositions [1,2].

On the other hand, an equilibrium approach can provide, in a general manner, a comprehension of redox processes occurring or expected in natural waters. Equilibrium computations give boundaries toward which one or more redox processes may be proceeding. Whenever differences are noted between

169

calculations and observations, some valuable information is obtained. It may be that more complex theories of redox processes are required.

An additional problem occurs when oxidation–reduction potential measurements are attempted in natural waters [1,2]. The nature and rates of the reactions at electrode surfaces influence the observed redox values. These data frequently are very difficult to interpret. For significant interpretation, components must behave electrochemically in a reversible manner at the electrode surface.

OXIDATION–REDUCTION EQUILIBRIA

Since oxidation–reduction equilibria involve the transfer of electrons (or, perhaps, of a group that carries the electron), an electron donor is labeled the reductant, whereas an electron acceptor is labeled the oxidant. Each oxidation is accompanied by a reduction and vice versa, because there are no free electrons. An example is as follows:

$$
\begin{array}{lll}
Cl_{2(g)} + 2e & = 2Cl^- & \text{reduction} \\
2Fe^{2+} & = 2Fe^{3+} + 2e & \text{oxidation} \\
\hline
Cl_{2(g)} + 2Fe^{2+} = 2Cl^- + 2Fe^{3+} & & \text{redox reaction}
\end{array}
$$

Please note the convention of writing the reduction half-reaction with the electron to the left of the = sign, whereas the oxidation has the electron to the right of the = sign. (More information is given below about redox conventions.)

There are changes in the oxidation states of the reactants and products resulting from electron transfer. Frequently, there are difficulties in assigning the electron loss or gain to a particular element. The oxidation state (or oxidation number) is an empirical concept; it is not synonymous with the number of bonds to an atom. "The oxidation number of an element in any chemical entity is the charge which would be present on an atom of the element if the electrons in each bond to that atom were assigned to the more electronegative atom" [3]. Consequently, there are rules to assign electrons to atoms not reproduced here, but the reader is referred to the literature [2,3] for the details. Roman numerals are used here to represent oxidation states, whereas arabic numbers indicate the charge on a free or coordinated ion. The concept of an oxidation number is an extremely useful tool in balancing redox reactions and in other aspects of chemistry.

Equilibrium Redox Potentials, Electron Activity and pE

The oxidizing intensity of a redox couple at equilibrium is represented by the equilibrium redox potential, i.e., the free energy change per mole of electrons for a specific reduction. Jörgensen [4] proposed the parameter pE to express these equilibrium potentials in various applications. pE is defined by

$$pE = \frac{F}{\ln 10\ RT}\ Eh \qquad (1)$$

where   pE = the negative logarithm of the relative electron activity
        F = the Faraday constant
        R = the gas constant
        T = the temperature ($^\circ$ Kelvin)
      Eh = the redox potential on the hydrogen scale in accord with the International Union's of Pure and Applied Chemistry conventions.

pE values in neutral waters range from about -10 to about $+14$. The larger, positive pE values represent strongly oxidizing conditions, whereas the negative values represent strongly reducing conditions.

From Equation 1, it may be derived:

$$pE = -\frac{1}{ne\ RT\ \ln 10}\ \Delta G \qquad (2)$$

$$pE^0 = -\frac{1}{ne\ RT\ \ln 10}\ \Delta G^0 \qquad (3)$$

where ne is the number of moles of electrons being transferred, pE is in a nonstandard-state value, and pE^0 and ΔG^0 are the respective standard state values. A general form may be written for the redox process:

$$\sum_i ni\ Ai + n_e e = 0 \qquad (4)$$

where Ai represents the redox species in the reaction and ni is their numerical coefficients, which is + for reactants and − for products. It follows:

$$pE = pE^0 + \frac{1}{ne} \log \prod_i (Ai)^{ni} \qquad (5)$$

and

$$pE^0 = \frac{1}{ne} \log K_{red} \qquad (6)$$

where pE^0 now is the relative electron activity when all species other than the electrons are at unit activity. (Please note that the application of pE and pE^0 below will use electron concentration [e] for convenience.) K_{red} is given by

$$K_{red} = \log \prod_i (Ai)^{-ni/ne} \qquad (7)$$

(The notation of K_{red} is explained below.)

Sign Conventions

Table I shows some reduction reactions commonly encountered in water chemistry. The sign conventions follow the International Union of Pure and Applied Chemistry (IUPAC) suggestions. The symbol V is used for the electrode potential. The reader is referred to the article by Licht and

Table I. Some Reduction Reactions and Their Equilibrium Constants[a]

Equation No.	Reaction	log K_{red}	pE^0	V^0 (volts)
(8)	$2H_2O + 2e = H_2 + 2OH^-$	−28	−14	−0.828
(9)	$Zn^{2+} + 2e = Zn_{(s)}$	−26	−13	−0.76
(10)	$Fe^{2+} + 2e = Fe_{(s)}$	−14.9	−7.5	−0.44
(11)	$2H^+ + 2e = H_{2(g)}$	0.0	0.0	0.00
(12)	$Cu^{2+} + 2e = Cu_{(s)}$	+12.0	+6.0	+0.34
(13)	$Fe^{3+} + e = Fe^{2+}$	+13.2	+13.2	+0.77
(14)	$O_{2(g)} + 4H^+ + 4e = 2H_2O$	+83.1	+20.75	+1.229
(15)	$Cl_{2(g)} + 2e = 2Cl^-$	+46	+23	+1.36
(16)	$Fe(OH)^{2+} + H^+ + e = Fe^{2+} + H_2O$	+15.2	−	
(17)	$Fe(OH)_3 + 3H^+ + e = Fe^{2+} + 3H_2O$	+17.8	−	
(18)	$Fe^{3+} + H_2O = Fe(OH)^{2+} + H^+$	−	−2.2[b]	
(19)	$Fe(OH)^{2+} + 2H_2O = Fe(OH)_{3(s)} + 2H^+$	−	−2.41[b]	

[a]Signs in accord with the IUPAC Convention, 25°C.
[b]Log K_{eq}.

deBethune [5] for a fuller explanation of the confusion surrounding electrode sign conventions. The IUPAC convention calls for all half-cell reactions to be written as reduction, with the pE^0 and V^0 values being sign invariant. All such reduction reactions are referred to or combined with the H_2 reduction reaction (Equation 11). If one wishes to combine two reactions to determine the spontaneity of the resultant redox reaction, the IUPAC convention utilizes the following equation:

$$pE^0_{rex} = pE^0_{red} - pE^0_{ox} \qquad (20)$$

This equation states that the pE^0 value of the final reaction is the algebraic difference between the pE^0 value of the reduction reaction and the pE^0 value of the oxidation reaction. Please note that the sign of the pE^0 value of the oxidation reaction does not change from its sign as a reduction reaction. An example illustrates this point:

$$
\begin{array}{lll}
Cl_{2(g)} + 2e = 2Cl^- & pE^0_{red} = +23.0 \\
2Fe^{2+} = 2Fe^{3+} + 2e & pE^0_{ox} = +13.2 \\
\hline
2Fe^{2+} + Cl_{2(g)} = 2Fe^{3+} + 2Cl^- & pE^0_{rex} = +9.8
\end{array}
$$

Since the pE^0 value for the final redox reaction is positive, it is feasible for the oxidation of ferrous iron to be effected by chlorine. Another example:

$$
\begin{array}{lll}
Cu^{2+} + 2e = Cu_{(s)} & pE^0_{red} = +6.0 \\
2Cl^- = Cl_{2(g)} + 2e & pE^0_{ox} = +23.0 \\
\hline
Cu^{2+} + 2Cl^- = Cl_{2(g)} + Cu_{(s)} & pE^0_{rex} = -17.0
\end{array}
$$

Since the pE^0 value for this redox reaction is negative, it is not feasible for oxidation of the chloride anion to be effected by a cupric cation. It should be noted from Table I that, by convention, the reduction reactions that lie above the hydrogen reaction carry a negative sign, whereas those that lie below hydrogen carry a positive sign.

Electron Energy Level Diagrams

Since Equations 2 and 3 show that the pE value represents an expression of the free energy change of a reduction reaction, an electron energy level diagram may be constructed. This is conceptually analogous to the proton energy level diagram shown in Chapter 3 on page 105. Figure 1 gives this

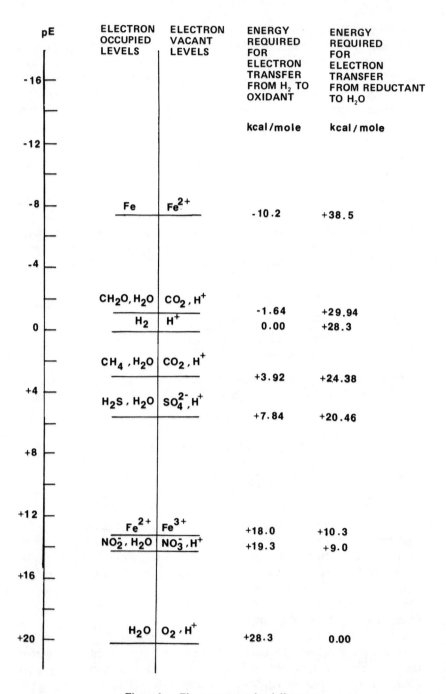

Figure 1.   Electron energy level diagram.

diagram for the transfer of one mole of electrons from reductant to oxidant. The diagram is constructed so that, by convention, the electrons are transferred from H_2 to the oxidant. The energy gained from this transfer increases as one proceeds from top to bottom. Above the H_2, H^+ line, the spontaneous reaction would be

$$Fe_{(s)} + 2H^+ = Fe^{2+} + H_{2(g)} \qquad (21)$$

whereas below the H_2, H^+ line, it would be

$$2Fe^{3+} + H_{2(g)} = Fe^{2+} + 2H^+ \qquad (22)$$

Another feature of this diagram shows the water-oxidized energy level with the H_2O, O_2, H^+ line. It is noted that the $Cl^-, Cl_{2(g)}$ line (not shown) lies below the water-oxidized line, which makes chlorine thermodynamically unstable in water. In the absence of such a reductant as Fe^{2+}, the H_2O molecule would serve as a source of electrons for $Cl_{2(g)}$. This accounts for the instability of standard chlorine solutions.

Another conceptual aspect of Figure 1 is that the left half of the diagram represents electron-occupied levels or reductants, whereas the right half represents electron-unoccupied levels or oxidants. Hypothetically, if one adds a reductant such as $Fe_{(s)}$ to a system containing several oxidants, the lowest unoccupied electron level would be populated first. For example, the $Fe_{(s)}$ would react first with $Cl_{2(g)}$, second with $O_{2(g)}$, third with Fe^{2+}, and so forth. It follows that an oxidant must be chosen from an electron-unoccupied level lower than the reductant for a redox reaction to be thermodynamically spontaneous.

Nernst-Peters Equation

A more commonly used expression for oxidation–reduction reactions is V, the electrode potential. (Symbolism in oxidation–reduction chemistry is confusing. The symbol E is used extensively in the technical literature. For all intents and purposes, this chapter considers E and V interchangeable.) In the IUPAC convention, the reduction standard free energy may be expressed as follows:

$$\Delta G^0 = -nFV^0 \qquad (23)$$

where   n = number of moles of electrons transferred
F = the Faraday (96,484 absolute coulombs)
V^0 = the electrode potential in volts.

The sign conventions for V are the same as for pE (sign invariant, + V denotes a spontaneous reaction, etc.). Substitution of Equation 23 into Equation 3 yields

$$pE^o = \frac{nFV^o}{2.3RT}$$ (24)

or, at 25°C,

$$pE^o = nV^o/0.059$$ (25)

which is the relation between pE^o and V^o.

Frequently, it is necessary to operate oxidation–reduction reactions at conditions other than the standard state. The nonstandard state values of V may be computed from the well-known Nernst-Peters equation:

$$V = V^o + \frac{2.3RT}{nF} \log \frac{(OXID)}{(RED)}$$ (26)

where (OXID) and (RED) are the activities of the oxidized and reduced species, respectively.

The concept of an electrode potential must be distinguished from the measurement of a potential. Lingane [6] was perhaps the first to indicate the importance of this distinction: "the potential of an actual electrode (observed physical quantity) is distinct from the thermodynamic concept of the potential of a half-reaction (defined quantity)." Consequently, there is no presumptive reason to equate thermodynamic redox potentials with measurable, or even measured, electrode potentials. The reader is referred to the literature [6-8] for greater detail and explanation of the foregoing aspects of redox chemistry.

pE–pH Diagrams

Many oxidation–reduction reactions are influenced by the [H+] or, as more commonly expressed, the pH value. This may be seen in the "water oxidation" reaction in Table I (Equation 14), where four moles of protons, four moles of electrons and one mole of $O_{2(g)}$ are involved. Thus, it may be readily seen that if one increases the [H+], there would be a concomitant increase in the [e] in order for the water molecules to be oxidized. Another aspect of this reaction is that it represents the oxidative stability boundary for water (illustrated below). The Fe(0)-Fe(+II)-Fe(+III) system may be cited as a classical example of the influence of pH values on pE values. Several (but not all) iron reduction reactions are cited in Table I.

pE–pH diagrams illustrate the effect of pH on the pE values of the iron system. Such a diagram is seen in Figure 2, where pE is plotted versus pH. Please note that the pE ordinate is compressed for illustrative purposes, whereas a correct plot would have one pE unit for every pH unit. Construction of the pE–pH diagram in Figure 2 goes something like this: First, the

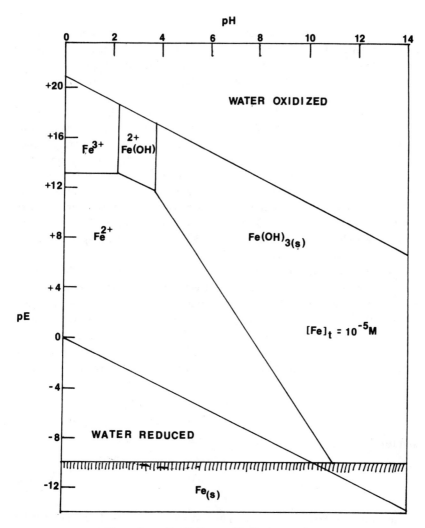

Figure 2.   pE–pH diagram for the iron system.

water-oxidized (Equation 14, Table I) and water-reduced (Equation 8, Table I) boundaries are drawn from their linear forms:

$$pE = 20.7 - pH \text{ (water-oxidized)} \tag{27}$$

and

$$pE = -pH \text{ (water-reduced)} \tag{28}$$

Water is stable toward oxidation–reduction at pH–pE values between these two boundaries. The Fe(0)-Fe(+II) boundary is established from Equation 10 (Table I) and from the assumption that $[Fe]_t$ of the soluble species is $10^{-5}M$. This boundary is set at a pE value of -9.95, below which $Fe_{(s)}$ is the stable species and above which $Fe^{2+}_{(aq)}$ is the stable species. Since this reaction is the oxidative portion of the corrosion reaction for metallic iron, it may be combined with Equation 8 (Table I) as the reductive reaction. This shows that metallic iron pipes are thermodynamically unstable in water, i.e., they have the inherent tendency to corrode. The Fe(+II)-Fe(+III) boundary is computed from Equation 13 (Table I) where the pE value is $+13.0$. At pE values $> +13.0$, Fe(+III) is the predominant species, whereas at pE values $< +13.0$, Fe(+II) is the predominant species. This reaction is independent of $[H^+]$ over the pH range of 0.0 to 2.2 Above this pH value of 2.2, Fe^{3+} enters into an acid–base equilibrium Equation 18 (Table I), whereupon $Fe(OH)^{2+}$ is the predominant Fe(+III) species. The pH boundary for Fe^{3+} and $Fe(OH)^{2+}$ may be readily computed from the equilibrium constant for this reaction. $Fe(OH)^{2+}$ is the predominant Fe(+III) species over the pH range 2.2 to 3.7, whereupon Equation 19 (Table I) becomes operative. The pH boundary for $Fe(OH)^{2+}$-$Fe(OH)_{3(s)}$ may be computed from the equilibrium constant for this reaction.

The influence of pH on pE of the Fe(+II)-Fe(+III) system may be seen from Equations 16 and 17 (Table I). The linear form of Equation 16 is

$$pE = 15.2 - pH \tag{29}$$

and for Equation 17 is

$$pE = 22.8 - 3pH \tag{30}$$

These two equations may be used to calculate the pH–pE boundaries for $Fe(+II)$-$Fe(OH)^{2+}$ and $Fe(+II)$-$Fe(OH)_{3(s)}$, respectively. The influence of pH value on the pE value of Equation 17 is seen readily from Equation 30 and Figure 2, where for every one-unit increase of pH there is a three-unit decrease in pE. This reaction is especially significant because it is employed in the removal of iron from natural waters. It is, of course, the oxidation of Fe^{2+} to Fe^{3+} with the subsequent precipitation of $Fe(OH)_{3(s)}$. This diagram also shows that the pE value required for the oxidation is lowered as the pH value is increased. This interpretation has very practical operational application in the removal of iron from natural waters.

pE LEVELS AND BIOTIC ACTIVITIES

In nature, there is a continuous "competition" between the photosynthetic process and the degradation process for electrons. This may be seen in a simple model:

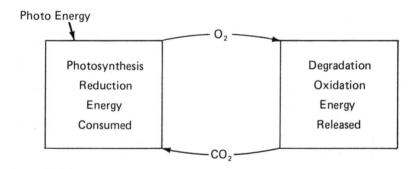

In this model, photosynthesis represents the reduction of $CO_{2(g)}$ to form organic compounds with consumption of energy. This process tends to disturb equilibrium created by oxidative reactions. On the other hand, degradation represents the oxidation of organic matter to form $CO_{2(g)}$, with the release of energy. This process tends toward a state of equilibrium. In nature, there may be a steady-state condition between the two processes. Disturbance of this condition with one process predominating over the other may be caused by "pollution." Often this disturbance is temporary and the steady state is resumed.

In the above model, nearly all the oxidation–reduction reactions occur in one form of a biological system or another. Any approach to equilibrium depends on the activities of various biota. For the most part, microbiological organisms, i.e., bacteria, yeast, fungi, algae, etc., are involved. These organisms provide the necessary mediation or catalysis for the reactions to occur. In many situations these organisms provide a biotic microenvironment whose pE level is quite different from the surrounding environment. This, in itself, provides a misleading aspect of the overall oxidation–reduction condition of the aquatic environment. In any event, it should be noted that these microorganisms also must obey the laws of thermodynamics. Equilibrium models are helpful to an overall comprehension of oxidation–reduction processes in natural waters. More importantly, these models indicate what may occur at a given pE level. Also, these models indicate a sequence of redox reactions. Again, it is extremely difficult for microorganisms to defy thermodynamics and to effect reactions against the feasible direction.

Only a few elements, C, N, O, S, Fe and Mn, for example, are involved in aquatic oxidation–reduction reactions. Since these elements have several oxidation states, it is possible for them to participate in biologically mediated processes. Some relevant half-cell reactions are given in Table II for aquatic conditions [1]. Application of these reactions is given below.

An unpolluted, natural water that is open to the atmosphere should be saturated with oxygen. In this case, this water has a well-defined pE level of +13.75 when the pH value is 7.0 and the temperature is 25°C, which is calculated from Equation 31 in Table II. This pE level establishes the oxidation–reduction conditions of the water and the oxidation states of the other elements: N, S, C, Fe and Mn. Carbon should be present as $CO_{2(g)}$, HCO_3^-, or CO_3^{2-}, with organic C virtually absent. Nitrogen should occur as NO_3^-, S as SO_4^{2-}, Fe as Fe(+III) and Mn as Mn(+IV). Also, a calculation may be made whereby atmospheric $N_{2(g)}$ should be oxidized to NO_3^-. Organic C and $N_{2(g)}$ are found in oxygen-saturated waters and in the atmosphere, respectively, which is indicative that a redox equilibrium is not reached. There may be localized environments where this may be observed, but oxidation–reduction processes are very "slow" in a kinetic sense. Nevertheless, the pE level defines the oxidation state of a specific element and a sequence of use of electron acceptors in natural aquatic environments. These two points are discussed below in the context of microbially mediated redox processes. For the purpose of presentation, it is assumed that there is active transformation of organic C to $CO_{2(g)}$, where the electron acceptors of O_2, NO_3^- and SO_4^{2-} are involved.

The Oxygen–Carbon pE "Ladder"

That oxygen establishes the pE level (at pH = 7.0) in aquatic environments has been cited above. A simple computation demonstrates the feasibility of oxygen to act as an electron acceptor in the microbially mediated degradation of organic C. Equations 31 and 48 are combined:

Table II. pE Values of Redox Processes Pertinent in Aquatic Conditions (25°C) [1]

Equation No.	Reaction	pE°	$pE_7^{\circ\,a}$
(31)	$\frac{1}{4} O_{2(g)} + H^+(W) + e = \frac{1}{2} H_2O$	+20.75	+13.75
(32)	$\frac{1}{5} NO_3^- + \frac{6}{5} H^+(W) + e = \frac{1}{10} N_{2(g)} + \frac{3}{5} H_2O$	+21.05	+12.65
(33)	$\frac{1}{2} NO_3^- + H^+(W) + e = \frac{1}{2} NO_2^- + \frac{1}{2} H_2O$	+14.15	+7.15
(34)	$\frac{1}{8} NO_3^- + \frac{5}{4} H^+(W) + e = \frac{1}{8} NH_4^+ + \frac{3}{8} H_2O$	+14.90	+6.15
(35)	$\frac{1}{6} NO_2^- + \frac{4}{3} H^+(W) + e = \frac{1}{6} NH_4^+ + \frac{1}{3} H_2O$	+15.14	+5.82
(36)	$\frac{1}{2} CH_3OH + H^+(W) + e = \frac{1}{2} CH_{4(g)} + \frac{1}{2} H_2O$	+9.88	+2.88
(37)	$\frac{1}{4} CH_2O + H^+(W) + e = \frac{1}{4} CH_{4(g)} + \frac{1}{4} H_2O$	+6.94	−0.06
(38)	$\frac{1}{2} CH_2O + H^+(W) + e = \frac{1}{2} CH_3OH$	+3.99	−3.01
(39)	$\frac{1}{6} SO_4^{2-} + \frac{4}{3} H^+(W) + e = \frac{1}{6} S_{(s)} + \frac{2}{3} H_2O$	+6.03	−3.30
(40)	$\frac{1}{8} SO_4^{2-} + \frac{5}{4} H^+(W) + e = \frac{1}{8} H_2S_{(g)} + \frac{1}{2} H_2O$	+5.75	−3.50
(41)	$\frac{1}{8} SO_4^{2-} + \frac{9}{8} H^+(W) + e = \frac{1}{8} HS^- + \frac{1}{2} H_2O$	+4.13	−3.75
(42)	$\frac{1}{2} S_{(s)} + H^+(W) + e = \frac{1}{2} H_2S_{(g)}$	+2.89	−4.11
(43)	$\frac{1}{8} CO_{2(g)} + H^+(W) + e = \frac{1}{8} CH_{4(g)} + \frac{1}{4} H_2O$	+2.87	−4.13
(44)	$\frac{1}{6} N_{2(g)} + \frac{4}{3} H^+(W) + e = \frac{1}{3} NH_4^+$	+4.68	−4.68
(45)	$\frac{1}{2} (NADP^+) + \frac{1}{2} H^+(W) + e = \frac{1}{2} (NADPH)$	−2.0	−5.5
(46)	$H^+(W) + e = \frac{1}{2} H_{2(g)}$	0.0	−7.00
(47)	Oxidized ferredoxin + e = reduced ferredoxin	−7.1	−7.1
(48)	$\frac{1}{4} CO_{2(g)} + H^+(W) + e = \frac{1}{24} (glucose) + \frac{1}{4} H_2O$	−0.20	−7.20
(49)	$\frac{1}{2} HCOO^- + \frac{3}{2} H^+(W) + e = \frac{1}{2} CH_2O + \frac{1}{2} H_2O$	+2.82	−7.68
(50)	$\frac{1}{4} CO_{2(g)} + H^+(W) + e = \frac{1}{4} CH_2O + \frac{1}{4} H_2O$	−1.20	−8.20
(51)	$\frac{1}{2} CO_{2(g)} + \frac{1}{2} H^+(W) + e = \frac{1}{2} HCOO^-$	−4.83	−8.73

[a] At pH = 7.0.

$$\frac{1}{4} O_{2(g)} + H^+ + e \qquad = \frac{1}{2} H_2O \qquad\qquad red \qquad\qquad (31)$$

$$\frac{1}{24} \text{(glucose)} + \frac{1}{4} H_2O \;\; = \frac{1}{4} CO_{2(g)} + H^+ + e \qquad ox \qquad\qquad (48)$$

$$\overline{\frac{1}{4} O_{2(g)} + \frac{1}{24} \text{(glucose)} = \frac{1}{4} H_2O + \frac{1}{4} CO_{2(g)}} \qquad rex \qquad\qquad (52)$$

The spontaneity of this reaction is

$$pE_{rex} = pE_{red} - pE_{ox} \qquad\qquad\qquad (20)$$
$$pE_{rex} = +13.75 - (-)\, 7.20 = +20.95$$

Since the pE value of the combined reaction is +, the left to right oxidation of glucose (representing organic C) is spontaneous. It should be noted also that this reaction occurs in an oxidizing environment (+pE values).

Two conditions should be indicated about Equation 31: (1) the decrease of P_{O_2} with decrease in pE value, and (2) the effect of $[H_3^+O]$ on the pE value. When Equation 31 is placed into a logarithmic form and rearranged (at pH 7.0),

$$\log P_{O_2} = 4pE - 55.1 \qquad\qquad\qquad (53)$$

This equation shows a four-unit decrease in the pE level for every one unit decrease in oxygen concentration. Thus, anaerobic conditions (absence of O_2) and a reducing environment result from removal of oxygen from the system (assuming no reaeration). Equation 31 may be arranged also (at constant P_{O_2} of 0.21 atm) to

$$pE = 20.75 - pH \qquad\qquad\qquad (54)$$

This equation indicates a unit decrease in the pE level with a unit increase in the pH value. Consequently, the pE level at which O_2 accepts an electron decreases as the $[H_3^+O]$ decreases. This cites one of the thermodynamic reasons why the $[H_3^+O]$ is a factor in many aerobic transformations of organic C in microbial systems.

The single C atom has five oxidation states for the purpose of modeling redox equilibria: $C^{+IV}(CO_2)$, $C^{+II}(HCOO^-)$, $C^0(CH_2O)$, $C^{-II}(CH_3OH)$ and $C^{-IV}(CH_4)$. There are appropriate reduction reactions in Table II for these carbon compounds. Consequently, carbon enters into a variety of biologically mediated transformations that may occur in aquatic environments: methane fermentation, photosynthesis, sewage sludge digestion, biological treatment of domestic sewage, etc. All of these transformations place C in one or more of its five oxidation states. Frequently, this is portrayed as a biochemical "cycle," with the attendant diagrams as shown in Figure 3 [9]. In a thermodynamic sense, this is not really a carbon "cycle"; rather, these reactions occur on different levels of the electron energy diagram where a "ladder" effect is observed. That is, the various oxidation states of C may be found on each "rung" (i.e., energy level) of a ladder. Consequently, as these transformations occur, carbon moves up or down the ladder seeking the appropriate energy level.

The oxygen–carbon pE ladder is seen in Figure 4, where four half-cell carbon reactions are given. Since each of these reactions lies below the energy level of the oxygen reduction reaction, the oxidation of organic carbon is thermodynamically feasible. That is, carbon is in a reduced valence

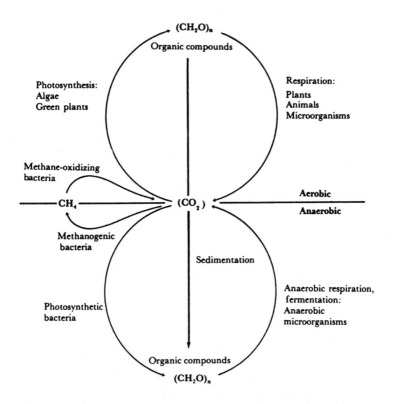

Figure 3.   Carbon cycle in a fresh water lake (reproduced from Brock [9] courtesy of Prentice-Hall, Inc.).

state and, as such, serves as a source of electrons for oxygen. The CH_2O-CO_2 reaction serves as the thermodynamic model for the biologically mediated oxidation of organic C. The $CH_{4(g)}$-$CO_{2(g)}$ couple serves as the model for the anaerobic fermentation of "sludge" in a sewage treatment plant. Also, it is interesting that the photosynthetic reaction, $C_6H_{12}O_6$ (glucose)-$CO_{2(g)}$,

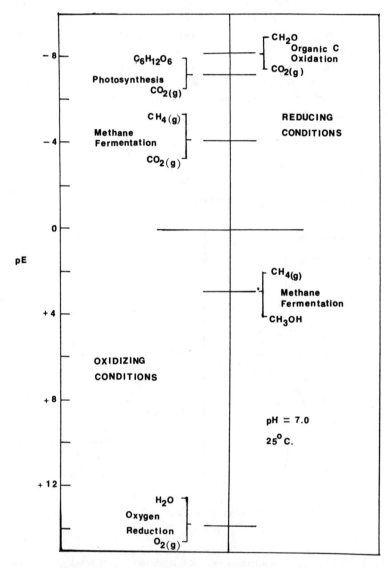

Figure 4.   Oxygen–carbon pE ladder for biological systems.

occurs in a reducing environment. In summary, Figure 4 shows the various energy levels for each oxidation state of carbon that serve as each "rung" in the O_2-C "ladder."

The Nitrogen pE "Ladder"

The biological transformations of nitrogen have been the focus of a considerable amount of research in the past century. There are several well-defined processes of "nitrification," "denitrification," "ammonification," etc., which may occur in aquatic environments as shown in Figure 5 [9]. The element nitrogen occurs in several oxidation states: $N^{+V}(NO_3^-)$, $N^{+III}(NO_2^-)$, $N^{+I}(N_2O)$, $N^0(N_2)$, $N^{-I}(HONH_2)$ and $N^{-III}(NH_3)$. Consequently, there are several "rungs" in the pE "ladder" of the nitrogen system. One of the most important transformations is the function of NO_3^- as an electron acceptor in the biologically mediated degradation of organic C. Combination of Equations 34 and 48 (Table II) yields (at pH = 7.0)

$$\frac{1}{8} NO_3^- + \frac{5}{4} H^+ + e \qquad = \frac{1}{8} NH_4^+ + \frac{3}{8} H_2O \qquad\qquad \text{red} \qquad (34)$$

$$\frac{1}{24} (\text{glucose}) + \frac{1}{4} H_2O \qquad = \frac{1}{4} CO_{2(g)} + H^+ + e \qquad\qquad \text{ox} \qquad (48)$$

$$\overline{\frac{1}{8} NO_3^- + \frac{1}{4} H^+ + \frac{1}{24} (\text{glucose}) = \frac{1}{8} NH_4^+ + \frac{1}{4} CO_{2(g)} + \frac{1}{8} H_2O} \qquad \text{rex} \qquad (55)$$

The spontaneity of this reaction is (at pH = 7.0)

$$pE_{rex} = pE_{red} - pE_{ox} \qquad\qquad (20)$$
$$pE_{rex} = +6.15 - (-) 7.20 = +13.35$$

The positive pE value indicates that it is feasible for NO_3^- to serve as an electron acceptor in the oxidation of organic C and in an oxidizing environment. However, it is noted that the pE_{rex} value of +13.35 is at a lower energy level than when oxygen is the electron acceptor.

The nitrogen pE ladder is seen in Figure 6 for several reactions. These models indicate that the microbially mediated processes of nitrification ($NH_4^+ \rightarrow NO_2^-$ and $NO_2^- \rightarrow NO_3^-$) occur in an oxidizing environment. In this case, O_2 serves as the electron acceptor. The "ammonification" reaction shows the production of NH_4^+ from glycine. This simulates the release of ammonia from the oxidation of proteins. For this one model, nitrogen is not the electron donor, whereas C is via this reaction:

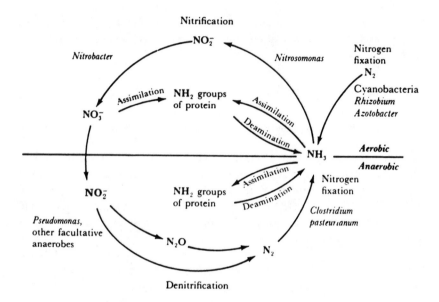

Figure 5. Microbiological processes in the nitrogen cycle in a fresh water lake (reproduced from Brock [9] courtesy of Prentice-Hall, Inc.).

$$\frac{1}{6} CO_2 + \frac{1}{6} HCO_3^- + \frac{1}{6} NH_4^+ + e = \frac{1}{6} CH_2NH_2COOH + \frac{1}{2} H_2O \qquad (56)$$

The pE value of Equation 56 is −6.17 at a pH value of 7.0. Two denitrification reactions are shown with $N_{2(g)}$ as one reductive product and NH_4^+ as the other. These reactions occur rather "high" on the nitrogen pE ladder. The biological fixation of atmospheric $N_{2(g)}$ to NH_4^+ occurs in a reducing environment and is close to the photosynthetic reaction (Equation 48, Table II). This suggests that these two processes may occur concurrently and may be mediated by such microorganisms as blue-green algae. It should be noted also that NH_4^+ is thermodynamically stable over a wide pE range, which extends from reducing conditions well into the oxidizing range. This is the thermodynamic reason why ammonia N accumulates in some heavily polluted rivers.

The Sulfur pE "Ladder"

Many inorganic and organic sulfur compounds are involved in biological transformations. The traditional biogeochemical S cycle is seen in Figure 7 [9]. Sulfur occurs in several oxidation states: $S^{-II}(H_2S_{(g)})$, $S^0(S^0_{(s)})$,

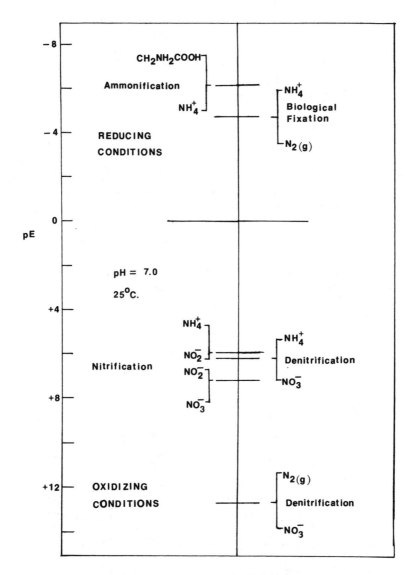

Figure 6.   pE nitrogen ladder for biological systems.

$S^{+II}(S_2O_3^{2-})$, $S^{+III}(S_2O_4^{2-})$, $S^{+IV}(SO_3^{2-})$, $S^{+V}(S_2O_6^{2-})$ and $S^{+VI}(SO_4^{2-})$. Thus, it is possible to have several "rungs" in the pE S ladder. There are occasions in aquatic environments when it is necessary for microorganisms to use SO_4^{2-} as an electron acceptor. That this is thermodynamically feasible may be seen from combination of Equations 40 and 48 (Table II) (at pH = 7.0):

$$\tfrac{1}{8} SO_4^{2-} + \tfrac{5}{4} H^+ + e \qquad = \tfrac{1}{8} H_2S_{(g)} + \tfrac{1}{2} H_2O \qquad\qquad red \qquad (40)$$

$$\tfrac{1}{24} (glucose) + \tfrac{1}{4} H_2O \qquad = \tfrac{1}{4} CO_{2(g)} + H^+ + e \qquad\qquad ox \qquad (48)$$

$$\tfrac{1}{8} SO_4^{2-} + \tfrac{1}{4} H^+ + \tfrac{1}{24} (glucose) = \tfrac{1}{8} H_2S_{(g)} + \tfrac{1}{4} CO_{2(g)} + \tfrac{1}{4} H_2O \qquad rex \qquad (57)$$

$$pE_{rex} = -3.50 - (-) 7.20 = +3.70$$

The positive pE_{rex} value indicates that it is feasible for SO_4^{2-} to serve as an electron acceptor in the oxidation of organic C and in an oxidizing environment. It should be noted that the energy level represented by the pE_{rex} value of +3.70 is lower than the levels of the electron acceptors $O_{2(g)}$ and NO_3^-. In this transformation, $H_2S_{(g)}$ is evolved whose characteristic odor serves to indicate, rather easily, the use of SO_4^{2-} as an electron acceptor by aquatic microorganisms.

The sulfur pE ladder is seen in Figure 8 for three reactions: the reduction of SO_4^{2-} and elemental $S_{(s)}^0$ to $H_2S_{(g)}$ and the oxidation of an organic S compound, cysteine, to SO_4^{2-}. The latter reaction resulted from

$$\tfrac{1}{18} SO_4^{2-} + \tfrac{1}{6} CO_{2(g)} + \tfrac{17}{18} H^+ + e = \tfrac{1}{18} C_3H_7O_2NS + \tfrac{2}{9} H_2O \qquad (58)$$

$$pE_{red} = -3.57 \text{ at pH} = 7.0$$

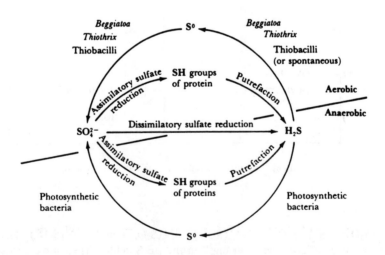

Figure 7.   Microbiological processes in the sulfur cycle in a fresh water lake (reproduced from Brock [9] courtesy of Prentice-Hall, Inc.).

A combination of Equation 58 and Equation 31 (Table II) yields a pE value of +17.32 (at pH = 7.0), which suggests that organic S compounds are thermodynamically unstable in aquatic environments. It should be noted also that the three "rungs" on the S pE ladder lie in the reducing portion of the environment. Transformation of S compounds do, indeed, occur under microbiologically anaerobic conditions.

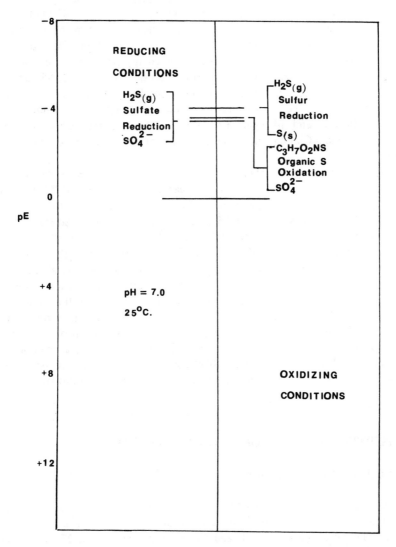

Figure 8.   pE sulfur ladder for biological systems.

Sequence of Electron Acceptor Reactions

In the microbially mediated oxidation of organic C compounds there is a well-defined sequence in which the four major electron acceptors—O_2, NO_3^-, SO_4^{2-} and CO_2—react. This may be demonstrated through the use of an electron acceptor diagram (Figure 9). A concentration of $10^{-4}M$ is assumed for each electron acceptor, with the exception of oxygen. When a surface water is saturated with oxygen (9.2 mg/l) at 20°C, the pE value is +13.75, which denotes an oxidizing environment. If an organic C compound (glucose, for example) is added now to this aquatic environment, there are four electron acceptors available for an oxidation reaction. The sequence of reactions will occur in accord with the energy level of the combined oxidation–reduction reaction. The energy levels of Equations 52, 55 and 57 are: −28.57, −18.21 and −5.05 kcal/mol, respectively. Carbon dioxide may be used also as an electron acceptor in the oxidation of glucose. Combination of Equations 43 and 48 yields a pE value of +3.07 and an energy level of −4.19 kcal/mol. Consequently, the sequence of use is O_2, NO_3^-, SO_4^{2-} and CO_2.

The electron acceptor diagram in Figure 9 shows that O_2 reacts over a very short range of pE values (assuming no reaeration) from +12.65 (1.0 mg/l) to +12.89 (9.2 mg/l). After depletion of oxygen, NO_3^- serves as the electron acceptor until a pE value of +6.15 is reached where the reduced state of N as NH_4^+ begins to predominate. Below the pE value of +6.15, SO_4^{2-} serves as the electron acceptor until the level of −3.50 is reached. Carbon dioxide serves as an electron acceptor concurrently with SO_4^{2-}. In the absence of SO_4^{2-}, CO_2 accepts electrons under reducing conditions and yields methane gas. The diagram in Figure 9 is, of course, simplified to present the sequential acceptance of electrons by O_2, NO_3^-, SO_4^{2-} and CO_2. There is a concurrent reaction, for example, of NO_3^- and O_2 as the concentration of the latter nears zero. Frequently, the pE level of NO_3^- is cited arbitrarily as the boundary between "aerobic" and "anaerobic" conditions in aquatic environments. The latter condition is characterized by the presence of odorous $H_2S_{(g)}$ and odorless $CH_{4(g)}$.

OBSERVED Eh VALUES OF NATURAL SYSTEMS

Limitations of Eh Measurement

There are many publications of Eh values of natural waters and other natural environments throughout the world of which some are reported here. The relation between pE and Eh is given in Equation 25. Since the work of Gillespie [10], the potential of the platinum electrode (Eh) has

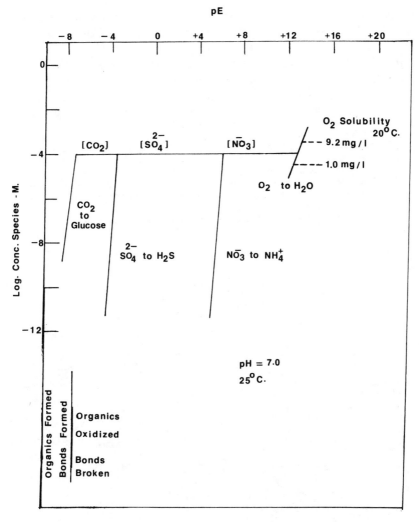

Figure 9.   Electron acceptor diagram for biological systems.

been used widely to characterize natural environments and occasionally make conclusions about the precise oxidation-reduction conditions within a particular system. Whitfield [11,12] has presented an excellent argument about the thermodynamic limitations on using the platinum electrode in Eh measurements. There is now sufficient evidence that such conclusions about the precise redox conditions are valid only in rather exceptional circumstances, such as the work by Hem [13], for example. However, considerable interest exists among aquatic chemists, biologists, geochemists, etc., in the use of Eh as an *operational* parameter for mapping variations in the levels of oxidation or reduction in situ.

Whitfield [11,12] summarizes the major physiochemical problems with use of Eh as an *operational* parameter (Figure 10)

1. There are difficulties associated with disturbance of the sample i.e., release or absorption of such gases as O_2, H_2S and reactions at the liquid junction of the reference electrode, i.e., precipitation of heavy metal sulfides, effect of suspended matter.
2. There are low exchange current densities at the platinum surface and the predominance of mixed potentials, which give rise to instability and poor reproducibility in the measurements.
3. The platinum electrode's response to changes in the environment depends, to a large extent, on the properties of the platinum surface and the presence of adherent surface coatings.
4. Another factor arises if the Eh values are interpreted as oxidation–reduction potentials.

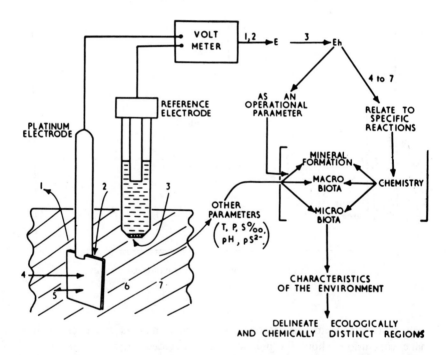

Figure 10.   Problems associated with Eh measurement: (1) release of gases (e.g., H_2S); (2) introduction of air; (3) liquid junction effects: (a) suspension, (b) precipitation of sulfides; (4) direct attack of platinum; (5) trace component controlling potential; (6) whole system out of equilibrium; and (7) microenvironments may be important. Reproduced from Whitfield [11] courtesy of the American Society of Limnology and Oceanography.

Whitfield [12] discusses in great detail the physicochemistry of the platinum electrode, which is assumed to be an inert sensor providing a site for electron exchange. The essentials of this discussion are presented here. Whitfield explores the consequences of direct attack of the platinum surface by substances dissolved in the water. This will alter the metal's properties and will display potentials characteristic of the electrode, rather than of the environment. Several reactions were given of which representative examples are as follows [12]:

	E^0 (V)	$\log K_{eq}$	
$PtO_{(s)} + 2H^+ + 2e = Pt^0_{(s)} + H_2O$	+0.9	34.0	(59)
$Pt^0_{(s)} + SO_4^{2-} + 8H^+ + 6e = PtS_{(s)} + H_2O$	–	52.05	(60)
$PtCl_4^{2-} + 2e = Pt^0_{(s)} + 4Cl^-$	–	24.7	(61)

Thus, the platinum oxides and hydroxides for the platinum–oxygen system, a sulfate reaction leading to formation of $PtS_{(s)}$, and a chloride complex of platinum are considered to be significant surface reactions.

A pE-pH diagram (Figure 11) was constructed by Whitfield to indicate the stability fields of $PtS_{(s)}, Pt^0_{(s)}, PtO_{(s)}, PtO_2 \cdot 2H_2O_{(s)}$ and $PtCl_6^{2-}$. "In well aerated environments, oxygen is the dominant electron acceptor and the formation of an oxide phase on the surface of the platinum electrode has a significant influence on its behavior. Oxygen rapidly becomes adsorbed onto a fresh platinum surface at moderate oxygen pressures and the 'derma-sorbed' layer, once formed, is difficult to remove." Furthermore, "most electro-chemical investigations have suggested that the stable phases are probably $PtO_{(s)}$, $PtO_{2(s)}$ or various admixtures of these components. These coatings are electronically conducting and, if the surface layer is intact, the platinum electrode should record the reversible oxygen potential (Equation 31, Table II). The electrode potential would then response to changes in the oxygen partial pressure." That is, if impurities are rigorously removed from the solution. Whitfield [12] cites evidence that, when the impurity concen-tration approaches $10^{-10}M$, the exchange current density is already being swamped. The implications of this statement for natural aquatic environ-ments are rather clear.

Field Measurements

Results follow of Eh measurements of several natural environments. How-ever, the reader should critically review the data in light of the above dis-cussion. For example, Hutchinson et al. [14] reported the "Eh values"

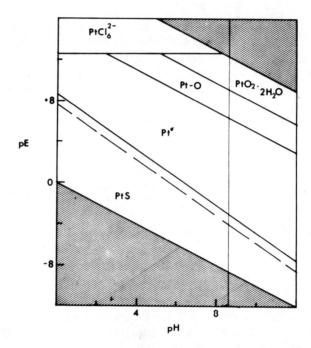

Figure 11.   Predominance area diagram for platinum compounds on the pH–pE plane for solutions where pCl = 0.45 and pSO_4 = 2.47. Broken line indicates the corresponding boundary in fresh water with pSO_4 = 3.92. The shaded area indicates conditions where water itself is unstable (reproduced from Whitfield [12] courtesy of the American Society of Limnology and Oceanography).

(quotation marks are added to question their validity) (pH = 7.0) for the surface waters of eight Connecticut Lakes ranging from 0.461 to 0.561 volts. These waters were presumed to be saturated, or nearly so, with oxygen. Allgeier et al. [15] reported in situ "Eh measurements" of the surface waters of 12 lakes in northern Wisconsin. The "Eh values" ranged from 0.39 to 0.50 V. (pH = 7.0). Hutchinson [14] also reports some "anomalously" low "Eh values" for two Wisconsin Lakes:

	pH	O_2 (mg/l)	Eh (V)	E_7 (V)
Nebish Lake				
Sample A	7.0	7.4	0.38	0.38
Sample B	7.0	–	0.45	0.45
Helmet Lake				
Sample A	6.0	5.3	0.42	0.37
Sample B	5.7	6.2	0.505	0.43

It must be assumed that these oxygen concentrations are less than the 100% saturation values. This would suggest that biological activity is responsible for the lower O_2 value and that the attendant oxidation–reduction reactions have decreased the Eh and E_7 values.

The vertical distribution of redox potentials in an impounded water has been reported by several investigators in the book by Hutchinson [14]. Where there is summer stratification in a lake, there should be an orthograde (i.e., uniform distribution) oxygen curve. This is observed when biological activity is minimum in the hypolimnion. Under these conditions, the Eh profile should be orthograde also and the redox potential values would decrease only when the mud surface is reached or penetrated. On the other hand, a clinograde (i.e., decreasing values with depth) oxygen curve does not imply a clinograde Eh curve. In natural waters, there may be redox processes other than the O_2, H_2O, H^+ (Equation 31) reaction in operation that would affect the Eh value. An example of an orthograde Eh curve and a clinograde oxygen curve is seen in Figure 12 for Lake Quossapaug, Connecticut [14]. In this example, the Eh value remains near 0.5 volt to about 19 meters of depth, whereas the $[O_2]$ decreases sharply at about 9 meters. A more classic example of clinograde oxygen and Eh curves is seen in Figure 13 for Linsley Pond, Connecticut [14]. There is a concomitant increase in Fe^{+2} with decreases in the Eh and oxygen values with depth.

Baas Becking et al. [16] have measured and reported Eh values from a variety of natural aquatic environments throughout the world. There were approximately 6200 sets of redox measurements reported by these authors. The four graphs in Figure 14 represent fresh water environments. Atmospheric waters appear to have positive Eh values between 0.4 and 0.8 volts and acidic pH values (Figure 14a). In all probability, these rain waters are saturated with oxygen, which controls the Eh values. The acidity of these waters has been discussed in Chapter 3. The Eh–pH characteristics of peat bogs are seen in Figure 14b. Surprisingly, Eh values of +0.7 volts have been measured in drained peat bogs. Most bogs, however, have values less than +0.4 and many have negative values due to sulfate reduction. The distribution of Eh and pH values for fresh waters is seen in Figure 14c. The range of Eh occurs from approximately –0.05 to +0.6 volts, with the pH value ranging from 4.0 to 9.0. Many geological as well as biological reasons may be cited for the wide spread of the two measurements. The Eh–pH characteristics of fresh-water sediments (Figure 14d) follow rather closely to fresh waters. This is, of course, an oversimplification of a very complex situation.

Hayes et al. [17] measured the redox potentials of mud cores to the 6-cm depth from seven lakes in Nova Scotia and Prince Edward Island. These measurements, E_7, are seen in Figure 15. Most of the E_7 values were positive, which reflect the observation that oxygen was present over the muds. In one lake, Punch Bowl, there was an oxygen depletion that led to lower E_7 values

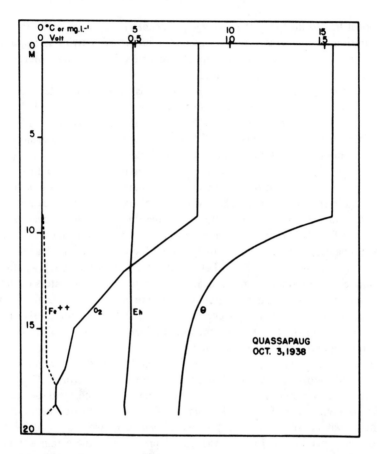

Figure 12.   Lake Quassapaug, Connecticut. Vertical distribution of temperature, oxygen, redox potential and ferrous iron late in the summer stagnation (reproduced from Hutchinson [14] courtesy of John Wiley & Sons, Inc.).

at depths below 1 cm. These authors were quite emphatic about the difficulties of obtaining accurate redox measurements. The lack of poise in the mud cores was cited as the reason for the scatter of readings to be ±50 mV.

There is considerable interest in the environmental measurement of the ferrous ion-ferric oxide redox potential. Several investigators [18-21] have attempted these field measurements with the primary objective of comparing in situ values to the calculated Nernst potential. For example, Doyle [18] obtained an experimental equation from Eh–pH measurements against the Pt electrode:

$$E = 0.965 - 0.177\ \text{pH} - 0.059\ \log\ (\text{Fe}^{2+}) \tag{62}$$

for a ferrous-ion and "a ferric species intermediate in stability between ferric hydroxide $(Fe(OH)_3)$ and hematite (Fe_2O_3)." The validity of Equation 62 was tested against Eh measurements in natural waters containing ferrous ion (sediments from the bottom of Linsley Pond, Connecticut). A measured potential of -190 mV was considerably more negative than the calculated Eh, -80 mV, from Equation 62 at the observed pH and (Fe^{2+}) values. Numerous other seasonal Eh measurements made in Linsley Pond by Doyle all fell between -150 mV and -200 mV. An attempt was made to explain this discrepancy by current-potential relationships in the natural sediment. In short, a precise explanation for the discrepancy between calculated and measured Eh values was not forthcoming. Doyle speculated that a true Fe^{2+}-Fe^{3+} potential cannot be measured either naturally or artificially due to oxidation of the Fe^{2+} ion and to absorbed oxygen on the Pt electrode.

Barnes and Back [19] and Langmuir [20] conducted field measurements of Eh and pH on iron-rich (up to 35 ppm) ground waters of southern Maryland and southern New Jersey, respectively. These authors confidently reported "significant" measurements of Eh and pH in natural waters. Their

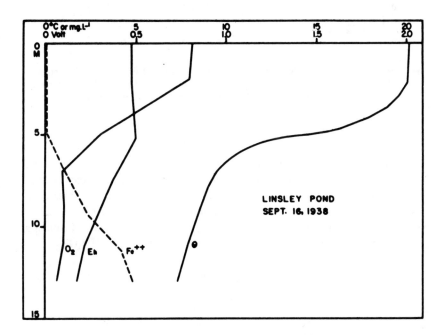

Figure 13.   Linsley Pond, Connecticut. Vertical distribution of temperature, oxygen, redox potential and ferrous iron (reproduced from Hutchinson [14] courtesy of John Wiley & Sons, Inc.).

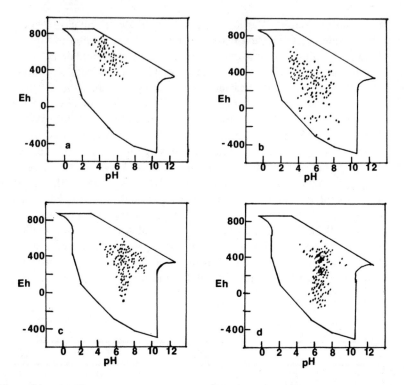

Figure 14.   (a) Eh–pH characteristics of meteoric waters; (b) Eh–pH characteristics of peat bogs; (c) Eh–pH characteristics of fresh waters; and (d) Eh–pH characteristics of fresh water sediments (reproduced from Baas-Becking et al. [16] courtesy of *J. Geol.*).

data were interpreted in terms of an equilibrium being obtained between the ground waters and ferric minerals intermediate in properties between freshly precipitated ferric hydroxide and hematite. Table III shows the measured and calculated values of Eh for the ground waters. The latter Eh values were obtained first from a calculation of $[Fe^{3+}]$ from Hem and Cropper's [22] values for the thermodynamic solubility of $Fe(OH)_{3(s)}$ and $Fe_2O_{3(s)}$:

$$Fe(OH)_{3(s)} + 3H^+ = Fe^{3+} + 3H_2O \tag{63}$$

from which

$$(Fe^{3+}) = K_{63}(10^{-pH}) \tag{64}$$

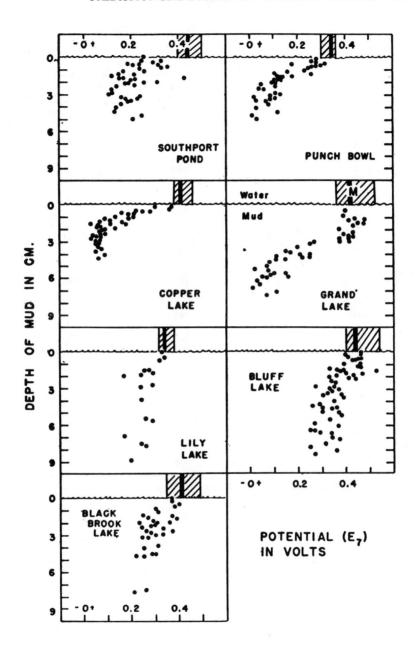

Figure 15.   The redox potential in volts at various depths in the mud of seven fresh water lakes (reproduced from Hayes et al. [17] courtesy of the American Society for Limnology and Oceanography).

Table III. Measured and Calculated Values of Eh[a]

Sample	Eh (mV)		
	Measured, Bright Pt Electrode	Calculated from Data for $Fe(OH)_{3(s)}$	Calculated from Data for $Fe_2O_{3(s)}$
15	471	660	311
16	455	597	246
17	397	565	220
18	384	700	348
19	375	628	279
20	345	515	170
21	306	271	−80
22	282	274	−73
23	243	627	279
24	242	260	−87
25	216	370	20
26	214	217	−124
27	190	224	−120
28	189	232	−120
30	185	394	45
31	164	405	48
32	158	204	−142
32a	153	699	352
33	145	392	49
34	123	774	425
35	113	306	−41
36	103	107	−240
37	82	438	90
38	55	86	−259
39	−7	213	−134
41	−20	198	−151

[a]Reproduced from Barnes and Back [19] courtesy of the *J. Geol.*

and

$$Fe_2O_{3(s)} + 6H^+ = 2Fe^{3+} + 3H_2O \tag{65}$$

from which

$$(Fe^{3+}) = K_{65}^{0.5}(10^{-3pH}) \tag{66}$$

The (Fe^{2+}) was calculated from the $[Fe^{2+}]$ times an activity coefficient for Fe^{2+}. Having thus obtained the values of (Fe^{2+}) and (Fe^{3+}), Eh values were calculated from

$$Fe^{2+} = Fe^{3+} + e \qquad\qquad (67)$$

The data in Table III indicated to Barnes and Back that Eh values are predicted more reliably from the behavior of $Fe(OH)_{3(s)}$. On the other hand, the $[Fe^{3+}]$ must be controlled by a mineral whose properties lie between those of freshly precipitated $Fe(OH)_{3(s)}$ and hematite. Given the limitations and difficulties of in situ Eh measurements, the observations of Barnes and Back are remarkable.

Interpretation

Whitfield [12] provides an interpretation of Eh values measured in the field against the pE–pH diagram for Pt compounds (Figure 11). For example, more than 90% of the data on natural aqueous environments of Baas-Becking et al. [16] lie within a fairly restricted area (Figure 16). At high Eh values (region A, Figure 16), the Pt electrode exhibits a pH-sensitive response that is described quite accurately by an equation summarizing the function of the Pt-O-coated electrode as a pH electrode:

$$E = E^0_{Pt/PtO} - 0.06 \text{ pH} \qquad\qquad (68)$$

where $E^0_{Pt/PtO}$ = 0.88 V. Most of the readings fall within ±50 mV of the line described by Equation 68 (line b, Figure 16).

Furthermore, the data points in region A, Figure 16, were taken entirely in environments where the P_{O_2} is high. These include: normal soils, shallow ground water, sea water and river water. Meteoric waters provided most of the data points on the acid side of the pH 6 boundary. According to Whitfield [12], the Pt electrode will not respond to changes in the P_{O_2}. Therefore, the Eh value will not indicate the redox status of the system unless other electroactive couples are present in sufficient concentrations ($>10^{-5}M$) to dominate the electrode potential. From an operational viewpoint, the consequence of the "oxide electrode" function of the Pt surface is to associate large, positive Eh values with well-aerated systems.

Redox readings in region B (Figure 16) reflect the behavior of the Pt electrode in stagnant environments where O_2 has been replaced by other electron acceptors. This region includes: fresh water lakes, evaporite basins, marginal marine and open-sea sediments. The Pt electrode is no longer responding primarily to pH in region B. Formation of the oxide coating is precluded either by the low pE values or the Pt surface is responding to other oxidation–reduction systems. Whitfield cites evidence that the Pt electrode responds to changes of the oxidation states of N, Fe and S. At low pE values

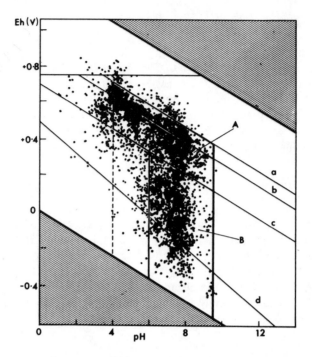

Figure 16.   Predominance area diagram for platinum compounds (Figure 11) super-
imposed on the observations of Baas-Becking et al. [16] for Eh–pH data in natural
waters. Lines a, b and d correspond to the boundaries shown in Figure 11. Line c
follows the "irreversible" oxygen potential. Vertical boundaries represent pH barriers, pCl = 0.45. Region A shows well-mixed environments where the oxygen partial
pressure is high. Region B shows stagnant environments where oxygen has been
replaced by other electron acceptors (reproduced from Whitfield [12] courtesy of
the American Society of Limnology and Oceanography).

(Figure 11), PtS becomes the stable phase, which may present a lower limit
to the usefulness of the Pt electrode. For the reaction

$$Pt^0 + S^{2-} = PtS_{(s)} + 2e \text{ (log K = 32.05)} \qquad (69)$$

for which

$$Eh = -0.96 + 0.0295 \text{ pS} \qquad (70)$$

these equations indicate that a relatively sharp drop in Eh would occur
when the Pt surface reacts with S^{2-}. Hayes et al. [17] reported a decrease

in Eh value of 0.36 V when a Pt electrode was left overnight in a sulfide-bearing mud.

In summary, Whitfield [12] indicates that the operational use of Eh measurements may be restricted to region B, Figure 16, for the reasons cited above.

Acid Mine Waters

The drainage waters from coal mines represent a unique water chemistry system. These waters usually have "very high concentrations" of iron and "very low" pH values due to sulfuric acid. The oxidative weathering of iron pyrite (FeS_2) is responsible for this unusual water quality [21] (see also Chapter 6). Sato [23] was able to demonstrate that the initial step was the release of ferrous ions and elemental sulfur:

$$FeS_{2(s)} = Fe^{2+} + S_2^0 + 2e \qquad (71)$$

In the next step, the sulfur is oxidized to sulfate and protons are produced:

$$S_2^0 + 8H_2O = 2SO_4^{2-} + 16H^+ + 12e \qquad (72)$$

The electron acceptor for these two reactions is dissolved oxygen. There is some evidence, however, that ferric iron is the primary oxidant that attacks the pyrite surface. Singer [24] has reported that the reaction rate depends on the availability of Fe^{3+} via reaction 13:

$$Fe^{2+} = Fe^{3+} + e \qquad (13)$$

Since the rate of this reaction proceeds very slowly under acidic conditions, it has been called the rate-determining step in the formation of acid mine drainage [24]. Such acidophilic iron-oxidizing bacteria as *Thiobacillus ferrooxidans* catalyze the oxidation rate by five or six orders of magnitude over the abiotic rate [21]. The geochemical formulation of this process is

$$FeS_{2(s)} + 14Fe^{3+} + 8H_2O = 15Fe^{2+} + 2SO_4^{2-} + 16H^+ \qquad (73)$$
$$\uparrow \overline{\text{regenerated by}} \mid$$
$$\textit{Thiobacillus ferrooxidans}$$

Nordstrom et al. [21] attempted to model the Eh values of acid mine waters in Shasta County, California by the $Fe^{2+}:Fe^{3+}$ activity ratio. The water quality analyses (temperature, pH, Eh, iron contents, etc.) from more than 100 samples were processed with the computer program WATEQ2 (see Chapter 6). Consequently, Eh values were calculated from the $Fe^{2+}:Fe^{3+}$ ratio by the appropriate Nernst-Peters Equation 26 and were compared with the measured Eh values. The iron activities were corrected for effects of complexing and ionic strength. Nordstrom et al. [21] were extremely confident that acid mine waters are well suited to reliable Eh measurements despite the limitations cited earlier in this chapter. Figure 17 shows the comparison of measured and calculated Eh values. The solid line represents perfect agreement. Of the measured values, 77% fell within ±30 mV, which is an anticipated uncertainty for most Eh measurements. According to Nordstrom et al. [21], "this excellent agreement suggests a simultaneous validation of the equilibrium conditions of acid mine waters, of the chemical model and its ability to represent that equilibrium, and of the accuracy of the Eh measurement."

Sulfur Species in Reducing Environments

Reducing conditions in aquatic environments are characterized frequently by the presence of hydrogen sulfide. In these environments there are several

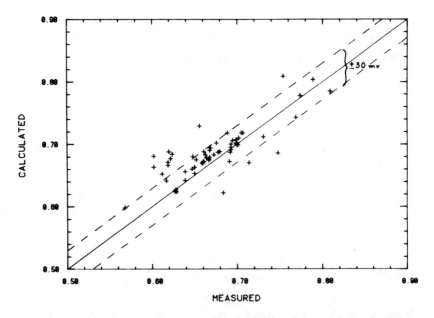

Figure 17.   Comparison of measured and calculated Eh values. Errors of ±30 mV are given for the uncertainty in the Eh measurements (reproduced from Nordstrom et al. [22] courtesy of the American Chemical Society).

factors, such as the slow diffusion of oxygen, the presence of Fe(+III) compounds and organic matter, which may result in the incomplete oxidation
of H_2S [25]. This may yield polysulfide (S_n^{2-}) and thiosulfate ($S_2O_3^{2-}$) ions
through these reactions:

$$2HS^- + O_2 = \frac{1}{4} S_8 + 2OH^- \tag{74}$$

$$HS^- + (n-1)/8 S_8 = S_n^{2-} + H^+ \tag{75}$$

$$2S_n^{2-} + O_2 + H_2O = S_2O_3^{2-} + S_{(n-1)}^{2-} + 2H^+ \tag{76}$$

$$2HS^- + 2O_2 = S_2O_3^{2-} + H_2O \tag{77}$$

According to Boulegue and Michard [25], "the above reactions proceed
easily and rapidly in the physicochemical conditions prevailing in natural
waters." Therefore, polysulfide ions should occur in reducing environments
where there is incomplete oxidation of H_2S.

Equation 75 shows the general process by which polysulfide ions are
produced. There is experimental evidence [25] that these polysulfide ions
are stable in the H_2S-S_8-H_2O system at sulfide concentrations in the 10^{-5}
to $10^{-6}M$ range. The major ions are probably S_5^{2-}, S_4^{2-} and S_6^{2-} over a "wide"
pH range. The following reactions are relevant:

	ΔG_{rex}^0 (kcal/mol)	$\log K_{eq}$	
$H_2S_{(aq)} = HS^- + H^+$	9.56	−7.01	(78)
$H_2S = S^0 + 2H^+ + 2e$	6.55	−4.80	(79)
$HS^- = S^0 + H^+ + 2e$	−3.01	2.21	(80)
$5HS^- = S_5^{2-} + 5H^+ + 8e$	0.79	−0.58	(81)
$4HS^- = S_4^{2-} + 4H^+ + 6e$	3.96	−2.90	(82)
$S_5^{2-} = 5S^0 + 2e$	−15.85	11.62	(83)
$5S_4^{2-} = 4S_5^{2-} + 2e$	−19.60	14.37	(84)

These reactions were used to draw the pE–pH diagram in Figure 18 for the
polysulfide ions.

According to Allen and Hickling [26], the sulfide/polysulfide couple is
electroactive and the reaction at the surface of the platinum electrode is
rapid. Thus, the potential measured in the H_2S-S_8-H_2O system must reflect
the equilibria between the sulfide and polysulfide species and the redox
potential is directly interpreted also in terms of these equilibria. Boulegue
and Michard [25] offer experimental evidence (not given here) that measured

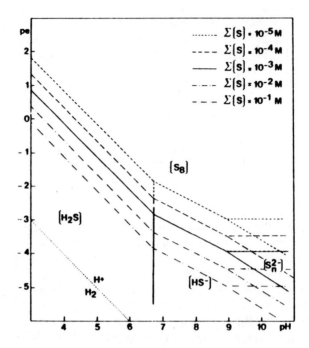

Figure 18.   pE–pH stability diagram of the system H_2S-S_8-H_2O-NaCl $(0.7M)$ for $\Sigma[S]$ = 10^{-X} g-atom/kg (reproduced from Boulegue and Michard [25] courtesy of the American Chemical Society).

and computed pE–pH values are good. Therefore, the reactions of the sulfide and polysulfide species appear to be chemically and electrochemically reversible under the appropriate reducing conditions.

REFERENCES

1. Morris, J. C., and W. Stumm. In: *Equilibrium Concepts in Natural Water Systems, Advances in Chemistry Series* 67 (1967), p. 270.
2. Stumm, W., and J. J. Morgan. *Aquatic Chemistry* (New York: John Wiley & Sons, Inc., 1970).
3. International Union of Pure and Applied Chemistry. *Nomenclature of Inorganic Chemistry*, 2nd ed. (London: Butterworths, 1970).
4. Jörgensen, H. *Redox-malinger* (Copenhagen: Gjellerup, 1965).
5. Licht, T. S., and A. J. deBethune. *J. Chem. Ed.* 34:433 (1957).
6. Lingane, J. J. *Electroanalytical Chemistry* (New York: Wiley-Interscience, 1953).
7. Pourbaix, M. *Atlas of Electrochemical Equilibria in Aqueous Solutions* (Elmsford, NY: Pergamon Press, Inc., 1966).

8. Latimer, W. M. *Oxidation Potentials* (Englewood Cliffs, NJ: Prentice-Hall, Inc., 1952).
9. Brock, T. D. *Biology of Microorganisms* (Englewood Cliffs, NJ: Prentice-Hall, Inc., 1978).
10. Gillespie, L. J. *Soil Sci.* 9:199 (1920).
11. Whitfield, M. *Limnol. Oceanog.* 14:547 (1969).
12. Whitfield, M. *Limnol. Oceanog.* 19:857 (1974).
13. Hem, J. D. *J. Am. Water Works Assoc.* 53:211 (1961).
14. Hutchinson, G. E. *A Treatise on Limnology*, Vol. 1, Parts 1 and 2 (New York: Wiley-Interscience, 1975).
15. Allgeier, R. J., et al. *Trans. Wis. Acad. Sci. Arts Lett.* 33:115 (1941).
16. Baas-Becking, L. G. M., et al. *J. Geol.* 68:243 (1960).
17. Hayes, F. R., et al. *Limnol. Oceanog.* 3:308 (1958).
18. Doyle, R. W. *Am. J. Sci.* 266:840 (1968).
19. Barnes, I., and W. Back. *J. Geol.* 72:435 (1964).
20. Langmuir, D. *U.S. Geol. Survey Prof. Paper 650-C* (1969), p. C224.
21. Nordstrom, D. K., et al. In: *Chemical Modeling in Aqueous Systems*, A.C.S. Symposium Series No. 93 (Washington, DC: American Chemical Society, 1979).
22. Hem, J. D., and W. H. Cropper. *U.S. Geol. Prof. Paper 1459A* (1959).
23. Sato, M. *Econ. Geol.* 55:1202 (1960).
24. Singer, P. *Science* 167:1121 (1970).
25. Boulegue, J., and G. Michard. In: *Chemical Modeling in Aqueous Systems*, E. A. Jenne, Ed., ACS Symposium Series No. 93 (Washington, DC: American Chemical Society, 1979).
26. Allen, P., and A. Hickling. *Trans. Faraday Soc.* 53:1626 (1957).

CHAPTER 5

CHEMICAL WEATHERING REACTIONS

INTRODUCTION

Most of the dissolved inorganic constituents that occur in natural waters have resulted from physical, chemical and biological weathering of geologic formations. Furthermore, these weathering interactions occur among the three phases of nature, namely, the atmosphere, the hydrosphere and the lithosphere. Consequently, natural waters, fresh and saline, acquire their chemical characteristics through reasonably well-defined processes. These waters vary considerably in their chemical composition, of course, which is a reflection of the type of geology contacted as water circulates through the hydrologic cycle. This chapter principally describes the chemical weathering processes by which natural waters obtain their dissolved inorganic constituents. Antropogenic sources are not considered here.

COMPOSITION OF THE EARTH'S CRUST

It is essential to have at least a working knowledge of the composition of the various rocks and minerals that are weathered to understand how the chemical composition of natural waters is derived. A brief discussion is given here with a suggestion to the reader to consult a geochemical textbook for terminology, definitions, etc. [1-4].

Geologically, the earth is divided into three zones: crust, mantle and core. Our concern is with the composition of the crust, which is separated from the mantle at an average depth below sea level ranging from 37 km (121,390 ft) under parts of the continents to about 11 km (36,090 ft) beneath the deep sea basins. The geochemical spheres—hydrosphere, atmosphere and biosphere—are included in the crust. The term "lithosphere" generally is applied to the lithic (pertaining to stone) crust of the earth.

The earth's crust is constituted principally of igneous rocks and contains only minor amounts of sedimentary and metamorphic rocks. Clarke and Washington [5] have indicated that the proportions of different classes of rocks in the crust are: 95% igneous (including metamorphic rocks) and 5% sedimentary rocks (consisting of 4% shale, 0.75% sandstone and 0.25% limestone). Although these estimates were published in 1924, they have not been significantly revised since.

The earth, in general, is composed of an iron-rich core surrounded by a thick mantle of magnesium and iron-rich silicates and a thin outer crust of the most common silicates and other types of minerals. The outer crust is, of course, the only portion that exerts a direct influence on the composition of natural waters. Any discussion on composition of this crust should be considered in terms of estimates and averages of available analyses, of which many have been attempted over the years. The data in Table I are from a computer analysis by Horn and Adams [6] of several publications since that of Clarke and Washington [5]. These data represent an elemental analysis and denote the abundance of each in the various rocks. Oxygen does not appear in the data but is, of course, the most abundant of all the elements in crustal rocks, constituting 466,000 ppm, or 46.6%, of the lithosphere's weight [7]. An alternate expression of the average composition of igneous and sedimentary rocks, for example, may be seen in Tables II, III and IV, in which several of the elements are expressed as their oxides [8]. Additional details about the composition of the earth's crust may be found in the excellent publication by Parker [9].

It should be obvious from the data in Tables I–IV that Si and Al are the two most abundant elements in the earth's crust. The concentrations of these constituents in natural waters are, in general, 100 mg/l and frequently in the order of 10-20 mg/l. This reflects the inverse relationship between elemental abundance in igneous and sedimentary rocks and concomitant concentrations in the water phase.

GENERAL ASPECTS OF WEATHERING REACTIONS

Mechanical and chemical processes weather rocks into smaller fragments and dissolved constituents. In general, these processes start with a solid rock, which must be broken into small fragments. In turn, these fragments are subjected to chemical weathering by ionization, dissolution, neutralization by carbon dioxide, hydrolysis and oxidation. All these processes provide the dissolved constituents that occur in natural waters (domestic and industrial pollution excepted).

Table I. Average Composition of Igneous and Some Types of Sedimentary Rocks [6][a]

Element	Igneous Rocks	Sedimentary Rocks		
		Resistates (sandstone)	Hydrolyzates (shale)	Precipitates (carbonates)
Si	285,000	359,000	260,000	34
Al	79,500	32,100	80,100	8,970
Fe	42,200	18,600	38,800	8,190
Ca	36,200	22,400	22,500	272,000
Na	28,100	3,870	4,850	393
K	25,700	13,200	24,900	2,390
Mg	17,600	8,100	16,400	45,300
Ti	4,830	1,950	4,440	377
P	1,100	539	733	281
Mn	937	392	575	842
F	715	220	560	112
Ba	595	193	250	30
S	410	945	1,850	4,550
Sr	368	28	290	617
C	320	13,800	15,300	113,500
Cl	305	15	170	305
Cr	198	120	423	7.1
Rb	166	197	243	46
Zr	160	204	142	18
V	149	20	101	13
Ce	130	55	45	11
Cu	97	15	45	4.4
Ni	94	2.6	29	13
Zn	80	16	130	16
Nd	56	24	18	8.0
La	48	19	28	9.4
N	46	–	600	–
Y	41	16	20	15
Li	32	15	46	5.2
Co	23	0.33	8.1	0.12
Nb	20	0.096	20	0.44
Ga	18	5.9	23	2.7
Pr	17	7.0	5.5	1.3
Pb	16	14	80	16
Sm	16	6.6	5.0	1.1
Sc	15	0.73	10	0.68
Th	11	3.9	13	0.20
Gd	9.9	4.4	4.1	0.77
Dy	9.8	3.1	4.2	0.53
B	7.5	90	194	16
Yb	4.8	1.6	1.6	0.20
Cs	4.3	2.2	6.2	0.77
Hf	3.9	3.0	3.1	0.23
Be	3.6	0.26	2.1	0.18
Er	3.6	0.88	1.8	0.45
U	2.8	1.0	4.5	2.2
Sn	2.5	0.12	4.1	0.17
Ho	2.4	1.1	0.82	0.18
Br	2.4	1.0	4.3	6.6

Table I, continued

Element	Igneous Rocks	Sedimentary Rocks		
		Resistates (sandstone)	Hydrolyzates (shale)	Precipitates (carbonates)
Eu	2.3	0.94	1.1	0.19
Ta	2.0	0.10	3.5	0.10
Tb	1.8	0.74	0.54	0.14
As	1.8	1.0	9.0	1.8
W	1.4	1.6	1.9	0.56
Ge	1.4	0.88	1.3	0.036
Mo	1.2	0.50	4.2	0.75
Lu	1.1	0.30	0.28	0.11
Tl	1.1	1.5	1.6	0.065
Tm	0.94	0.30	0.29	0.075
Sb	0.51	0.014	0.81	0.20
I	0.45	4.4	3.8	1.6
Hg	0.33	0.057	0.27	0.046
Cd	0.19	0.020	0.18	0.048
In	0.19	0.13	0.22	0.068
Ag	0.15	0.12	0.27	0.19
Se	0.050	0.52	0.60	0.32
Au	0.0036	0.0046	0.0034	0.0018

[a]Concentration in ppm by weight.

The Rock Cycle

Mechanical or physical weathering may be described as a cycle. Igneous rocks undergo metamorphic processes (heat, pressure and migrating fluids), which, in turn, are transformed into sedimentary rocks via uplifting and erosive processes. To complete the cycle, sediments undergo deposition and cementation processes that eventually lead to metamorphic and igneous rocks. This so-called rock cycle is seen in Figure 1.

A major portion of the dissolved constituents is derived from weathering of sediments since small rock fragments provide greater surface areas. This does not except the larger igneous and metamorphic from dissolution reactions. However, the rate at which these latter rocks are weathered is somewhat slower. In the weathering of sediments, the hydrosphere acts as a conveyor of these fragments in suspended and dissolved form and as a reactant in various chemical reactions. In turn, the atmosphere provides two chemical reactants: acids from volcanic emanations, HCl, CO_2, H_2S and S, and the oxidant, O_2.

Table II. Approximate Chemical Composition of Common Minerals
of Igneous Rocks (% wt)[a]

	SiO_2	Al_2O_3	$MgO + FeO$	CaO	$Na_2O + K_2O$
Ferromagnesian Minerals					
Olivine[b]	40	–	60	–	–
Augite[c]	50	3	23	20	–
Hornblende[c]	40	10	30	12	trace
Biotite[c]	36	15	30	1	10
Plagioclase Series					
Calcic plagioclase[d]	54	29	–	12	5
Sodic plagioclase (albite)	68	20	–	–	12
Alkali Feldspars					
Potash feldspar (orthoclase, microcline etc.)	65	18	–	–	17
Feldspathoids					
Nepheline	42	36	–	–	22
Leucite	55	23	–	–	22
Silica Minerals					
Quartz, etc.	100	–	–	–	–

[a]Reproduced from Cox [8] courtesy of The M.I.T. Press.
[b]Composition of the common olivine of basic igneous rocks, containing about 25% of Fe_2SiO_4 and 75% of Mg_2SiO_4.
[c]In addition to the constituents shown, augite, hornblende and biotite contain significant amounts of TiO_2 and Fe_2O_3. Hornblende and biotite usually also contain about 2% of water.
[d]Composition of the common plagioclase of basic igneous rocks, containing about 60% of $CaAl_2Si_2O_8$ and 40% of $NaAlSi_3O_8$.

Agents of Chemical Weathering

Dissolution and Hydration

The role of water in chemical weathering processes is extremely important and varied. Moisture accelerates the rate at which weathering occurs. In addition, water acts as a conveyor of the dissolved constituents and of such weathering agents as CO_2 and O_2. On occasion, water itself enters into these processes through hydration and dehydration reactions, of which an example may be the conversion of gypsum to anhydrite, and vice versa:

$$CaSO_4 \cdot 2H_2O_{(s)} = CaSO_{4(s)} + 2H_2O$$
$$\text{Gypsum} \qquad \text{Anhydrite}$$

(1)

Table III. Approximate Average Compositions of Some Common Igneous Rocks[a,b]

Rock Type	Granite (Rhyolite)	Diorite (Andesite)	Gabbro (Basalt)	Syenodiorite (Trachyandesite)	Syenite (Trachyte)	Nepheline Syenite (Phonolite)	Ijolite (Nephelinite)	Syenogabbro (Trachybasalt)
SiO_2	70.8	57.6	49.0	56.7	62.5	55.4	43.2	49.3
TiO_2	0.4	0.9	1.0	1.1	0.6	0.9	1.6	1.9
Al_2O_3	14.6	16.9	18.2	17.1	17.6	20.2	19.1	18.2
Fe_2O_3	1.6	3.2	3.2	3.0	2.1	3.4	3.9	4.4
FeO	1.8	4.5	6.0	4.1	2.7	2.2	4.9	5.7
MgO	0.9	4.2	7.6	3.3	0.9	0.9	3.2	4.1
CaO	2.0	6.8	11.2	6.6	2.3	2.5	10.6	9.0
Na_2O	3.5	3.4	2.6	3.7	5.9	8.4	9.7	4.4
K_2O	4.2	2.2	0.9	3.8	5.2	5.5	2.3	2.3

[a] Analyses given are those of the plutonic rocks. Names of the approximately equivalent volcanics are shown in brackets.
[b] Reproduced from Cox [8] courtesy of The M.I.T. Press.

Table IV. Chemical Composition of Some Typical Sedimentary Rocks[a]

	1	2	3	4	5	6
SiO_2	74.14	78.14	99.14	58.10	55.02	7.61
TiO_2	0.15	–	0.03	0.65	1.00	0.14
Al_2O_3	10.17	11.75	0.40	16.40	22.17	1.55
Fe_2O_3	0.56	1.23	0.12	4.02	8.00	0.70
FeO	4.15	–	–	2.45		1.20
MgO	1.43	0.19	nil	2.44	1.45	2.70
CaO	1.49	0.15	0.29	3.11	0.15	45.44
Na_2O	3.56	2.50	0.01	1.30	0.17	0.15
K_2O	1.36	5.27	0.15	3.24	2.32	0.25
H_2O^+	2.66	0.64	0.17	5.00	7.76	0.38
H_2O^-					2.10	0.30
CO_2	0.14	0.19	–	2.63	–	39.27
C	–	–	–	0.80	–	0.09
Total	99.81	100.06	100.31	100.14	100.14	99.78

1. Typical greywacke.
2. Typical arkose. Torridonian, Scotland.
3. Devonian orthoquartzite.
4. Silty clay.
5. Residual clay from weathering of gneiss.
6. Allochemical limestone.

[a]Reproduced from Cox [8] courtesy of The M.I.T. Press.

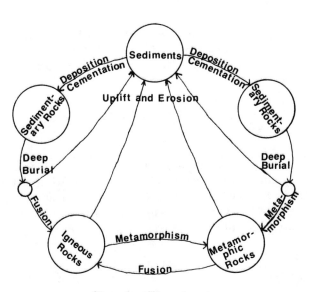

Figure 1.   The rock cycle.

Another example may be the instability of freshly precipitated ferric hydroxide:

$$2Fe(OH)_{3(s)} \quad = Fe_2O_{3(s)} + 3H_2O$$
$$\text{Ferric hydroxide} \quad \text{Hematite}$$

(2)

Dissociation and Acidic Dissolution

The dissolving of various carbonate minerals is particularly significant because these reactions supply the hardness constituents to water, namely, Ca^{2+} and Mg^{2+}. Carbonates may be weathered through simple dissociation:

$$CaCO_{3(s)} = Ca^{2+} + CO_3^{2-}$$
$$\text{Calcite}$$

(3)

or through an acidic reaction:

$$CaCO_{3(s)} + CO_{2(g)} + H_2O = Ca^{2+} + 2HCO_3^-$$

(4)

The chemical weathering of carbonates is discussed in some detail in Chapter 6.

Oxidative Reactions

Many of the constituents in natural waters occur from oxidative reactions of various rocks and minerals where they reside in reduced valence states. For example, a commonly occurring reaction in nature is the oxidation of iron pyrite:

$$FeS_{2(s)} + 7.5\ O_2 + H_2O = Fe^{2+} + 2SO_4^{2-} + 2H^+$$
$$\text{Pyrite}$$

(5)

This is a rather unique reaction since it may be mediated in biological systems and produces sulfuric acid leading to very low pH values in natural waters (see Chapter 4). It should be noted that the sulfur atom in the pyrite undergoes the oxidation and not the ferrous atom, as is erronously reported on occasion. Some other oxidative reactions are the following:

$$2FeCO_{3(s)} + 0.5\ O_2 + 2H_2O = Fe_2O_3 + 2H_2CO_3$$
$$\text{Siderite}$$

(6)

$$MnCO_{3(s)} \quad + 0.5\ O_2 + H_2O = MnO_{2(s)} + H_2CO_3 \tag{7}$$
$$\text{Rhodochrosite} \qquad\qquad\qquad \text{Pyrolusite}$$

These three reactions represent only the elements of sulfur, iron and manganese in common rocks, for which oxidation is an important part of weathering.

Hydrolytic Reactions of the Silicates

The chemical weathering of silicates is, perhaps, the predominant reaction in nature and is the major source of many dissolved constituents. Hydrolytic reactions account for most of the weathering of silicates. Representative samples are as follows:

$$Mg_2SiO_{4(s)} + 4H_2O = 2Mg^{2+} + 4OH^- + H_4SiO_4 \tag{8}$$
forsterite

$$SiO_{2(s)} + 2H_2O = H_4SiO_4 \tag{9}$$
quartz

$$4KAlSi_3O_{8(s)} + 22H_2O = 4K^+ + 4OH^- + 8H_4SiO_4 + Al_4Si_4O_{10}(OH)_{8(s)} \tag{10}$$
microcline $\qquad\qquad\qquad\qquad\qquad\qquad$ kaolinite

These reactons are, of course, oversimplified in their presentation and suggest straightforward hydrolysis. Undoubtedly these reactions require several sequential steps before the listed products are formed. Also, they are very slow processes and may require several years for completion or to reach equilibrium. Many have been studied in the laboratory but almost always at elevated temperatures circa 200°C.

Frequently, silicates may undergo an acidic reaction:

$$Mg_2SiO_{4(s)} + 4H_2CO_3 = 2Mg^{2+} + 4HCO_3^- + H_4SiO_4 \tag{11}$$

This reaction suggests that the carbonic acid attack on forsterite and similar silicates may be a source of bicarbonate alkalinity in water.

There are three unique features of the hydrolytic weathering of silicates. Firstly, dissolved silica is represented by the very weak silicic acid, H_4SiO_4 ($K_1 = 10^{-9.5}$), which is usually written as a neutral species. Secondly, prolonged contact of water with the silicate minerals eventually leads to alkaline solutions and reduction of acidity in natural waters. The extent of the latter depends, of course, on the nature of the silicate under attack. Thirdly, clay minerals, as seen in Equation 10, are frequently residual solids.

CHEMICAL WEATHERING REACTIONS

Standard Free Energy of Formation of Common Minerals

A compilation of the standard free energy of formation values for several commonly occurring minerals is given in Table V. These data have been selected mostly from the *Geological Survey Bulletin* of Robie and Waldbaum [10]. The reader will note from this source and other publications that thermodynamic data vary considerably for many of these solid phases. Additional data may be found in the geochemistry text by Krauskopf [1] and the publication by Ball et al. [16].

Dissolution of Carbonates

Carbonate rocks and minerals weather chemically by acidic dissolution which occurs by reaction with hydrogen ions or with dissolved CO_2 (carbonic acid, H_2CO_3). Several acidic weathering reactions may be written for various carbonate minerals found in nature: calcite, aragonite, magnesite and dolomite. Table VI shows 15 reactions that were written in an attempt to provide equilibrium models for naturally occurring carbonate waters [17]. Reactions 12 through 21 may occur in nature, whereas Reactions 22 through 26, as written, may be considered hypothetical and may not occur naturally. These carbonate reactions are discussed in greater detail in Chapter 6.

Some miscellaneous carbonate weathering reactions are presented in Table VII that have some significance in water quality chemistry. For example, Reaction 27 (reading right to left) has been proposed to be one of the reactions leading to the formation of dolomite from calcite. If this reaction does, indeed, occur naturally, then the Ca^{2+} and Mg^{2+} contents in the aqueous phase will be affected. Reaction 31 shows the acidic weathering of malachite, the blue-green corrosion product of copper pipes. Number 32 (reading left to right) is also a "formation" reaction where dolomitic limestone rocks have been found "contaminated" with a 4 mol% excess of calcite. It is an attempt to provide a more accurate representation of carbonate weathering reactions as they occur in nature.

Dissolution of Silicates and Clay Minerals

Al_2O_3-SiO_2-H_2O Systems

Reactions 35 and 36 in Table VIII represent the dissolution and solubilities of $SiO_{2(s)}$ in water [21,22]. Note that the amorphous form of silica is more soluble in water than is quartz. In nature, however, amorphous silica is metastable with respect to quartz. Reactions 37 and 41 show the laboratory

Table V. Standard Free Energy of Formation of Some Minerals[a]

Group	Mineral Name	Formula	ΔG_f^0 (kcal/GFW)
Silicon Oxides	α-Quartz	SiO_2	−204.646
	Silica glass	SiO_2	−203.298
	α-Tridymite	SiO_2	−204.076
Aluminum Oxides	Corundum	Al_2O_3	−378.082
	Boehmite	$AlO(OH)$	−217.674
	Gibbsite	$Al(OH)_3$	−273.486
	Gibbsite	$Al_2O_3 \cdot 3H_2O$	−554.6
Clay Minerals			
a. Sheet structure	Kaolinite	$Al_2Si_2O_5(OH)_4$	−902.868
	Halloysite	$Al_2Si_2O_5(OH)_4$	−898.419
	Muscovite	$KAl_2[AlSi_3O_{10}](OH)_2$	−1330.103
	Talc	$Mg_3Si_4O_{10}(OH)_2$	−1324.486
b. Framework structure	Anorthite	$CaAl_2Si_2O_8$	−955.626
	Phillipsite	$Ca_2Al_4Si_8O_{24} \cdot 9H_2O$	−3260.328[b]
	Microcline	$KAlSi_3O_8$	−892.817
	Low albite	$NaAlSi_3O_8$	−883.988
	High albite	$NaAlSi_3O_8$	−882.687
	Anacline	$NaAlSi_2O_6 \cdot H_2O$	−734.262
c. Chain and band structure	Diopside	$CaMg(SiO_3)_2$	−725.784
	Clinoenstatite	$MgSiO_3$	−349.394
	Rhodonite	$MnSiO_3$	−297.390
	Wollastonite	$CaSiO_3$	−370.313
d. *Ortho* and ring structure	Sillimanite	Al_2SiO_5	−583.600
	Fayalite	Fe_2SiO_4	−329.668
	Forsterite	Mg_2SiO_4	−491.938
	Iron metasilicate	$FeSiO_3$	−257.0[c]
Sulfides and Arsenides	Realgar	AsS	−16.806
	Orpiment	As_2S_3	−40.250
	Troilite	FeS	−24.219
	Pyrrhotite	$FeS_{Hex.}$	−24.3
	Pyrite	FeS_2	−38.296
	Alabandite	MnS	−52.140
	Galena	PbS	−22.962
	Wurtzite	ZnS	−45.760
Oxides and Hydroxides	Arsenolite	As_2O_3	−137.731
	Lime	CaO	−144.352
	Portlandite	$Ca(OH)_2$	−214.673
	Tenorite	CuO	−30.498
	Cuprite	Cu_2O	−35.022
	Ferrous oxide	FeO	−60.097
	Ferrous hydroxide	$Fe(OH)_2$	−115.57
	Hematite	Fe_2O_3	−177.728
	Magnetite	Fe_3O_4	−243.094
	Ferric hydroxide	$Fe(OH)_3$	−166.0
	Periclase	MgO	−136.087
	Brucite	$Mg(OH)_2$	−199.460
	Manganosite	MnO	−86.720
	Pyrolusite	β-MnO_2	−111.342
	Pyrochroite	$Mn(OH)_2$	−147.14[d]

Table V, continued

Group	Mineral Name	Formula	ΔG_f^0 (kcal/GFW)
	Manganite	γ-MnOOH	−133.3
	Bixbyite	Mn_2O_3	−210.097
	Hausmanite	Mn_3O_4	−306.313
	Sodium oxide	Na_2O	−90.161
	Phosphorus pentoxide	P_2O_5	−649.968
	Eskolaite	Cr_2O_3	−253.203
Halides	Lawrencite	$FeCl_2$	−72.273
	Molysite	$FeCl_3$	−79.827
	Sylvite	KCl	−97.693
	Chloromagnesite	$MgCl_2$	−141.521
	Scacchite	$MnCl_2$	−105.295
	Halite	NaCl	−91.807
	Villiaumite	NaF	−129.812
	Fluorite	CaF_2	−277.799
	Cryolite	Na_3AlF_6	−745.4[e]
Carbonates	Aragonite	$CaCO_3$	−269.678
	Calcite	$CaCO_3$	−269.908
	Dolomite	$CaMg(CO_3)_2$	−518.734
	Malachite	$Cu_2(OH)_2CO_3$	−216.440
	Azurite	$Cu_3(OH)_2(CO_3)_2$	−343.730
	Siderite	$FeCO_3$	−161.030
	Magnesite	$MgCO_3$	−246.112
	Rhodochrosite	$MnCO_3$	−195.045
	Hydromagnesite	$MgCO_3 \cdot Mg(OH)_2 \cdot 3H_2O$	−1100.100
	Huntite	$Mg_3Ca(CO_3)_4$	−1007.700
	Nesquehonite	$MgCO_3 \cdot 3H_2O$	−412.660[f]
	Otavite	$CdCO_3$	−159.964
	Cerussite	$PbCO_3$	−150.325
	Strontianite	$SrCO_3$	−275.450
	Smithsonite	$ZnCO_3$	−174.786
Sulfates	Barite	$BaSO_4$	−325.30
	Anhydrite	$CaSO_4$	−316.475
	Thenardite	Na_2SO_4	−303.400
	Celestite	$SrSO_4$	−319.83
	Gypsum	$CaSO_4 \cdot 2H_2O$	−430.137
	Celestite	$SrSO_4$	−319.830
Phosphates	Berlinite	$AlPO_4$	−388.0
	Whitlockite	$Ca_3(PO_4)_2$	−932.8
	Hydroxylapatite	$Ca_5(PO_4)_3OH$	−1506.3[b]
	Strengite	$Fe(PO_4) \cdot 2H_2O$	−397.700
	Vivianite	$Fe_3(PO_4)_2$	−590.9[g]
	Fluorapatite	$Ca_5(PO_4)_3F$	−1545.2[b]
	−	$CaHPO_4$	−401.5

[a]From Robie and Waldbaum [10].
[b]Calculated.
[c]From Latimer [11].
[d]From Bricker [12].
[e]From Roberson and Hem [13].
[f]From Langmuir [14].
[g]From Chen and Faust [15].

Table VI. Carbonate Weathering Reactions by Acidic Dissolution at 25°C[a,b]

Reaction No.	Carbonate	Reaction	ΔG^0_{Rex} (kcal/mol)	log K_{eq}
(12)	Calcite	$CaCO_3(s) + H^+ = Ca^{2+} + HCO_3^-$	-2.532	1.856
(13)	Calcite	$CaCO_3(s) + H_2CO_3 = Ca^{2+} + 2HCO_3^-$	6.148	-4.507
(14)	Calcite	$2CaCO_3(s) + H^+ + H_2CO_3 = 2Ca^{2+} + 3HCO_3^-$	3.616	-2.651
(15)	Aragonite	$CaCO_3(s) + H^+ = Ca^{2+} + HCO_3^-$	-2.762	2.025
(16)	Magnesite	$MgCO_3(s) + H^+ = Mg^{2+} + HCO_3^-$	-3.048	2.235
(17)	Magnesite	$MgCO_3(s) + H_2CO_3 = Mg^{2+} + 2HCO_3^-$	5.632	-4.129
(18)	Magnesite	$2MgCO_3(s) + H^+ + H_2CO_3 = 2Mg^{2+} + 3HCO_3^-$	2.584	-1.894
(19)	Dolomite	$CaMg(CO_3)_2(s) + 2H^+ = Ca^{2+} + Mg^{2+} + 2HCO_3^-$	-2.866	2.101
(20)	Dolomite	$CaMg(CO_3)_2(s) + 2H_2CO_3 = Ca^{2+} + Mg^{2+} + 4HCO_3^-$	14.494	-10.626
(21)	Dolomite	$CaMg(CO_3)_2(s) + H^+ + H_2CO_3 = Ca^{2+} + Mg^{2+} + 3HCO_3^-$	5.814	-4.262
(22)	Mixed-1	$CaCO_3(s) + MgCO_3(s) + 2H^+ = Ca^{2+} + Mg^{2+} + 2HCO_3^-$	-5.580	4.091
(23)	Mixed-2	$2CaCO_3(s) + MgCO_3(s) + 2H^+ + H_2CO_3 = 2Ca^{2+} + Mg^{2+} + 4HCO_3^-$	0.568	-0.416
(24)	Mixed-3	$CaCO_3(s) + 2MgCO_3(s) + 2H^+ + H_2CO_3 = Ca^{2+} + 2Mg^{2+} + 4HCO_3^-$	0.052	-0.038
(25)	Mixed-4	$CaMg(CO_3)_2(s) + CaCO_3(s) + 3H^+ = 2Ca^{2+} + Mg^{2+} + 3HCO_3^-$	-5.398	3.957
(26)	Mixed-5	$CaMg(CO_3)_2(s) + MgCO_3(s) + 3H^+ = Ca^{2+} + 2Mg^{2+} + 3HCO_3^-$	-5.914	4.336

[a]Solid phases are denoted with (s). Remainder of constituents are considered to be dissolved in the aqueous phase.
[b]From Faust [17].

Table VII. Miscellaneous Carbonate Weathering Reactions at 25°C

Reaction No.	Reaction	ΔG^0_{Rex} (kcal/mol)	log K_{eq}	Reference
(27)	$CaMg(CO_3)2(s) + Ca^{2+} = Mg^{2+} + CaCO_3(s)$ Calcite	2.198	−1.61	18
(28)	$3MgCO_3 \cdot Mg(OH)_2 \cdot 3H_2O = 4Mg^{2+} + 3CO_3^{2-} + 2OH^- + 3H_2O$ Hydromagnesite	41.197	−30.2	14
(29)	$MnCO_3(s) + H^+ = Mn^{2+} + HCO_3^-$ Rhodochrosite	1.385	−1.02	Calculated
(30)	$FeCO_3(s) + H^+ = Fe^{2+} + HCO_3^-$ Siderite	0.47	−0.34	Calculated
(31)	$Cu_2(OH)_2CO_3(s) + 3H^+ = 2Cu^{2+} + HCO_3^- + 2H_2O$ Malachite	−6.142	4.50	Calculated
(32)	$CaCO_3(s) + 0.46Mg^{2+} = Ca_{0.54}Mg_{0.46}CO_3(s) + 0.46Ca^{2+}$	−0.061	0.045	19
(33)	$MgCO_3 \cdot 3H_2O(s) = Mg^{2+} + CO_3^{2-} + 3H_2O$ Nesquehonite	7.487	−5.49	1
(34)	$Zn^{2+} + CO_2(g) + H_2O = ZnCO_3(s) + 2H^+$ Smithsonite	11.346	−8.32	20

Table VIII. Chemical Weathering Reactions of Al_2O_3-SiO_2-H_2O Systems at 25°C

Reaction No.	Reaction	ΔG^0_{Rex} (kcal/mol)	log Keq	Reference
(35)	$SiO_2(s) + 2H_2O = H_4SiO_4$ Silica glass	3.78	−2.77	21, 22
(36)	$SiO_2(s) + 2H_2O = H_4SiO_4$ Quartz	5.13	−3.76	21, 22
(37)	$SiO_2(s) + H_4SiO_4 + 2Al^{3+} + 3H_2O = 6H^+ + Al_2Si_2O_5(OH)4(s)$ Synthetic halloysite Quartz	22.6	−16.6	23
(38)	$Al_2Si_2O_5(OH)4(s) + 5H_2O = 2Al(OH)3(s) + 2H_4SiO_4$ Kaolinite Gibbsite	13.6	−9.93	25
(39)	$0.5Al_2Si_2O_5(OH)4(s) + 3H^+ = SiO_2(s) + 2.5H_2O + Al^{3+}$	−10.9	8.02	26
(40)	$Al_2O_3 \cdot 3H_2O(s) + 2SiO_2(s) = Al_2Si_2O_5(OH)4(s) + H_2O$ Gibbsite	4.33	−3.18	27
(41)	$2Al^{3+} + 2H_4SiO_4 + H_2O = 6H^+ + Al_2Si_2O_5(OH)4(s)$ Synthetic halloysite	17.5	−12.8	23
(41a)	$Al_2Si_2O_5(OH)4(s) + 6H^+ = 2Al^{3+} + 2H_4SiO_4^0 + H_2O$	−8.13	5.96	24

synthesis of halloysite (similar to kaolinite) by Hem and co-workers [23]. Reaction 38 shows the hydrolytic weathering of kaolinite into gibbsite and dissolved silica. If this reaction is read from left to right, the thermodynamically feasible synthesis of kaolinite is indicated. Reaction 39 is the acidic weathering of kaolinite to quartz and Al^{3+} ions. Reaction 40 indicates that the formation of kaolinite from two solid phases, gibbsite and quartz, is not thermodynamically feasible, but does occur to some extent.

An extensive review of the literature provided Bassett et al. [24] with sufficient information and justification to recompute the standard free energy of formation for kaolinite. An average value of -907.7 kcal/mol was obtained, which is slightly higher than the value in Table V. Reaction 41a in Table VIII is an acidic weathering reaction for kaolinite that used the recomputed free energy value.

Na_2O-Al_2O_3-SiO_2-H_2O Systems

The clay mineral, albite ($NaAlSi_3O_8$), weathers spontaneously under acidic conditions to yield Na^+ ions, dissolved silica, kaolinite, gibbsite and quartz. This is seen in Reactions 42, 45 and 51 of Table IX. When carbonic acid is the chemical agent, the weathering reaction yields bicarbonate anions. Albite also undergoes hydrolysis (Reaction 44) to kaolinite but to a lesser extent than by acidic dissolution. Sodium montmorillonite also dissolves under acidic conditions, Reaction 43, in an incongruent manner to yield kaolinite. Analcite ($NaAlSi_2O_6 \cdot H_2O$) is a common authigenic silicate [27] that undergoes acidic weathering in accord with Reactions 47 and 48. Analcite may also react with K^+ ions and quartz, Reaction 46, to yield microcline ($KAlSi_3O_8$) and Na^+ ions. This is, in effect, an ion exchange type of reaction. Reaction 50 is the laboratory synthesis of halloysite from cryolite (Na_3AlF_6) by Hem [23]. This, or similar reactions, may be sources of $\overline{F}$ ions in nature.

K_2O-Al_2O_3-SiO_2-H_2O Systems

Many clay mineral weathering reactions involve the K^+ ion, as seen in Table X. Microcline (K-feldspar, $KAlSi_3O_8$) undergoes incongruent hydrolytic and acidic dissolution in accord with Reactions 52, 53, 57, 58, 62, 63 and 64. Another K-clay mineral, muscovite ($KAl_3Si_3O_{10}(OH)_2$), undergoes these same reactions: 54 and 55. Reaction 56 suggests that microcline and albite may be in equilibrium through the exchange of Na^+ and K^+ ions. That analcite may be produced, to some extent, from muscovite may be seen in Reaction 59, in which concomittant control of the equilibrium concentration of Na^+, K^+ and H^+ may be accomplished. The acidic weathering of microcline in the presence of Na^+ ions yields Na-montmorillonite, as seen in Reaction 60. The synthesis of muscovite from gibbsite, quartz and K^+ ions is given in

（CHEMICAL WEATHERING REACTIONS 225）

Table IX. Chemical Weathering Reactions of Na_2O-Al_2O_3-SiO_2-H_2O Systems at 25°C

Reaction No.	Reaction	ΔG^0_{Rex} (kcal/mol)	$\log K_{eq}$	Reference
(42)	$NaAlSi_3O_8{}_{(s)} + H^+ + 4.5H_2O = Na^+ + 2H_4SiO_4 + 0.5Al_2Si_2O_5(OH)_4{}_{(s)}$ Albite	−0.68	0.495	28
(43)	$3Na\text{-Mont-}_{(s)} + H^+ + 11.5H_2O = Na^+ + 4H_4SiO_4 + 3.5Al_2Si_2O_5(OH)_4{}_{(s)}$ Montmorillonite[a]	12.4	−9.1	28
(44)	$NaAlSi_3O_8{}_{(s)} + 5.5H_2O = Na^+ + OH^- + 2H_4SiO_4 + 0.5Al_2Si_2O_5(OH)_4{}_{(s)}$	18.4	−13.5	18
(45)	$NaAlSi_3O_8{}_{(s)} + H_2CO_3 + 4.5H_2O = Na^+ + HCO_3^- + 2H_4SiO_4 + 0.5Al_2Si_2O_5(OH)_4{}_{(s)}$	8.00	−5.87	18
(46)	$NaAlSi_2O_6 \cdot H_2O_{(s)} + K^+ + SiO_2{}_{(s)} = Na^+ + H_2O + KAlSi_3O_8{}_{(s)}$ Analcite Microcline	−5.44	3.99	27
(47)	$2NaAlSi_2O_6 \cdot H_2O_{(s)} + 2H^+ = Al_2O_3 \cdot 3H_2O + 2Na^+ + 4SiO_2{}_{(s)}$ Gibbsite Quartz	−29.7	21.8	27
(48)	$2NaAlSi_2O_6 \cdot H_2O_{(s)} + 2H^+ = 2Na^+ + 2SiO_2{}_{(s)} + H_2O + Al_2Si_2O_5(OH)_4{}_{(s)}$	−25.4	18.6	27
(49)	$2NaAlSi_3O_8{}_{(s)} + KAlSi_3O_8{}_{(s)} + 2H^+ = KAl_3Si_3O_{10}(OH)_2 + 2Na^+ + 6SiO_2{}_{(s)}$ Muscovite	−22.3	16.3	29
(50)	$2Na_3AlF_6{}_{(s)} + 2H_4SiO_4 + H_2O = 6Na^+ + 12F^- + 6H^+ + Al_2Si_2O_5(OH)_4{}_{(s)}$ Cryolite Synthetic halloysite	108.1	−79.3	23
(51)	$2NaAlSi_3O_8{}_{(s)} + 2CO_2{}_{(g)} + 3H_2O = 2Na^+ + 2HCO_3^- + 4SiO_2{}_{(s)} + Al_2Si_2O_5(OH)_4{}_{(s)}$	−0.49	0.36	30

[a] Na-Mont. = $Na_{0.33}Al_{2.33}Si_{3.67}O_{10}(OH)_2$ whose ΔG^0_f is −1278.21 kcal/mol by calculation.

Table X. Chemical Weathering Reactions of $K_2O-Al_2O_3-SiO_2-H_2O$ Systems at 25°C

Reaction No.	Reaction	ΔG^0_{Rex} (kcal/mol)	log Keq	Reference
(52)	$3KAlSi_3O_8(s) + 12H_2O + 2H^+ = 2K^+ + 6H_4SiO_4 + KAl_3Si_3O_{10}(OH)_2(s)$ Microcline Muscovite	15.8	-11.6	28
(53)	$2KAlSi_3O_8(s) + 2H^+ + 9H_2O = 2K^+ + 4H_4SiO_4 + Al_2Si_2O_5(OH)_4(s)$ Kaolinite	5.99	-4.39	28
(54)	$KAl_3Si_3O_{10}(OH)_2(s) + H^+ + 9H_2O = 3Al(OH)_3(s) + K^+ + 3H_4SiO_4$ Muscovite Gibbsite	13.4	-9.86	28
(55)	$2KAl_3Si_3O_{10}(OH)_2(s) + 2H^+ + 3H_2O = 2K^+ + 3Al_2Si_2O_5(OH)_4(s)$	-13.7	10.1	28
(56)	$KAlSi_3O_8(s) + Na^+ = K^+ + NaAlSi_3O_8(s)$ Albite	3.67	-2.69	28
(57)	$KAlSi_3O_8(s) + 5.5H_2O = K^+ + OH^- + 2H_4SiO_4 + Al_2Si_2O_5(OH)_4(s)$	22.1	-16.1	29
(58)	$1.5KAlSi_3O_8(s) + H^+ = K^+ + 3SiO_2(s) + 0.5KAl_3Si_3O_{10}(OH)_2$	-7.46	5.47	26
(59)	$KAl_3Si_3O_{10}(OH)_2(s) + 3Na^+ + 3SiO_2(s) + 3H_2O = 3NaAlSi_2O_6·H_2O(s) + K^+ + H^+$ Analcite	31.2	-22.9	27
(60)	$7KAlSi_3O_8(s) + Na^+ + 6H^+ = 3Na-Mont. + 7K^+ + 10SiO_2(s)$	-42.7	31.3	27
(61)	$3Al_2O_3·3H_2O(s) + 2K^+ + 6SiO_2(s) = 2KAl_3Si_3O_{10}(OH)_2(s) + 2H^+ + 6H_2O$	26.7	-19.6	27
(62)	$KAlSi_3O_8(s) + H^+ + 0.5H_2O = K^+ + 2SiO_2(s) + 0.5Al_2Si_2O_5(OH)_4(s)$	-7.26	5.32	31
(63)	$KAlSi_3O_8(s) + 8H_2O = K^+ + Al(OH)_4^- + 3H_4SiO_4$	28.2	-20.7	29
(64)	$KAlSi_3O_8(s) + 8H_2O = Al(OH)_3(s) + K^+ + OH^- + 3H_4SiO_4$ Gibbsite	28.9	-21.2	29
(65)	$KAlSi_3O_8(s) + NaAlSi_3O_8(s) + 16H_2O = 2Al(OH)_4^- + K^+ + Na^+ + 6H_4SiO_4$	52.8	-38.7	29

Reaction 61. The cohydrolysis of microcline and albite is suggested in Reaction 65. However, the standard free energy change of this reaction of 52.8 kcal/mol indicates that very little hydrolysis would occur.

CaO- and MgO-Al_2O_3-SiO_2-H_2O Systems

That Ca^{2+} and Mg^{2+} ions are associated with the chemical weathering of clay minerals is seen in Table XI. Anorthite ($CaAl_2Si_2O_8$) may hydrolyze (Reaction 66) or undergo acidic dissolution (Reactions 67 and 68) to yield kaolinite. Ca-montmorillonite also yields kaolinite by an acidic hydrolysis (Reaction 69). Forsterite (Mg_2SiO_4) also is hydrolyzed (Reaction 70) or dissolves under acidic conditions (Reactions 71 and 72), to form soluble constituents.

Bassett et al. [24] recomputed from several literature sources a standard free energy of formation value for sepiolite, a magnesium silicate clay mineral: -1105.6 kcal/mol. Sepiolite is not a major clay mineral but is commonly associated with deep-sea sediments, salt formations, etc. Reaction 72a, Table XI, is a hydrolytic weathering reaction of sepiolite.

Dissolution of Phosphates

The occurrence of phosphates in surface and ground waters has caused some concern, which is centered on the role of phosphorus in the biological process of eutrophication. In the presence of Ca^{2+}, Al^{3+} and Fe^{3+}, phosphate forms solid phases. The significance of phosphate precipitation in surface waters lies, perhaps, not in the removal of this element from the aqueous phase, which would lower the soluble concentration; rather, there would be a reservoir of phosphorus in the bottom sediments. As soluble phosphorus is removed from the aqueous phase by photosynthetic activities, there would be a tendency to dissolve previously precipitated phosphates. This would serve as a continuous source of this nutrient to eutrophic processes.

Leckie and Stumm [33] have published a comprehensive paper on the precipitation of phosphates. Table XII gives some of the forms of phosphorus that may have some significance in natural waters and their dissolution equilibria. There is considerable scientific controversy surrounding the precipitation reactions of phosphorus in natural and artificial aqueous environments. Therefore, the dissolution equilibria in Table XII should be interpreted as a qualitative indication of what occurs in nature. There are, however, some well-defined phosphate solid phases: $Ca_{10}(PO_4)_6(OH)_{2(s)}$ (hydroxylapatite), $AlPO_4 \cdot 2H_2O$ (variscite), and $FePO_4 \cdot 2H_2O$ (strengite). Figure 2 shows the solubility diagrams for six phosphate solid phases as a function of pH. It would appear that Fe^{3+} and Al^{3+} would tend to control

Table XI. Chemical Weathering Reactions in the CaO- and MgO-Al$_2$O$_3$-SiO$_2$-H$_2$O Systems at 25°C

Reaction No.	Reaction	ΔG^0_{Rex} (kcal/mol)	log K_{eq}	Reference
(66)	$CaAl_2Si_2O_8(s) + 3H_2O = Ca^{2+} + 2OH^- + Al_2Si_2O_5(OH)_4(s)$ Anorthite Kaolinite	15.5	-11.3	—
(67)	$CaAl_2Si_2O_8(s) + 2H_2CO_3 + H_2O = Ca^{2+} + 2HCO_3^- + Al_2Si_2O_5(OH)_4(s)$	-5.99	4.39	—
(68)	$CaAl_2Si_2O_8(s) + 2H^+ + H_2O = Ca^{2+} + Al_2Si_2O_5(OH)_4(s)$	-23.4	17.1	18
(69)	$3Ca\text{-}Mont. + 2H^+ + 23H_2O = Ca^{2+} + 8H_4SiO_4 + 7Al_2Si_2O_5(OH)_4$ Montmorillonite[a]	21.0	-15.4	18
(69a)	$Ca_2Al_4Si_8O_{24}(s)\cdot9H_2O + 4H^+ = 4SiO_2(s) + 2Al_2Si_2O_5(OH)_4 + 2Ca^{2+} + 7H_2O$ Phillipsite	-26.4	19.4	32
(70)	$Mg_2SiO_4(s) + 4H_2O = 2Mg^{2+} + 4OH^- + H_4SiO_4$ Forsterite	37.6	-27.6	1
(71)	$Mg_2SiO_4(s) + 4H_2CO_3 = 2Mg^{2+} + 4HCO_3^- + H_4SiO_4$	-4.04	2.96	1
(72)	$Mg_2SiO_4(s) + 4H^+ = 2Mg^{2+} + H_4SiO_4$	-38.8	28.4	1
(72a)	$Mg_2Si_3O_{7.5}(OH)\cdot3H_2O(s) + 4.5H_2O = 2Mg^{2+} + 3H_4SiO_4 + 4OH^-$ Sepiolite	55.1	-40.4	24

[a] Ca-Mont. = $Ca_{0.33}Al_{4.67}Si_{7.33}O_{20}(OH)_4$ whose ΔG^0_f is -2557.73 kcal/mol by calculation.

Table XII. Dissolution of Some Phosphatic Minerals at 25°C

Reaction No.	Reaction	ΔG^0_{Rex} (kcal/mol)	log K_{eq}	Reference
(73)	$FePO_4 \cdot 2H_2O_{(s)} = Fe^{+3} + PO_4^{-3} + 2H_2O$ Strengite	37.8	−27.7	33
(74)	$AlPO_4 = Al^{+3} + PO_4^{-3}$ Berlinite	28.0	−20.5	a
(75)	$CaHPO_{4(s)} = Ca^{+2} + HPO_4^{-2}$	7.82	−5.73	a
(76)	$Ca_4H(PO_4)_3 = 4Ca^{+2} + 3PO_4^{-3} + H^+$	64.0	−46.9	33
(77)	$Ca_5(PO_4)_3(OH)_{(s)} = 5Ca^{+2} + 3PO_4^{-3} + OH^-$ Hydroxylapatite	75.8	−55.6	a
(78)	$Ca_{10}(PO_4)_6F_{2(s)} = 10Ca^{+2} + 6PO_4^{-3} + 2F^-$ Fluoroapatite	161.0	−118.0	a
(79)	$Ca_{10}(PO_4)_6(OH)2_{(s)} + 6H_2O = 4[Ca_2(HPO_4)(OH)_2] + 2Ca^{+2} + 2HPO_4^-$	23.2	−17.0	a
(80)	$CaHAl(PO_4)2_{(s)} = Ca^{+2} + Al^{+3} + H^+ + 2PO_4^{-3}$	53.2	−39.0	a
(81)	$MgNH_4PO_{4(s)} = Mg^{+2} + NH_4^+ + PO_4^{-3}$	17.2	−12.6	a
(82)	$Fe_3(PO_4)2_{(s)} + 2H^+ = 3Fe^{2+} + 2HPO_4^{2-}$ Vivianite	7.0	−5.13	15, 34

aCalculated and are considered to be approximations.

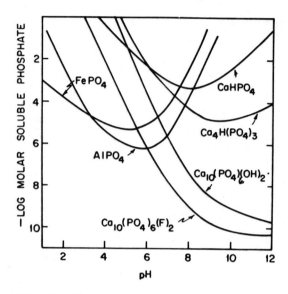

Figure 2.   Solubility diagrams for some solid phosphates (reproduced from Leckie and Stumm [33] courtesy of the University of Texas Press).

the aqueous phosphate concentration under acid conditions, whereas the Ca^{2+} compounds would be the controlling phases under alkaline conditions. Reaction 77 may give some qualitative indication of the phosphate concentration at a pH of 7.0 and a calcium content of 40 mg/l ($10^{-3}M$). This yields a soluble phosphate concentration of $6.3 \times 10^{-12}M$, or 0.60 ng/l as PO_4^{3-} This suggests that such solid phases as hydroxylapatite would produce extremely low concentrations of soluble phosphate in aquatic environments.

In all probability, phosphatic minerals weather by acidic dissolution, of which an example may be Reaction 82. This indicates the dissolution of vivianite (ferrous phosphate) by reaction with hydrogen ions. That the log K_{eq} is -5.13 suggests that this dissolution occurs to a small extent. An increase in the hydrogen ion content will, of course, increase the extent to which vivianite is dissolved.

Dissolution of Halide and Sulfate Minerals

Geologic formations composed of halide and sulfate minerals are, perhaps, the most soluble solid substances and contribute substantial quantities of soluble constituents to the water phase. These formations undoubtedly originated from salt deposits called "evaporites" [1]. Table XIII lists some of

Table XIII. Dissolution Reactions of Some Halide and Sulfate Minerals at 25°C

Reaction No.	Reaction	$\Delta G^0_{Rex}{}^a$ (kcal/mol)	$\log K_{eq}{}^a$
(83)	$NaCl_{(s)} = Na^+ + Cl^-$ Halite	−2.10	1.54
(84)	$KCl_{(s)} = K^+ + Cl^-$ Sylvite	−1.38	1.01
(85)	$MgCl_{2(s)} = Mg^{2+} + 2Cl^-$ Chloromagnesite	−30.1	22.1
(86)	$CaF_{2(s)} = Ca^{2+} + 2F^-$ Fluorite	12.3	−9.05
(87)	$Na_3AlF_{6(s)} = 3Na^+ + Al^{3+} + 6F^-$ Cryolite	41.9	−30.8
(88)	$NaF_{(s)} = Na^+ + F^-$ Villiaumite	0.63	−0.46
(89)	$CaSO_{4(s)} = Ca^{2+} + SO_4^{2-}$ Anhydrite	6.96	−5.1
(90)	$CaSO_4 \cdot 2H_2O_{(s)} = Ca^{2+} + SO_4^{2-} + 2H_2O$ Gypsum	7.24	−5.3
(91)	$BaSO_{4(s)} = Ba^{2+} + SO_4^{2-}$ Barite	13.96	−10.2
(92)	$SrSO_{4(s)} = Sr^{2+} + SO_4^{2-}$ Celestite	9.29	−5.8
(93)	$Na_2SO_{4(s)} = 2Na^+ + SO_4^{2-}$ Thenardite	0.98	−0.72

[a]Calculated values.

the dissolution reactions of single species of several halide and sulfate minerals. These reactions represent a simplification of natural events since many of the evaporites are double salts and hydrates. Dissolution of these minerals may lead to unusually high concentrations of ionic constituents in natural waters. For example, brine waters associated with oil fields may reach sodium contents as high as 63,900 mg/l, and chloride contents as high as 124,000 mg/l [35]. Occasionally, cold to moderately thermal spring waters will have saline contents similar to the oil field brines [35].

The chloride minerals—halite, sylvite and chloromagnesite—are extremely soluble in water, which is indicated by negative standard free energy values of their dissolution reactions (Table XIII). The fluoride minerals—fluorite, cryolite, villiaumite and fluoroapatite (Table XII)—exhibit low solubilities in water through their positive dissolution standard free energy values. The sulfate minerals—anhydrite, gypsum, barite, celestite and thenardite—also exhibit low solubilities in water.

Dissolution of Iron and Manganese Minerals

Iron and manganese may be considered ubiquitous, trace constituents of natural waters. There are numerous rocks containing iron and manganese minerals that weather chemically through oxidation–reduction reactions. These reactions have received much laboratory and field study in recent years. Consequently, numerous reactions have been proposed to account for the natural water chemistry of iron and manganese.

Table XIV shows 60 weathering reactions for various iron minerals. Reactions 94 through 110 are for the iron oxides; Reactions 111 through 119 are for the carbonates; Reactions 120 through 150 are for the iron sulfides; and Reactions 151 through 153 are for the iron silicates. All ΔG^0_{Rex} and log K_{eq} values have been recomputed from the thermodynamic data of Robie and Waldbaum [10]. The extent to which these reactions occur in nature is somewhat speculative because of the difficulties inherent in the thermodynamic data and of the uncertain existence of some of the iron minerals. Certainly not all of these reactions are operative in nature, but they do indicate the possibilities and the complexities of aqueous iron chemistry.

Table XV shows some 25 weathering reactions for various manganese minerals. Reactions 154 through 165 are for the manganese oxides; Reactions 166 through 171 are for the manganese carbonates; and Reactions 172 through 178 are for the manganese sulfides. The aqueous chemistry of manganese has been studied somewhat less than iron. However, similar uncertainties surround the chemical weathering of manganese minerals under natural conditions. These reactions, therefore, serve as indicators of what may occur in natural waters.

Dissolution of Trace Minerals

Trace concentrations of such metals ions as Li^+, Ni^{2+}, Be^{2+}, Cd^{2+}, etc., may occur in natural waters from the chemical oxidation of their sulfide or oxide ores, or from the acidic dissolution of their carbonates. Table XVI lists several of these chemical weathering reactions of various naturally occurring trace minerals. The various solid phases are those most likely to occur for the specific trace constituent. Weathering of the sulfide ore is assumed to be oxidation with gaseous oxygen as the electron acceptor, which, in all cases, is thermodynamically feasible. Reaction 181 shows a

Table XIV. Chemical Oxidation-Reduction Weathering Reactions of Iron Minerals at 25°C

Reaction No.	Reaction	ΔG_{Rex}^{0} [a] (kcal/mol)	log K_{eq} [a]	Reference
	Iron Oxides			
(94)	$3Fe_2O_3(s) + 2H^+ + 2e = 2Fe_3O_4(s) + H_2O$ Hematite / Magnetite	-9.695	7.11	36
(95)	$Fe_2O_3(s) + 6H^+ + 2e = 2Fe^{2+} + 3H_2O$	-32.945	24.2	37
(96)	$Fe_3O_4(s) + 8H^+ + 2e = 3Fe^{2+} + 4H_2O$	-44.6	32.7	37
(97)	$Fe_3O_4(s) + 8H^+ + 8e = 3Fe^{0}(s) + 4H_2O$ Iron	16.33	-11.97	38
(98)	$Fe_2O_3(s) + 6H^+ + 6e = 3Fe^{0}(s) + 3H_2O$	7.65	-5.61	38
(99)	$Fe_3O_4(s) + 2H^+ + 2e = 3FeO(s) + H_2O$ Ferrous Oxide	6.11	-4.48	38
(100)	$Fe_2O_3(s) + 3H^+ + 2e = 3FeO(s) + H_2O$	0.84	-0.62	38
(101)	$2Fe(OH)_3(s) + 6H^+ + 2e = 2Fe^{2+} + 6H_2O$ Ferric Hydroxide	-48.7	35.7	38
(102)	$Fe(OH)_2(s) + 2H^+ + 2e = Fe^{0}(s) + 2H_2O$ Ferrous Hydroxide	2.19	-1.60	39
(103)	$Fe(OH)_3(s) + H^+ + e = Fe(OH)_2(s) + H_2O$	-6.26	4.59	39
(104)	$3Fe^{3+} + 4H_2O + e = Fe_3O_4(s) + 8H^+$	-8.77	6.43	39
(105)	$3Fe(OH)^{2+} + H_2O + e = Fe_3O_4(s) + 5H^+$	-18.7	13.7	39
(106)	$3Fe(OH^+)_2 + e = Fe_3O_4(s) + 2H_2O + 2H^+$	-37.9	27.8	39
(107)	$Fe_2O_3(s) + 4H^+ + 2e = 2Fe(OH)^+ + H_2O$	-9.36	6.86	39

Table XIV, continued

Reaction No.	Reaction	ΔG^0_{Rex} [a] (kcal/mol)	log K_{eq} [a]	Reference
	Iron Oxides			
(108)	$Fe_3O_4(s) + 5H^+ + 2e = 3Fe(OH^+) + H_2O$	−9.2	6.74	39
(109)	$Fe_2O_3(s) + H_2O + 2e = 2HFeO_2^-$	54.0	−39.6	39
(110)	$Fe_3O_4(s) + 2H_2O + 2e = 3HFeO_2^- + H^+$	85.9	−63.0	39
	Iron Carbonates			
(111)	$Fe_3O_4(s) + 3CO_2(g) + 2H^+ + 2e = 3FeCO_3(s) + H_2O$ Siderite	−13.9	10.2	36
(112)	$Fe_2O_3(s) + 2CO_2(g) + 2H^+ + 2e = 2FeCO_3(s) + H_2O$	−12.5	9.2	36
(113)	$Fe(OH)_3(s) + CO_2(g) + H^+ + e = FeCO_3(s) + 2H_2O$	−14.2	10.4	40
(114)	$Fe_3O_4(s) + 3H_2CO_3 + 2H^+ + 2e = 3FeCO_3(s) + 4H_2O$	−19.9	14.6	39
(115)	$Fe_3O_4(s) + 3HCO_3^- + 5H^+ + 2e = 3FeCO_3(s) + 4H_2O$	−46.0	33.7	39
(116)	$Fe_3O_4(s) + 3CO_3^{2-} + 8H^+ + 2e = 3FeCO_3(s) + 4H_2O$	−88.3	64.7	39
(117)	$Fe_2O_3(s) + 2H_2CO_3 + 2H^+ + 2e = 2FeCO_3(s) + 3H_2O$	−16.5	12.1	39
(118)	$Fe_2O_3(s) + 2HCO_3^- + 4H^+ + 2e = 2FeCO_3(s) + 3H_2O$	−33.9	24.8	39
(119)	$Fe_2O_3(s) + 2CO_3^{2-} + 6H^+ + 2e = 2FeCO_3(s) + 3H_2O$	−62.1	45.5	39
	Iron Sulfides			
(120)	$FeS_2(s) + 3.5O_2 + H_2O = Fe^{3+} + 2SO_4^{2-} + 2H^+$ Pyrite	−262.1	192.2	

(121)	$2FeS_2(s) + 4H_2O + 7.5O_2 = Fe_2O_3(s) + 4SO_4^{2-} + 8H^+$	-583.8	427.9	—
(122)	$FeS(s) + CO_2(g) + H_2O = FeCO_3(s) + 2H^+ + S^{2-}$	34.7	-25.5	36
(123)	$FeS_2(s) + 2e = FeS(s) + S^{2-}$	34.5	-25.3	36
(124)	$Fe_3O_4(s) + 8H^+ + 3S^{2-} + 2e = 3FeS(s) + 4H_2O$	-118.1	86.6	36
(125)	$Fe_2O_3(s) + 6H^+ + 2S^{2-} + 2e = 2FeS(s) + 3H_2O$	-81.9	60.1	36
(126)	$3FeS_2(s) + 4H_2O + 4e = Fe_3O_4(s) + 8H^+ + 6S^{2-}$	221.6	-162.4	36
(127)	$FeS_2(s) + H_2O + CO_2(g) + 2e = FeCO_3(s) + 2H^+ + 2S^{2-}$	69.2	-50.7	36
(128)	$2FeS_2(s) + 3H_2O + 2e = Fe_2O_3(s) + 6H^+ + 4S^{2-}$	150.9	-110.7	36
(129)	$FeS_2(s) + 2H^+ + 2e = FeS(s) + H_2S(aq)$	7.46	-5.47	41
(130)	$FeS_2(s) + H^+ + 2e = FeS(s) + HS^-$	17.0	-12.5	41
(131)	$Fe^{2+} + 2SO_4^{2-} + 16H^+ + 14e = FeS_2(s) + 8H_2O$	-116.8	85.7	41
(132)	$Fe(OH)_2(s) + 2SO_4^{2-} + 18H^+ + 14e = FeS_2(s) + 10H_2O$	-135.0	98.9	41
(133)	$Fe^{2+} + 2S^0(s) + 2e = FeS_2(s)$	-18.0	13.2	41

Sulfur

(134)	$FeS_2(s) + H_2O + 3.5O_2 = Fe^{2+} + 2H^+ + SO_4^{2-}$	-102.7	75.3	42
(135)	$Fe_3O_4(s) + 1.5S_2(g) + 8H^+ + 8e = 3FeS(s) + 4H_2O$	-85.3	62.5	39
(136)	$Fe_2O_3(s) + S_2(g) + 6H^+ + 6e = 2FeS(s) + 3H_2O$	-60.1	44.0	39
(137)	$Fe_3O_4(s) + 3S_2(g) + 8H^+ + 8e = 3FeS_2(s) + 4H_2O$	-155.9	114.3	39
(138)	$Fe_2O_3(s) + 2S_2(g) + 6H^+ + 6e = 2FeS_2(s) + 3H_2O$	-107.2	146.2	39
(139)	$Fe_3O_4(s) + 3H_2S(aq) + 2H^+ + 2e = 3FeS(s) + 4H_2O$	-36.95	27.1	39
(140)	$3FeS_2(s) + 4H_2O + 4H^+ + 4e = Fe_3O_4(s) + 6H_2S(aq)$	59.3	-43.5	39

Table XIV, continued

Reaction No.	Reaction	ΔG^0_{Rex} [a] (kcal/mol)	$\log K_{eq}$ [a]	Reference
	Iron Sulfides			
(141)	$2FeS_2(s) + 3H_2O + 2H^+ + 2e = Fe_2O_3(s) + 4H_2S(aq)$	42.8	-31.4	39
(142)	$Fe_2O_3(s) + 4SO_4^{2-} + 38H^+ + 30e = 2FeS_2(s) + 19H_2O$	-266.6	195.5	39
(143)	$Fe_3O_4(s) + 6SO_4^{2-} + 56H^+ + 44e = 3FeS_2(s) + 28H_2O$	-395.1	289.7	39
(144)	$4HSO_4^- + Fe_2O_3(s) + 34H^+ + 30e = 2FeS_2(s) + 19H_2O$	-256.2	187.9	39
(145)	$FeS_2(s) + 4H^+ + 2e = 2H_2S(aq) + Fe^{2+}$	4.92	-3.6	39
(146)	$2HSO_4^- + Fe^{2+} + 14H^+ + 14e = FeS_2(s) + 8H_2O$	-111.6	81.9	39
(147)	$2SO_4^{2-} + Fe^{2+} + 16H^+ + 14e = FeS_2(s) + 8H_2O$	-116.8	85.7	39
(148)	$FeS_2(s) + 2H^+ + 2e = 2HS^- + Fe^{2+}$	24.0	-17.6	39
(149)	$FeCO_3(s) + 2SO_4^{2-} + 17H^+ + 14e = FeS_2(s) + 8H_2O + HCO_3^-$	-116.4	85.3	43
(150)	$HFeO_2^- + 2SO_4^{2-} + 19H^+ + 14e = FeS_2(s) + 10H_2O$	-160.3	117.5	43
	Iron Silicates			
(151)	$2Fe(OH)_3(s) + SiO_2(s) + 2H^+ + 2e = Fe_2SiO_4(s) + 4H_2O$ Amorphous Fayalite	-21.1	15.5	40
(152)	$Fe_3O_4(s) + 3SiO_2(s) + 2H^+ + 2e = 3FeSiO_3(s) + H_2O$ Iron Metasilicate	25.3	-18.5	39
(153)	$Fe_2O_3(s) + 2SiO_2(s) + 2H^+ + 2e = 2FeSiO_3(s) + H_2O$	13.6	-9.99	39

[a]Calculated values.

clay mineral weathering of lithium aluminum silicate. The positive standard free energy change of this reaction suggests there should be a very small extent of occurrence of lithium ions in natural waters. Nickel carbonate is reasonably soluble in water, whereas the carbonates of strontium, cobalt and lead are only slightly soluble. Additional information about the chemical weathering of ores may be found in the papers of Sato [40], Garrels [46], Mason [47] and Bostrom [48].

An additional source of thermochemical data and chemical weathering reactions is found in the publication by Ball et al. [16], which is concerned with a computer program for trace major element speciation and mineral equilibria in natural waters. Greater detail is given in Chapter 6.

REFERENCES

1. Krauskopf, K. *Introduction to Geochemistry* (New York: McGraw-Hill Book Co., 1967).
2. Rankama, K., and T. G. Sahama. *Geochemistry* (Chicago, IL: University of Chicago Press, 1950).
3. Degens, E. T. *Geochemistry of Sediments* (Englewood Cliffs, NJ: Prentice-Hall, Inc., 1965).
4. Grim, R. E. *Clay Mineralogy* (New York: McGraw-Hill Book Co., 1953).
5. Clarke, F. W., and H. S. Washington. "The Composition of the Earth's Crust," *U.S. Geol. Survey Prof. Paper 127*, Washington, DC (1924).
6. Horn, M. K., and J. A. S. Adams. *Geochim. Cosmochim. Acta* 30:279 (1966).
7. Goldschmidt, V. M. *Geochemistry*, Alex Muir, Ed. (Oxford, England: Clarendon Press, 1954).
8. Cox, K. "Minerals and Rocks" in *Understanding the Earth*, 2nd ed., I. G. Gass, et al., Eds. (Cambridge, MA: The M.I.T. Press, 1972).
9. Parker, R. L. "Composition of the Earth's Crust," Chapter D "Data of Geochemistry," *U.S. Geol. Survey Prof. Paper 440D*, Washington, DC (1967).
10. Robie, R. A., and D. R. Waldbaum. "Thermodynamic Properties of Minerals and Related Substances at 298.15°K (25°C.) and One Atmosphere (1.013 Bars) Pressure and at Higher Temperatures," *Geol. Survey Bull.* 1259, Washington, DC (1968); *Geol. Survey Bull.* 1452 (1979).
11. Latimer, W. M. *The Oxidation States of the Elements and Their Potentials in Aqueous Solutions*, 2nd ed. (Englewood Cliffs, NJ: Prentice-Hall, Inc., 1952).
12. Bricker, O. *Am. Min.* 50:1296 (1965).
13. Roberson, C. E., and J. D. Hem. "Solubility of Aluminum in the Presence of Hydroxide, Fluoride, and Sulfate," *Geol. Survey Water Supply Paper 1827-C*, Washington, DC (1969).
14. Langmuir, D. *J. Geol.* 73(5):730 (1965).
15. Chen, P. J., and S. D. Faust. *Environ. Lett.* 6(4):287 (1974).
16. Ball, J. W., et al. "Additional and Revised Thermochemical Data and Computer Code for WATEQ2," U.S. Geological Survey, Water Res. Invest. WRI 78-116, Menlo Park, CA (1980).

Table XV. Chemical Oxidation–Reduction Weathering Reactions of Manganese Minerals at 25°C

Reaction No.	Reaction	ΔG^0_{Rex} [a] (kcal/mol)	log K_{eq} [a]	Reference
	Manganese Oxides			
(154)	$Mn_3O_4(s) + 8H^+ + 2e = 3Mn^{2+} + 4H_2O$ Hausmanite	−80.7	59.1	12
(155)	$MnOOH(s) + 3H^+ + e = Mn^{2+} + 2H_2O$ Manganite	−33.5	24.5	12
(156)	$MnO_2(s) + 4H^+ + e = Mn^{2+} + 2H_2O$ Pyrolusite	−55.4	40.6	12
(157)	$Mn_3O_4(s) + 2H^+ + 2H_2O + 2e = 3Mn(OH)2(s)$ Pyrochroite	−23.8	17.4	44
(158)	$Mn_3O_4(s) + 5H^+ + 2e = 3MnOH^+ + H_2O$	−40.2	29.5	44
(159)	$Mn_3O_4(s) + 2H_2O + 2e = 3HMnO_2^- + H^+$	55.0	−40.3	44
(160)	$Mn_2O_3(s) + 6H^+ + 2e = 2Mn^{2+} + 3H_2O$	−66.8	48.96	44
(161)	$MnO_2(s) + H^+ + e = MnOOH(s)$	−22.0	16.1	45
(162)	$Mn_3O_4(s) + 5H_2O + 2e = 3Mn(OH)_3^- + H^+$	57.0	−41.8	45
(163)	$3MnOOH(s) + H^+ + e = Mn_3O_4(s) + 2H_2O$	−19.8	14.5	45
(164)	$MnO_4^- + 4H^+ + 3e = MnO_2(s) + 2H_2O$	−117.3	86.0	45
(165)	$MnO_4^{2-} + 4H^+ + 2e = MnO_2(s) + 2H_2O$	−104.3	76.4	45

Manganese Carbonates

(166)	$MnCO_3(s) + 0.5O_2 + H_2O = MnO_2(s) + H_2CO_3$ Rhodochrosite	-8.55	6.26	—
(167)	$Mn_3O_4(s) + 3HCO_3^- + 5H^+ + 2e = 3MnCO_3(s) + 4H_2O$	-84.8	62.2	44
(168)	$Mn_2O_3(s) + 2HCO_3^- + 4H^+ + 2e = 2MnCO_3(s) + 3H_2O$	-69.5	51.0	44
(169)	$MnO_2(s) + HCO_3^- + 4H^+ + 2e = MnHCO_3^+ + 2H_2O$	-58.9	43.2	44
(170)	$MnCO_3(s) + H^+ + 2e = Mn(s) + HCO_3^-$ Manganese	131.3	-96.2	45
(171)	$MnOOH(s) + HCO_3^- + 2H^+ + e = MnCO_3(s) + 2H_2O$	-34.9	25.6	45

Manganese Sulfides

(172)	$Mn(OH)_2(s) + SO_4^{2-} + 10H^+ + 8e = MnS(s) + 6H_2O$ Alabandite	-67.8	49.7	44
(173)	$MnCO_3(s) + SO_4^{2-} + 8H^+ + 8e = MnS(s) + CO_3^{2-} + 4H_2O$	-32.7	24.0	44
(174)	$Mn^{2+} + SO_4^{2-} + 8H^+ + 8e = MnS(s) + 4H_2O$	-48.2	35.3	44
(175)	$HMnO_2^- + SO_4^{2-} + 11H^+ + 8e = MnS(s) + 6H_2O$	-94.0	69.0	44
(176)	$MnSO_4(aq) + 8H^+ + 6e = S(s) + Mn^{2+} + 4H_2O$	-45.4	33.3	44
(177)	$MnO_2(s) + SO_4^{2-} + 4H^+ + 2e = MnSO_4(aq) + 2H_2O$	-59.5	43.6	44
(178)	$S(s) + Mn^{2+} + H_2CO_3(aq) + 2e = MnCO_3(s) + H_2S(aq)$	0.76	-0.55	44

[a]Calculated values.

Table XVI. Chemical Weathering Reactions of Various Trace Minerals at 25°C

Reaction No.	Reaction	ΔG^0_{Rex}[a] (kcal/mol)	$\log K_{eq}$[a]
(179)	$2AsS_{(s)} + 5.5O_{2(g)} + H_2O = 2SO_4^{2-} + 2HAsO_2$ Realgar	−456.9	335.0
(180)	$As_2S_3{(s)} + 7.5O_{2(g)} + H_2O = 3SO_4^{2-} + 2HAsO_2$ Orpiment	−627.6	460.1
(181)	$LiAl(SiO_3)_2 + H^+ + 2.5H_2O = Li^+ + H_4SiO_4 + 0.5Al_2Si_2O_5(OH)_{4(s)}$ β-Spodumene	28.2	−20.7
(182)	$SrCO_3{(s)} + H^+ = Sr^{2+} + HCO_3^-$ Strontianite	1.99	−1.46
(183)	$Cr_2O_3{(s)} + 6H^+ + 2e = 2Cr^{2+} + 3H_2O$ Eskolaite	−1.07	0.78
(184)	$CoCO_3{(s)} + H^+ = Co^{2+} + HCO_3^-$ Cobalt carbonate	3.01	−2.2
(185)	$NiS_{(s)} + 1.5O_2 + H_2O = SO_4^{2-} + Ni^{2+} + 2H^+$ Millerite	−111.1	81.5
(186)	$NiCO_3{(s)} + H^+ = Ni^{2+} + HCO_3^-$ Nickel carbonate	−4.36	3.2

(187)	$CuS_{(s)} + 1.5O_2 + H_2O = SO_4^{2-} + Cu^{2+} + 2H^+$ Covellite	-93.4	68.5
(188)	$Cu_2S_{(s)} + 1.5O_2 + H_2O = SO_4^{2-} + 2Cu^+ + 2H^+$ Chalcocite	-87.9	64.5
(189)	$ZnS_{(s)} + 1.5O_2 + H_2O = SO_4^{2-} + Zn^{2+} + 2H^+$ Sphalerite	-107.2	78.6
(190)	$CdS_{(s)} + 1.5O_{2(g)} + H_2O = SO_4^{2-} + Cd^{2+} + 2H^+$ Greennockite	-104.4	76.6
(191)	$CdCO_{3(s)} + H^+ = Cd^{2+} + HCO_3^-$ Otavite	1.12	-0.82
(192)	$HgS_{(s)} + 1.5O_{2(g)} + H_2O = SO_4^{2-} + Hg^{2+} + 2H^+$ Cinnabar	-69.2	50.7
(193)	$PbS_{(s)} + 1.5O_{2(g)} + H_2O = SO_4^{2-} + Pb^{2+} + 2H^+$ Galena	-103.5	75.9
(194)	$PbCO_{3(s)} + H^+ = Pb^{2+} + HCO_3^-$ Cerussite	4.24	-3.1
(195)	$BeS_{(s)} + 1.5O_{2(g)} + H_2O = SO_4^{2-} + Be^{2+} + 2H^+$ Beryllium Sulfide	-150.0	110.0

[a] Calculated values.

17. Faust, S. D. (Unpublished results).
18. Stumm, W., and J. J. Morgan. *Aquatic Chemistry* (New York: John Wiley & Sons, Inc., 1970), p. 193.
19. Langmuir, D. *Geochim. Cosmochim. Acta* 35:1023 (1971).
20. Schindler, P. W. In: *Equilibrium Concepts in Natural Water Systems*, Advances in Chemistry Series No. 67 (Washington, DC: American Chemical Society, 1967).
21. Siever, R. *J. Geol.* 70:127 (1962).
22. Siever, R. *Am. Min.* 42:821 (1957).
23. Hem, J. D., et al. "Chemical Interactions of Aluminum with Aqueous Silica at 25°C," *Geol. Survey Water Supply Paper 1827-E*, Washington, DC (1973).
24. Bassett, R. L., et al. In: *Chemical Modeling in Aqueous Systems*, ACS Symposium Series, No. 93 (Washington, DC: American Chemical Society, 1979).
25. Garrels, R. M. *Am. Min.* 42:789 (1957).
26. Hemley, J. J., and W. R. Jones. *Econ. Geol.* 59:538 (1964).
27. Hess, P. C. *Am. J. Sci.* 264:289 (1966).
28. Feth, J. H., et al. "Sources of Mineral Constituents in Water from Granitic Rocks," *Geol. Survey Water Supply Paper 1535-I*, Washington, DC (1964).
29. Helgeson, H. C., et al. *Geochim. Cosmochim. Acta* 33:455 (1969).
30. Garrels, R. M. In: *Researches in Geochemistry*, Vol. 2, P. H. Abelson, Ed. (New York: John Wiley & Sons, Inc., 1967), p. 405.
31. Helgeson, H. C. *Geochim. Cosmochim Acta* 32:853 (1968).
32. Sillén, L. G. *Science* 156(3779):1189 (1967).
33. Leckie, J., and W. Stumm. "Phosphate Precipitation," in *Water Quality Improvement by Physical and Chemical Processes*, Water Resources Symposium No. 3. E. F. Gloyna, and W. W. Eckenfelder, Eds. (Austin, TX: University of Texas Press, 1970), p. 237.
34. Singer, P. C. *J. Water Poll. Control Fed.* 44(4):663 (1972).
35. White, D. E., et al. "Chemical Composition of Subsurface Waters," Chapter F, "Data of Geochemistry" *U.S. Geol. Survey Prof. Paper 440-F*, Washington, DC (1963).
36. Berner, R. S. *J. Geol.* 72:826 (1964).
37. Doyle, R. W. S. "Eh and Thermodynamic Equilibrium in Environments Containing Dissolved Ferrous Iron," PhD. Thesis, Yale University (1968).
38. Pourbaix, J. *Atlas of Electrochemical Equilibria in Aqueous Solutions* Elmsford, NY: Pergamon Press, Inc., 1966).
39. Garrels, R. M., and C. L. Christ. *Solutions, Minerals, and Equilibria*, (New York: Harper & Row Publishers, Inc., 1965).
40. Sato, M. *Econ. Geol.* 55:928 (1960).
41. Hem, J. D. "Some Chemical Relationships Among Sulfur Species and Dissolved Ferrous Iron," *Geol. Survey Water Supply Paper 1459-C*, Washington, DC (1960).
42. Langmuir, D. *U.S. Geol. Survey Prof. Paper 650-C* (1969), p. C224.
43. Hem. J. D. In: *Principles and Applications of Water Chemistry*, S. D. Faust and J. V. Hunter, Eds. (New York: John Wiley & Sons, Inc., 1967), p. 625.
44. Hem, J. D. "Chemical Equilibria and Rates of Manganese Oxidation," *Geol. Survey Water Supply Paper 1667-A*, Washington, DC (1963).

45. Morgan, J. J. In: *Principles and Applications of Water Chemistry*, S. D. Faust and J. V. Hunter. Eds. (New York: John Wiley & Sons, Inc., 1967), p. 561.
46. Garrels, R. M. *Geochim. Cosmochim. Acta* 5:153 (1954).
47. Mason, B. *J. Geol.* 57:62 (1949).
48. Boström, K. In: *Equilibrium Concepts in Natural Water Systems*, Advances in Chemistry Series #67 (Washington, DC: American Chemical Society, 1962).

CHAPTER 6

CHEMICAL EQUILIBRIUM MODELS

CONSTRUCTION OF MODELS

Concentrations of the naturally occurring constituents of fresh and saline waters are controlled by many of the weathering reactions cited in Chapter 5. Consequently, many attempts have been reported in the literature to interpret or to predict these concentrations through the use of chemical equilibrium models. Some of these attempts have been fairly successful whereas others have not. This chapter shows how to construct, apply and interpret these models. Limitations are discussed also.

Approaches

There are several approaches for the construction of equilibrium models for chemical constituents in natural waters. In general, the complexity of the system is chosen first, i.e., simple versus complex, single or several constituents, one or several solid phases, open or closed to the atmosphere, etc. Next, the chemical weathering reactions with their appropriate equilibrium constants are chosen. The third step consists of computing the equilibrium concentrations of the various dissolved constituents for the model selected. The last step compares the actual concentrations of the constituents with the computed equilibrium concentrations. Then an interpretation is made. Much trial and error labor is involved in these models. The degree of success is somewhat dependent on the value of the equilibrium constant and the analytical accuracy of the available water quality data. The many difficulties with equilibrium models are discussed later in this chapter. Methods for the computation of the equilibrium concentrations are discussed next using a closed aqueous solution of $CaCO_{3(s)}$ as an example.

Charge Balance

Equilibrium concentrations may be computed from the relationship between the oppositely charged ionic species in water. In the case of a closed system of pure $CaCO_{3(s)}$ in contact with pure water (a closed system to the atmosphere negates CO_2 as a gaseous phase and dissolved H_2CO_3 as a constituent), the charge balance is

$$2[Ca^{2+}] + [H^+] = [HCO_3^-] + 2[CO_3^{2-}] + [OH^-] \qquad (1)$$

For this system, dissolved Ca^{2+} is considered equal to the mass balance or the total concentration of the dissolved carbonate species. If $[H_2CO_3]$ is neglected, then

$$[Ca^{2+}] = [HCO_3^-] + [CO_3^{2-}] \qquad (2)$$

For this system, the $[HCO_3^-]$ is equal to the $[OH^-]$, which may be seen by multiplication of the terms in Equation 2 by 2 and then subtraction from Equation 1 ($[H^+]$ also neglected). This yields

$$[HCO_3^-] = [OH^-] = \frac{K_w}{[H^+]} \qquad (3)$$

Before computation of the equilibrium concentrations of the various ionic species, let us look at the Phase Rule constraints on this system. There are two phases: solid $(CaCO_{3(s)})$ and liquid (H_2O). There are three components: H_2O, CO_2 and CaO. Consequently, $F = 3 - 2 + 2 = 3$, or three degrees of freedom. The constraints of temperature ($25°C$), pressure (1 atm) and $[Ca^{2+}]$ = mass balance of carbonate species are placed on the system. Subsequently, the concentrations of H^+, HCO_3^-, CO_3^{2-} and H_2CO_3 are fixed and cannot vary.

The equilibrium concentrations are computed from Equation 1 ($[H^+]$ neglected):

$$2[Ca^{2+}] = [HCO_3^-] + 2[CO_3^{2-}] + [OH^-] \qquad (4)$$

Substitution for these terms is made from: $[Ca^{2+}] = K_s/[CO_3^{2-}]$, $[HCO_3^-]$ from Equation 3, $[CO_3^{2-}] = K_2[HCO_3^-]/[H^+] = K_2K_w/[H^+]^2$, and $[OH^-] = K_w/[H^+]$. Whereupon

$$\frac{2K_s[H^+]^2}{K_2K_w} = \frac{K_w}{[H^+]} + \frac{2K_2K_w}{[H^+]^2} + \frac{K_w}{[H^+]} \qquad (5)$$

Gathering terms and multiplication through by $[H^+]^2$ yields

$$\frac{K_S[H^+]^4}{K_2K_W} - K_W[H^+] - K_2K_W = 0 \qquad (6)$$

Trial and error solution of Equation 6 gives a value of $10^{-9.95}$ for the $[H^+]$, which is the equilibrium concentration for this system. The appropriate calculations for the other constituents yield: $[CO_3^{2-}] = 3.72 \times 10^{-5}M$ or 2.2 mg/l; $[Ca^{2+}] = 1.23 \times 10^{-4}M$ or 4.9 mg/l; $[HCO_3^-] = 8.91 \times 10^{-5}M$ or 5.4 mg/l; and $[OH^-] = 8.91 \times 10^{-5}M$ or 1.5 mg/l as their equilibrium concentrations.

Diagrams

Diagrams that depict areas of "stability" may be employed to establish concentrations of various ionic species.

In this case of $CaCO_{3(s)}$ in water, the stability diagram is a plot of log $[Ca^{2+}]$ versus pH. From the solubility equilibrium for $CaCO_{3(s)}$,

$$K_S = [Ca^{2+}][CO_3^{2-}] \qquad (7)$$

Substituting for $[CO_3^{2-}]$ (Equation 27, Chapter 3),

$$K_S = [Ca^{2+}]C_t\,\alpha_{C^{2-}} \qquad (8)$$

Since $[Ca^{2+}] = C_t$,

$$K_S = [Ca^{2+}]^2\alpha_{C^{2-}} \qquad (9)$$

To construct a stability diagram for $CaCO_{3(s)}$, it is convenient to divide the pH scale into three regions: 0.00–6.35(I); 6.35–10.33(II); and 10.33–14.0(III). In Region I, $\alpha_{C^{2-}} \approx K_1K_2/[H^+]^2$ (Equation 33, Chapter 3), whereupon

$$K_S = \frac{[Ca^{2+}]^2K_1K_2}{[H^+]^2} \qquad (10)$$

or

$$\log[Ca^{2+}] = -pH + 0.5(pK_1 + pK_2 - pK_S) \qquad (11)$$

In Region II, $\alpha_{C^{2-}} \approx K_2/[H^+]$ (Equation 36, Chapter 3), whereupon

$$K_S = \frac{[Ca^{2+}]^2K_2}{[H^+]} \qquad (12)$$

or

$$\log[Ca^{2+}] = -0.5pH + 0.5(pK_2 - pK_S) \qquad (13)$$

In Region III, $\alpha_{C^{2-}} \approx K_1K_2/K_1K_2 = 1.0$ (Equation 39, Chapter 3), whereupon

$$K_S = [Ca^{2+}]^2 \qquad (14)$$

or

$$\log[Ca^{2+}] = -0.5pK_S \qquad (15)$$

Equations 11, 13 and 15 are the linear relations between $[Ca^{2+}]$ and pH that represent the equilibrium boundary conditions between soluble and insoluble $CaCO_{3(s)}$.

The appropriate solution of Equations 11, 13 and 15 leads to the stability diagram for aqueous $CaCO_{3(s)}$ in a closed system, as seen in Figure 1. Three areas exist in this diagram for $CaCO_{3(s)}$: (1) a soluble region to the left of the flagged line; (2) an insoluble region to the right of the flagged line; and (3) an equilibrium region on the flagged line. From this diagram, the concentrations of Ca^{2+} and C_t (mass balance on carbonate species) may be determined at any pH value. Pragmatic illustration of this stability diagram is given below.

Free Energy

The free energy change, ΔG_{Rex}, of a reaction may be used to denote a condition of equilibrium for dissolved constituents. Combination of Equations 41 and 47 of Chapter 2 leads to:

$$\Delta G_{Rex} = \Delta G^0_{Rex} + RT \ln \frac{(Products)}{(Reactants)} \qquad (16)$$

where ΔG_{Rex} is the free energy difference of the reaction under nonstandard-state conditions, ΔG^0_{Rex} is the free energy difference of the reaction under standard state conditions, and ( ) are the activities of reactants and products, respectively. In the application of this equation, the actual activities (concentrations corrected for ionic strength) are computed for the dissolved constituents. When ΔG_{Rex} is calculated to be 0.0, the system is in equilibrium and the attendant dissolved species are at their equilibrium concentrations.

This method has the advantage of relating the free energy difference in

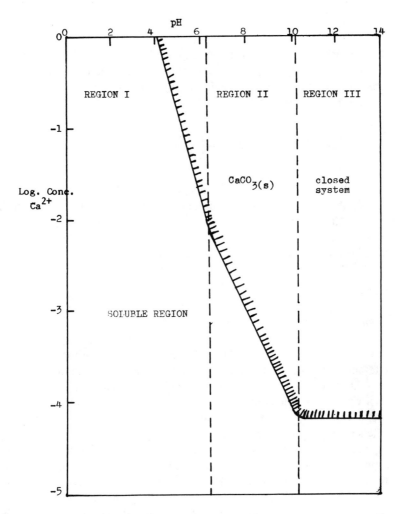

Figure 1.   Solubility diagram for calcium carbonate in a closed system at 25°C.

kcal/mol to the actual concentrations of substances in natural waters. An example may be cited for the closed $CaCO_{3(s)}$ system. A natural ground water in a limestone geology had these contents: $[Ca^{2+}]$ = 48 mg/l, $[Mg^{2+}]$ = 3.6 mg/l, $[HCO_3^-]$ = 152 mg/l, and a pH value of 7.5 [5]. The ΔG_{Rex} for Equation 10 was computed to be +0.006 kcal/mol. This suggests that this natural water may be in equilibrium with $CaCO_{3(s)}$ in a closed system.

Ratio

Many geochemists use a mathematical ratio of an empirical solubility product constant obtained from the ionic activities in the system under examination to the thermodynamic constant. For example, Langmuir [2] uses a saturation index of calcite:

$$SI_c = \log\left(\frac{IAP_c}{K_c}\right) \tag{17}$$

where IAP_c is the ion activity product of (Ca^{2+}) (CO_3^{2-}) in solution. When $SI_c = 0$, the ground water is saturated with respect to the carbonate in question. Undersaturation is denoted by negative SI values, whereas positive indices indicate supersaturation. This method is limited somewhat to systems where only one solid phase is present.

Examples

Carbonate Systems

Many ground waters in the United States contact the carbonate minerals calcite and dolomite. Consequently, several investigators have attempted to fit equilibrium models to the calcium and magnesium contents of these waters. Conflicting conclusions have been reported concerning the solubilities of these two carbonate minerals under natural conditions. Some of these conflicts are reported below.

An example of an excellent fit to an equilibrium model is provided by Langmuir [2], who examined the geochemistry of some carbonate ground waters in central Pennsylvania. The field values of pH and the contents of Ca^{2+}, Mg^{2+} and HCO_3^- were measured in 29 spring waters and 29 well waters in folded and faulted Paleozoic carbonate rocks. Most of the springs originated in limestone, whereas most of the well waters came from dolomite. Langmuir claimed that these natural carbonate rocks were "relatively pure." The question of solid phase purity is emphasized below. A saturation index was employed as the criterion of equilibrium for the calcitic and dolomitic waters: $SI_c = \log(IAP_c/K_c)$ and $SI_d = \log(IAP_d/K_d)^{1/2}$, where SI_c is the saturation index for calcite, IAP_c is the ion activity product (Ca^{2+}) (CO_3^{2-}) in solution (i.e., the natural waters), and K_c is the solubility product constant for calcite. Similar designations are carried for dolomite. It is important to note that Langmuir employed the log K_c value of -8.400 ($25°C$) for calcite and the log K_d value of -17.0 ($25°C$) for dolomite.

Difficulties with the variability of thermodynamic data of carbonate species are discussed below.

A summary of Langmuir's results is given in Table I. The average SI_c and SI_d values were negative, which led to these statements: "None of the spring or well waters are significantly supersaturated with calcite or dolomite. Based on the uncertainty of ±0.1 units in SI_c and SI_d, 3 spring waters are saturated with both minerals. Similarly, 12 well waters are just saturated with calcite and 7 just saturated with dolomite. These results show that the theoretical solubilities of stoichiometric calcite and dolomite represent real limits on the relative concentrations of Ca^{2+}, Mg^{2+}, HCO_3^-, and H^+ in the natural ground water. This is true even in well waters which are polluted, having specific conductances above 600-700 $u\Omega^{-1}$, and SO_4^{2-}, Cl^-, and NO_3^- ion concentrations in excess of 10-15 ppm."

This excellent study of Langmuir also raises a common observation in the chemistry of natural carbonate waters. This is, the partial pressure of $CO_{2(g)}$ (P_{CO_2}) in well and spring waters is about one order of magnitude ($10^{-2.2}$ atm) above that of the atmosphere ($10^{-3.5}$ atm). This suggests some extraordinary source of $CO_{2(g)}$, which, if not constantly supplied, will affect the pH value and the contents of Ca^{2+}, HCO_3^- and CO_3^{2-}. This may be seen in Reaction 13 (Table VI, Chapter 5), from which an equilibrium model is derived that reflects the change in $[HCO_3^-]$ and pH value when the P_{CO_2} is increased or decreased:

$$[HCO_3^-] = \frac{K_{H(CO_2)} \times K_1 \times P_{CO_2} \times 10^{pH}}{\gamma_{HCO_3^-}} \qquad (18)$$

Table I. Ranges and Averages of Some Chemical Quality
Parameters for the Spring and Wellwaters[a]

Parameter	Spring Waters		Well Waters	
	Range	Average	Range	Average
μ	180–475	347	173–945	499
DO (% sat)	42–100	71	30–117	82
Ca^{2+}/Mg^{2+} (molar)	0.8–11	3.4	0.6–3.9	1.3
pH	7.11–7.80	7.37	7.06–8.15	7.46
P_{CO_2} (atm)	$10^{-1.92}$–$10^{-2.60}$	$10^{-2.18}$	$10^{-1.63}$–$10^{-3.23}$	$10^{-1.99}$
SI_c	−1.17 to −0.04	−0.41	−0.38 to +0.04	−0.15
SI_d	−1.48 to −0.06	−0.63	−0.39 to +0.02	−0.18

[a]Reproduced from Langmuir [2] courtesy of *Geochim. Cosmochim. Acta.*

Furthermore, it was postulated that a ground water at or near saturation with carbonate rocks will change in bicarbonate content and pH value up or down an equilibrium boundary line. The appropriate equation for calcite saturation is

$$-\log[HCO_3^-] = \frac{1}{2}\left[pH - \log\left(\frac{2K_c}{K_2}\right) + \log(\gamma_{Ca}\gamma_{HCO_3^-})\right] \qquad (19)$$

For dolomite saturation, it is

$$-\log[HCO_3^-] = \frac{1}{2}\left[pH - \log\left(\frac{4K_d^{1/2}}{K_2}\right) + \frac{1}{2}\log(\gamma_{Ca}\gamma_{Mg}\gamma_{HCO_3^-}^2)\right] \qquad (20)$$

These two equations are seen plotted in Figure 2 for the pH range of 7.0 to 8.2. When $CO_{2(g)}$ is exsolved, carbonates may precipitate with the attendant water chemistry moving diagonally down the equilibrium line to the right. When the $CO_{2(g)}$ content is increased, more carbonate rocks dissolve and the water chemistry moves up the equilibrium boundary line to higher HCO_3^- contents and lower pH values. By the same token, an increase in pH value, Ca^{2+}, HCO_3^- or CO_3^{2-} can cause precipitation of calcite with a subsequent increase in P_{CO_2}. The pH values and HCO_3^- contents of the 29 well waters in Langmuir's study are seen plotted in Figure 2. That the points fell on or near the equilibrium boundary lines was considered evidence for carbonate saturation or just saturation of these waters. This is also strong evidence that idealized thermodynamic equilibrium models are obeyed under natural field conditions.

An example of a lack of fit to carbonate equilibria is provided by Hanshaw et al. [3]. In this case, the ground waters of the carbonate peninsula of Florida were examined for their equilibria with calcite, aragonite and dolomite. The criterion of equilibrium was the ratio of the $K_{iap}:K_s$. When the ratio was equal to one, the water was considered to be saturated, less than one designated undersaturation, and greater than one indicated supersaturation. In the study area, the ground water was undersaturated with respect to aragonite, but was supersaturated with respect to calcite in all but one sample and with respect to dolomite in four of eight samples. Hanshaw's data appear in Table II. No acceptable explanation has been advanced for this apparent permanent supersaturation of the ground waters with respect to calcite and dolomite.

Several carbonate models were developed by the author to determine the equilibrium conditions of various ground waters in the United States. These models are listed in Table VI of Chapter 5. All ΔG_{Rex}^0 values were computed from the free energy of formation data supplied by Robie and

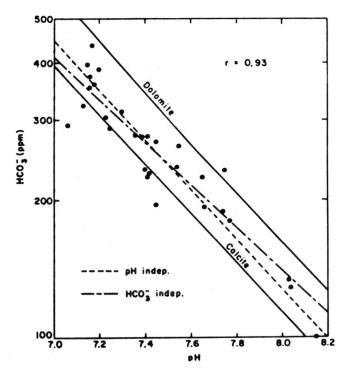

Figure 2. Plot of pH vs HCO_3^- (ppm) for the well waters. Regression lines are drawn assuming HCO_3^- or pH as the independent variable. The theoretical solubilities of calcite and dolomite in pure water are shown also (reproduced from Langmuir [2] courtesy of *Geochim. Cosmochim. Acta.*).

Table II. A Comparison of K_{eq} and K_{iap} for Aragonite, Calcite and Dolomite from the Ground Waters of Florida [3]

Well Location	Aragonite $\log K_{eq} = -8.14$		Calcite $\log K_{eq} = -8.34$		Dolomite $\log K_{eq} = -17.0$	
	$\log K_{iap}$	% Saturation	$\log K_{iap}$	% Saturation	$\log K_{iap}$	% Saturation
Ocala 4	−8.22	82	−8.22	129	−17.02	96
Wildwood 2	−8.40	52	−8.40	83	−17.88	13
Groveland	−8.29	67	−8.29	106	−17.38	42
Polk City	−8.24	77	−8.24	122	−17.04	92
Fort Meade	−8.18	94	−8.18	149	−16.66	220
Wauchula	−8.20	87	−8.20	138	−16.53	293
Arcadia	−8.21	87	−8.21	138	−16.44	362
Cleveland	−8.33	68	−8.32	107	−16.55	279

Waldbaum [4]. Some of these carbonate reactions admittedly are not strictly applicable to the real world, but this does represent a concerted effort to provide an equilibrium model for a given set of water quality data. Activities of the dissolved constituents were taken from various published sources that were either field or laboratory data. The two exceptions, (H_2CO_3) and P_{CO2}, were computed from the carbonate mass balance.

Some ground waters are apparently in equilibrium with more than one carbonate mineral. Several sets of water quality data (mostly U.S. Geological Survey data [5-7]) were examined in an attempt to find a fit to one or more of the 15 carbonate models. Table III shows five representative computations. All were ground waters from various types of limestone (the only available geological data). In most cases, the water quality data were applied to all of the 15 carbonate models. It was coincidental that the water quality analyses in Table III were performed in the laboratory.

The results from this data-fitting exercise are most revealing. Water #3 from Hem [6] was found to be in equilibrium (i.e., $\Delta G_{Rex} = 0.0 \pm 0.100$ kcal/mol) with the three calcite models. This water was undersaturated with respect to magnesite and oversaturated with respect to dolomite. This leads to these conclusions: (1) the water is, indeed, in an apparent equilibrium with calcite in an "open" system (because P_{CO2} was calculated to be greater than atmospheric CO_2); (2) magnesite, if present,

Table III. Carbonate Waters Showing

Water[a,b]		pH	Ca^{2+}	Mg^{2+}	HCO_3^-	H_2CO_3	P_{CO_2} (atm $\times 10^3$)
#3	mg/l	–	48	3.6	152	10.3	–
	log Act.	7.5	–3.035	–3.943	–2.604	–3.780	4.8
#5	mg/l	–	100	19	281	–	–
	log Act.	7.2	–2.777	–3.281	–2.383	–3.232	17.0
#3	mg/l	–	40	24	204	–	–
	log Act.	7.9	–3.137	–3.141	–2.513	–4.062	2.5
#7	mg/l	–	76	1.3	240	–	–
	log Act.	7.1	–2.848	–4.398	–2.437	–3.186	18.9
#4	mg/l	–	58	3.2	164	–	–
	log Act.	7.5	–2.965	–4.006	–2.602	–3.752	5.14

[a]All are ground waters from various types of limestone. Water # conforms to the
[b]All water quality analyses are laboratory.
[c]Superscripts eq, u and s are: equilibrium, undersaturated and supersaturated,

is being dissolved; and (3) dolomite should be forming. The fallacies of these conclusions are discussed below. Water #5 from Hem [6] is slightly oversaturated with respect to calcite, in an apparent equilibrium with the metastable aragonite; undersaturated with respect to magnesite; and over-saturated with respect to dolomite. Although water #3 from Back [7] was derived from a limestone geology, it appears to be in equilibrium with magnesite. This water was calculated to be undersaturated with calcite and oversaturated to dolomite despite the equilibrium weight ratio of 0.6 for $Mg^{2+}:Ca^{2+}$. Water #7 from White et al. [5] provides a rather peculiar example of carbonate equilibria. Three models were fitted by the water quality data: calcite, dolomite and the mixed dolomite-calcite! It was undersaturated with respect to magnesite. It is extremely doubtful that calcite and dolomite are present in the proportions suggested by Reaction 25 of Table VI (Chapter 5). Water #4 from White et al. [5] fits very closely to the calcite and aragonite models with an undersaturation to magnesite and an oversaturation to dolomite.

Many examples of ground waters apparently not in equilibrium are given in Table IV, where five representative computations of carbonate waters are seen. The Polk City and Venus waters probably provide the most accurate data since the analyses were performed in the field, and the geology is defined better than in the other reports [8]. These two waters were oversaturated with respect to calcite, aragonite and dolomite and

an Apparent Equilibrium

			Models[c]				
No.	ΔG_{Rex} (kcal/mol)	No.	ΔG_{Rex} (kcal/mol)	No.	ΔG_{Rex} (kcal/mol)	Reference
12	+0.006[eq]	14	+0.067[eq]	19	+0.972[s]	6
13	+0.060[eq]	16	−1.748[u]			
12	+0.251[s]	15	+0.021[eq]	19	+2.012[s]	6
13	+0.268[s]	16	−0.953[u]			
12	+0.537[s]	16	+0.015[eq]	19	+3.257[s]	7
15	+0.307[s]	17	+0.032[eq]			
12	−0.056[eq]	14	−0.095[eq]	19	−0.029[eq]	5
13	−0.083[eq]	16	−2.687[u]	25	−0.085[eq]	
12	+0.105[eq]	15	−0.125[eq]	19	+0.987[s]	5
13	+0.228[s]	16	−1.830[u]			

original publication.

respectively (Table VI, Chapter 5).

undersaturated with respect to magnesite. Water #4 from Back [7] shows a $Mg^{2+}:Ca^{2+}$ weight ratio of 0.58, which approaches the equilibrium ratio of 0.61 for dolomite; however, this water was 4.540 kcal/mol oversaturated, which suggests that dolomite should be forming. Water #4 also showed oversaturation to calcite, aragonite and magnesite. Water #3 from Back [7] also showed oversaturation to the four carbonate phases. These two waters were derived from dolomitic formations and should be in equilibrium with this carbonate. Water #2 from Back [7] was derived from a limestone formation and exhibited undersaturation to calcite, aragonite, magnesite and dolomite.

Many other investigators have employed carbonate equilibrium models with varying degrees of success: Hostetler [9] examined the degree of saturation of magnesium and calcium carbonate minerals in natural waters; Holland et al. [10] and Thrailkill [11] looked at the chemical evolution of cave waters through carbonate equilibria; Barnes [12] was concerned with the deposition of calcite in Birch Creek, Inyo County, California; Shuster and White [13] looked at several carbonate springs in the Nittany Valley in central Pennsylvania through the Saturation Index Method of Langmuir [2]; and Hsu [14] reported a solubility product constant for dolomite from the chemical composition of Florida ground waters. Some of the major

Table IV. Carbonate Waters

Water[a]		pH	Ca^{2+}	Mg^{2+}	HCO_3^-	H_2CO_3	P_{CO_2} (atm $\times$ 10^3)
Polk	mg/l[c]	–	34	5.6	124	–	–
City	log Act.	8.0	–3.169	–3.724	–2.719	–4.369	1.24
Venus	mg/l[c]	–	80	44	108	–	–
	log Act.	7.83	–2.893	–2.928	–2.863	–4.302	1.45
#4	mg/l[d]	–	67	39	390	–	–
	log Act.	7.9	–2.943	–2.955	–2.236	–3.785	0.16
#3	mg/l[d]	–	35	33	241	–	–
	log Act.	8.2	–3.214	–3.013	–2.431	–4.288	0.05
#2	mg/l[d]	–	43	1.2	133	–	–
	log Act.	7.3	–3.068	–4.404	–2.684	–3.633	6.75

[a]All are ground waters from the designated geological source. Water # conforms to
[b]Superscripts u and s are: undersaturated and supersaturated.
[c]Field data.
[d]Laboratory data.

difficulties with the application of idealized calcite models to natural systems are discussed below.

Silicate Systems

  Stability Diagrams. Stability may be constructed for several silicate systems whose chemical weathering reactions may control the concentrations of Na^+, K^+, Ca^{2+}, H^+, Mg^{2+} and H_4SiO_4 in natural waters. The major naturally occurring silicates are amorphous silica, quartz, kaolinite and illite (considered by some to be a mixture of muscovite and montmorillonite). Several idealized chemical weathering reactions are given in Tables VIII, IX, X and XI of Chapter 5. Almost all of these reactions involve the incongruent dissolution of one or more silicate solid phases, which is accompanied by a reaction with one or more of the dissolved constituents. Whenever these reactions occur, the Phase Rule constraints bring these constituents to their equilibrium concentrations.

  The foundation of a silicate stability diagram lies in the relation between kaolinite, gibbsite and silica (Reactions 38 and 40, Table VIII, Chapter 5). These two reactions show the incongruent dissolution of kaolinite in pure water into gibbsite and either dissolved silica or solid silica. The clay

Apparently not in Equilibrium

						Models[b]		
No.	ΔG_{Rex} (kcal/mol)	No.	ΔG_{Rex} (kcal/mol)	No.	ΔG_{Rex} (kcal/mol)	Geological Source	Reference
12	+0.349[S]	16	−0.924[u]			Limestone	8
15	+0.119[S]	19	+2.138[S]			and dolomite	
12	+0.493[S]	16	−0.267[u]	14	+3.041[S]	Limestone	8
15	+0.263[S]	19	+2.744[S]			and gypsum	
12	+1.179[S]	16	+0.647[S]			Dolomite	6
15	+0.949[S]	17	+4.540[S]				
12	+0.953[S]	16	+0.711[S]			Dolomite	6
15	+0.723[S]	19	+4.378[S]				
12	−0.421[u]	16	−2.759[u]			Limestone	6
15	−0.651[u]	19	−0.465[u]				

original publication.

mineral, kaolinite, is apparently the pivotal solid phase because it precedes the final dissolution of solids in the weathering process. Also, kaolinitic soils are very common in temperate climates with moderate to heavy rainfall, whereas the montmorillonite and illite minerals are predominant in soils of semiarid regions [15].

Silicate stability diagrams are log-log plots of dissolved silica concentration versus the algebraic difference between $[H^+]$ and $[Na^+]$ or $[Ca^{2+}]$ or $[K^+]$. The diagrams depicted here vary slightly from those of Garrels and Christ [16] and Stumm and Morgan [17] because of the abscissa and ordinate origin of 0.0 (1 M) for the concentrations of the dissolved constituents. This graphic technique is preferred because it yields the proper intercept values on the axis from the various straight line equations. In addition, two-dimensional diagrams are employed here instead of the three-dimensional technique preferred by Garrels and Christ [16].

Figure 3 shows the stability diagram for the Na_2O-Al_2O_3-SiO_2-H_2O system in which these solid phases are considered: amorphous silica, quartz, kaolinite, gibbsite, albite and Na montmorillonite. This diagram was kept simple for didactic purposes. A much more complex one could have been constructed from all of the reactions listed in Table IX of Chapter 5. Dissolved silica concentration is controlled by either Reaction 35 (amorphous) or 36 (quartz) (Chapter 5). These equilibrium concentrations are $10^{-2.77}M$ and $10^{-3.76}M$, respectively, and are independent of pH. Most diagrams show only amorphous silica, which is metastable to quartz. Siever [18] indicated that the rate of crystallization of quartz is extremely slow at low temperatures. Consequently, amorphous silica tends to control the solubility of dissolved silica, which is shown in Figure 3 at the equilibrium boundary line of $pH_4SiO_4 = 2.77$. The kaolinite–gibbsite domains are designated by Reaction 38 (Chapter 5) with the equilibrium position located at a pH_4SiO_4 value of 4.97 ($K_{eq\,38} = 10^{-9.93} = [H_4SiO_4]^2$). This reaction is also independent of pH. Whenever the concentration of dissolved silica falls below $10^{-4.97}M$, kaolinite tends to dissolve incongruently to leave a residuum of gibbsite. Whenever the concentration of dissolved silica falls between $10^{-2.77}$ and $10^{-4.97}M$, the formation of kaolinite is favored. However, when the concentration exceeds $10^{-2.77}M$, dissolved silica would precipitate to form amorphous silica. There are now two domains for the solid phases of amorphous silica and kaolinite, whose dissolution stability is controlled by the concentration of dissolved silica. The dissolution stability of gibbsite is not affected by dissolved silica but rather by its own solubility product constant ($10^{-32.65}$ [19]). It is assumed, however, that solid phase gibbsite is formed from the incongruent dissolution of kaolinite.

If kaolinite is in contact with water containing H^+ and Na^+ ions and dissolved silica, then incongruent solution may occur with the formation

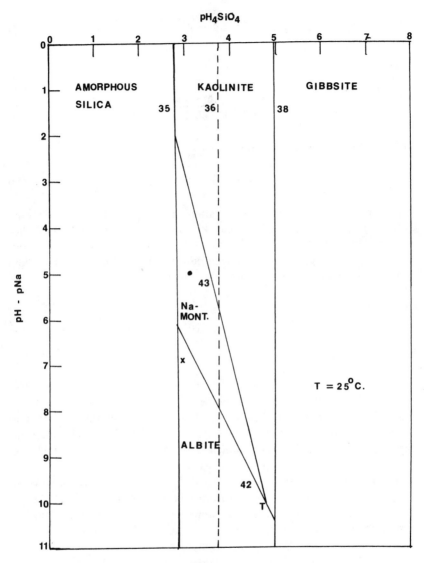

Figure 3.   Stability diagram for the $Na_2O-Al_2O_3-SiO_2-H_2O$ system.

of albite (Na feldspar) and sodium montmorillonite (Reactions 42 and 43, Table IX, Chapter 5). These reactions also have the consequence of exchanging H^+ for Na^+ with the concomitant control of their equilibrium concentrations. The stability domain of albite in relation to kaolinite is seen from the equilibrium expression for Reaction 42:

$$K_{eq42} = 10^{0.495} = \frac{[Na^+][H_4SiO_4]^2}{[H^+]} \tag{21}$$

Rearranging,

$$\frac{[Na^+]}{[H^+]} = \frac{10^{0.495}}{[H_4SiO_4]^2} \tag{22}$$

or in the negative logarithm form,

$$pH - pNa = 0.495 + 2pH_4SiO_4 \tag{23}$$

Equation 23 is a straight line equation that represents the equilibrium boundary line between albite and kaolinite, where the dissolved silica concentrations range from $10^{-2.77}$ to $10^{-4.96}M$ and the $[Na^+]:[H^+]$ ratios of 6.05 to 10.4. Similarly, the equilibrium boundary line between sodium montmorillonite and kaolinite comes from Reaction 43, Chapter 5:

$$K_{eq43} = 10^{-9.1} = \frac{[Na^+][H_4SiO_4]^4}{[H^+]} \tag{24}$$

or,

$$pH - pNa = -9.1 + 4pH_4SiO_4 \tag{25}$$

Note the interception of the albite and the sodium montmorillonite lines at a pH − pNa value of 9.95 and a pH_4SiO_4 value of 4.8. This intercept represents a triple point (T) at which the three silicates are in equilibrium and separates the stability domains for each.

Similar diagrams may be constructed for the systems of K_2O-Al_2O_3-SiO_2-H_2O and CaO-Al_2O_3-SiO_2-H_2O, in which the principal cations are K^+ and Ca^{2+}, respectively. The appropriate reactions for the potassium system shown in Figure 4 are as follows:

$$K_{eq53} = 10^{-4.39} = \frac{[K^+]^2[H_4SiO_4]^4}{[H^+]^2} \tag{26}$$

or,

$$pH - pK = -2.2 + 2pH_4SiO_4 \tag{27}$$

This reaction shows the equilibrium relationship between kaolinite and microline. The incongruent dissolution of microcline to muscovite is seen from Reaction 52 of Chapter 5:

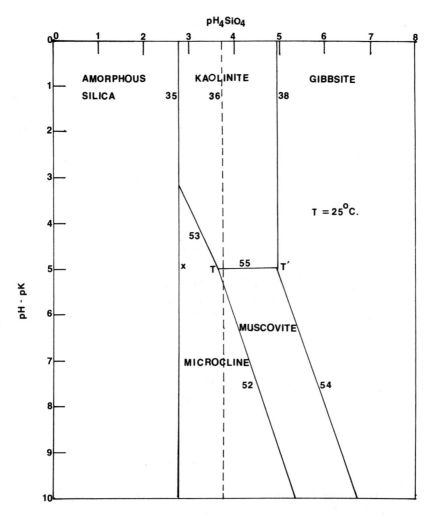

Figure 4.   Stability diagram for the K_2O-Al_2O_3-SiO_2-H_2O system.

$$K_{eq_{52}} = 10^{-11.6} = \frac{[K^+]^2[H_4SiO_4]^6}{[H^+]^2} \tag{28}$$

or,

$$pH - pK = -5.8 + 3pH_4SiO_4 \tag{29}$$

The reaction between muscovite and kaolinite is independent of dissolved silica concentration:

$$K_{eq_{55}} = 10^{10.1} = \frac{[K^+]^2}{[H^+]^2} \tag{30}$$

or,

$$pH - pK = 5.05 \tag{31}$$

Muscovite may also undergo incongruent dissolution to gibbsite via Reaction 54 (Table X, Chapter 5):

$$K_{eq_{54}} = 10^{-9.86} = \frac{[K^+][H_4SiO_4]^3}{[H^+]} \tag{32}$$

or,

$$pH - pK = -9.86 + 3pH_4SiO_4 \tag{33}$$

Two triple points are seen in this diagram: one separates the domains of kaolinite, microcline and muscovite (T), whereas the other separates the domains of kaolinite, muscovite and gibbsite (T').

The appropriate reactions for the calcium system shown in Figure 5 are as follows:

$$K_{eq_{69}} = 10^{-15.4} = \frac{[Ca^{2+}][H_4SiO_4]^8}{[H^+]^2} \tag{34}$$

or,

$$2pH - pCa = -15.4 + 8pH_4SiO_4 \tag{35}$$

Anorthite undergoes incongruent dissolution to kaolinite via Reaction 68 (Chapter 5), which is independent of dissolved silica concentration:

$$K_{eq_{68}} = 10^{17.1} = \frac{[Ca^{2+}]}{[H^+]^2} \tag{36}$$

or,

$$2pH - pCa = 17.1 \tag{37}$$

The equilibrium boundary line between anorthite and calcium montmorillonite is obtained by combination of Equations 68 and 69 (Chapter 5), which yields

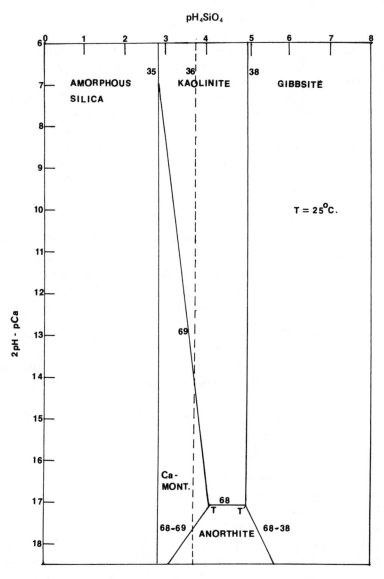

Figure 5. Stability diagram for the CaO-Al_2O_3-SiO_2-H_2O system.

$$2pH - pCa = 22.5 - \frac{4}{3} pH_4SiO_4 \qquad (38)$$

Likewise, the equilibrium boundary line between anorthite and gibbsite is obtained by combination of Equations 38 and 68 (Chapter 5), which yields

$$2pH - pCa = 7.17 + 2pH_4SiO_4 \qquad (39)$$

Two triple points are given in this diagram also: one separates the domains of anorthite, kaolinite and calcium montmorillonite (T), whereas the other separates the domains of anorthite, kaolinite and gibbsite (T').

These idealized silicate stability diagrams may be employed to gain some insight into chemical weathering processes under natural conditions. If one is given the appropriate chemical water quality data, a plot of this information would indicate which silicate(s) is controlling the observed concentrations of H^+, Na^+, K^+, Ca^{2+} and H_4SiO_4. Another use would be to gain some inference about the mineral geology of a particular region from chemical water quality data. Two didactic examples are given. A ground water in Owyhee County, Idaho [5] yielded these contents: pH = 9.2, $[Na^+]$ = 100 mg/l, $[K^+]$ = 2.9 mg/l, and $[H_4SiO_4]$ = 99 mg/l. The appropriate computations and plots (X) in Figures 3 and 4 fall into the stability domains of albite (Reaction 42, Chapter 5) and microcline (Reaction 52, Chapter 5), respectively. This strongly suggests that a kaolinite weathering reaction to the silicate $AlSi_3O_{8(s)}$ is controlling the concentrations of H^+, Na^+, K^+ and H_4SiO_4. Also note that the dissolved silica is saturated with respect to amorphous silica. In another ground water from Umatilla County, Oregon [5], these contents were observed: pH = 7.8, $[Na^+]$ = 30 mg/l, $[Ca^{2+}]$ = 32 mg/l, and $[H_4SiO_4]$ = 49 mg/l. The appropriate plots (•) in Figures 3 and 5 fall into the stability domains of sodium and calcium montmorillonite, respectively. This strongly suggests that kaolinite (if present) is being weathered to montmorillonite, with the concomitant control of the aqueous concentrations of Na^+, Ca^{2+}, H^+ and H_4SiO_4. In this example, dissolved silica is undersaturated with respect to amorphous silica.

Applications of Stability Diagrams. Several researchers have used the silicate stability diagrams in attempts to explain natural water chemistry or to interpret geologic formations. A few examples are cited here.

As part of a wastewater reclamation study, the U.S. Geological Survey and the Nassau County government conducted a series of artificial recharge experiments at Bay Park, New York [20]. These experiments were designed to obtain some of the scientific and the economic data needed to evaluate the feasibility of injecting highly treated sewage into a proposed network of barrier-recharge wells in an area of potential salt water encroachment. Data were collected to show the change in water quality as it moves through the aquifer [21].

The chemical composition of the native water in the recharge area is influenced by reactions involving the minerals of the Magothy formation. At Bay Park, the Magothy lies between 104 and 593 feet below land surfaces. This formation is composed mostly of gray, very fine to medium quartzose sand, and some silt. Layers of clay occur throughout the formation. An X-ray analysis of a clay sample from 66 feet above the base of the Magothy (527 feet below land surface) showed that the sample was composed of equal amounts of kaolinite and illite. Kaolinite is a common constituent of the Magothy with lignite, pyrite-marcasite and muscovite occurring as minor constituents.

Chemically, water from the Magothy aquifer may be characterized as low in dissolved solids (25 mg/l). The principal cation is Na^+ (3.8 mg/l), but none of the anions (Cl^-, SO_4^{2-}, HCO_3^-) predominates in concentration. Dissolved silica, which is a nonionic constituent, constitutes roughly one-third of the dissolved solids by weight (8 mg/l). The pH value of the native water is between 5 and 6. The major chemical constituents of the reclaimed water were: $[Na^+]$ = 70 mg/l, $[NH_4^+]$ = 32 mg/l, $[Ca^{2+}]$ = 16 mg/l, $[HCO_3^-]$ = 86 mg/l, $[SO_4^{2-}]$ = 140 mg/l, $[Cl^-]$ = 69 mg/l and $[H_4SiO_4]$ = 14 mg/l. The pH values ranged from 6.1 to 6.6 over several injection tests. Phosphate contents ranged from 1.3 to 4.0 mg/l as PO_4^{3-}.

The effect of reclaimed water on native water quality was examined through the use of clay–mineral stability field diagrams. Samples obtained from an observation well (20 feet from point of injection) on or about the terminal day of each injection test were considered as mixed water. The pH–pNa versus pH_4SiO_4 points for the mixed waters from three tests plotted well below those for the native water quality. Similar observations were made for the Ca^{2+} and K^+ ions. All points, however, were within the kaolinite stability domain. Ion exchange reactions probably controlled the equilibrium concentrations of Na^+, K^+, Ca^{2+}, Mg^{2+}, NH_4^+ and H^+ ions in the mixed waters.

Sutherland [22] applied the equilibrium concepts that involve the silicate minerals to water quality data from the North Channel and Lakes Erie, Ontario and Huron of the Great Lakes system. The major concern was the solubility control of dissolved silica. In general terms, the open waters of the Great Lakes are relatively low in dissolved solids with ionic strengths less than 0.005. The major cations are Ca^{2+} and Mg^{2+} in concentrations of less than 50 mg/l. Dissolved silica concentrations are less than 5.0 mg/l. pH values are mostly between 7 and 8. These slightly alkaline values are influenced greatly by the limestone bedrock of the area. Several silicate minerals were identified in the sediments: quartz, muscovite, K feldspar (microcline variety), NaCa feldspar, amphiboles, chlorite, kaolinite, montmorillonite and structurally intermediate clay minerals. Figures 6 and 7 show the $K-H-SiO_2$ and $Ca-H-SiO_2$ stability diagrams, respectively, for the open water quality data. Some sediment interstitial water chemical data are included also. All of the water quality data are plotted in the kaolinite domain. The chemical weathering of kaolinite (Reaction 38, Table VIII,

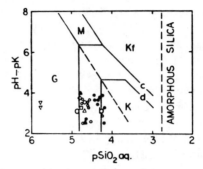

Figure 6. K-H-SiO$_2$ mineral stability diagram for open waters with boundaries at 20°C (reproduced from Sutherland [22] courtesy of the American Chemical Society). G = gibbsite, K = kaolinite, M = muscovite, Kf = K feldspar. Lines a and b join field of gibbsite with kaolinite of excellent and fair crystallinity, respectively. Lines c and d join field of K feldspar with kaolinite of excellent and fair crystallinity, respectively. ▲ = L. Huron, o = N. Channel, □ = Croker Island (granite), ♦ = L. Erie, ▽ = L. Ontario. Smaller symbols are open water analyses; the open symbols = surface; and the dark symbols = bottom.

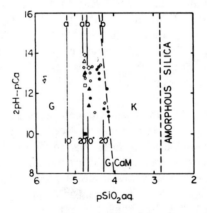

Figure 7. Ca-H-SiO$_2$ mineral stability for open waters. Lines a and b are as in Figure 6. Slightly inclined dashed line joins gibbsite with Ca montmorillonite (CaM) (reproduced from Sutherland [22] courtesy of the American Chemical Society).

Chapter 5) yields gibbsite and an equilibrium concentration of $10^{-4.8}M$ for dissolved silica at 20°C. All measured concentrations of silica in the surface waters of the Great Lakes lie below this value, which suggested to Sutherland that kaolinite was slowly yielding gibbsite and dissolved silica.

In cooler bottom waters, the concentrations of dissolved silica were higher, which shifted the points away from the kaolinite-gibbsite boundary line. Sutherland interpreted this shift as making kaolinite more stable with respect to gibbsite. It may be observed also that muscovite (Figure 6) and Ca-montmorillonite (Figure 7) are unstable with respect to gibbsite. These two silicates may lose their silica and cations at their layer edges, leaving the Al-rich gibbsite layers intact. Sutherland felt that "the equilibrium model perspective is useful in attempting to identify and understand chemical self-regulation in the Great Lakes."

The spring waters of the Sierra Nevada mountains in California and Nevada provide an almost classic example for the application of silicate stability diagrams. These waters emerge from contact with igneous rock minerals, which are attacked by soil waters high in dissolved CO_2. Also, this system is apparently "closed" in the sense that there is little loss or gain of water or carbon dioxide. Feth et al. [23] and Garrels and Mackenzie [24] studied rather extensively this system whose geologic composition is reasonably well known. Feldspars and quartz are the major minerals with hornblende (a Ca-Mg-Fe silicate) and biotite (a mica mineral) as accessories. K-feldspar and plagioclase (a mixture of albite and anorthite) feldspars are about equally abundant. Andesine ($Na_2CaAl_4Si_8O_{24}$) is the dominant plagioclase with traces of microcline. After the CO_2 "attack" on these igneous rocks, an aluminosilicate residue is left, traces of gibbsite, mica and montmorillonite are found. Kaolinite, however, was identified as the major alteration product.

Table V shows the mean values for the chemical composition of the ephemeral and perennial spring waters of the Sierra Nevada. These waters may be classified as low in dissolved solids and slightly acid, with Ca^{2+} and Na^+ as the principal cations. That HCO_3^- is the principal anion may be seen from the reaction of the CO_2 (calculated pressure of $10^{-1.8}$ atm) with the igneous rocks. Reaction 51, Table IX, Chapter 5 is a typical example of this type of acidic weathering.

Three major points resulted from the study of Garrels and Mackenzie [24]. First, an earlier conclusion of Feth et al. [23] was reemphasized: "the silica in the water came from the breakdown of the silicates and an insignificant amount from the direct solution of quartz." Second, the rock minerals react with the dissolved CO_2-forming kaolinite continuously in the system with the consequence that kaolinite controls the water composition by its presence. Third, about 80% of the dissolved constituents in the ephemeral springs can be attributed to the decomposition of the plagioclase alone. Support for these conclusions came from deductive reconstruction of the original minerals through the water quality data and from a $Na_2O-Al_2O_3-SiO_2-H_2O$ stability diagram (Figure 8 [23]) showing points from the chemical weathering of siliceous rocks fall mainly into the kaolinite domain.

Table V. Mean Values for Compositions of Ephemeral and
Perennial Springs of the Sierra Nevada[a]

	Ephemeral Springs		Perennial Springs	
	ppm	molality $\times$ 10^4	ppm	molality $\times$ 10^4
SiO_2	16.4	2.73	24.6	4.1
Al	0.03	–	0.018	–
Fe	0.03	–	0.031	–
Ca	3.11	0.78	10.4	2.6
Mg	0.70	0.29	1.70	0.71
Na	3.03	1.34	5.95	2.59
K	1.09	0.28	1.57	0.40
HCO_3	20.0	3.28	54.6	8.95
SO_4	1.00	0.10	2.38	0.25
Cl	0.50	0.14	1.06	0.30
F	0.07	–	0.09	–
NO_3	0.02	–	0.28	–
Dissolved Solids	36.0		75.0	
pH	6.2		6.8	

[a]Reproduced from Garrels and Mackenzie [24] courtesy of the American Chemical Society.

Garrels and Mackenzie [24] also suggested criteria for an idealized water that is derived from the CO_2 attack of siliceous rocks to produce only kaolinite. In terms of molar concentration, these criteria are as follows [24]:

(1) HCO_3^- should be the only anion, except for small concentrations of Cl^- and SO_4^{2-} from fluid inclusions in the minerals, oxidation of pyrite, and other minor sources.
(2) Na^+ and Ca^{2+} should be the chief cations, and the ratio of Na^+ to Ca^{2+} should be the same as that in the plagioclase of the rock.
(3) The total of Mg^{2+} and K^+ should be less than about 20% of the total of Na^+ and Ca^{2+}. The ratio of Mg^{2+} to K^+ should range around 1 to 1, with higher values related to higher percentages of mafic minerals, and lower values to higher ratios of K-feldspar to mafic minerals.
(4) The ratio of SiO_2 to Na^+ should be about 2 to 1, with somewhat higher ratios from rocks unusually high in K-spar and/or mafic minerals.
(5) The ratio of Na^+ to K^+ should be 5 to 1 or higher because of the abundance and high weathering rate of plagioclase, as opposed to the low weathering rate of K-feldspar and the generally low abundance of micas or other K-bearing phases.

It appears that the water quality of the ephemeral and perennial springs of the Sierra Nevada fits these criteria reasonably well.

Some other applications of silicate stability diagrams may be found in the publications of Sutherland et al. [25] (Lake Ontario of the Great Lakes System), Bricker et al. [26] (Pond Branch of Baisman Run, Baltimore County, Maryland) and Helgeson et al. [27] (an excellent discussion of a

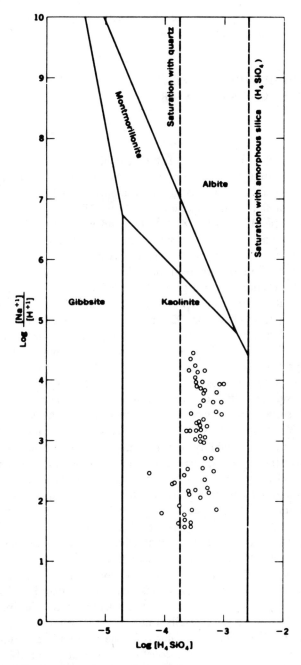

Figure 8.   Stability relations of phases in the system Na_2O-Al_2O_3-SiO_2-H_2O at 25°C and 1 atm total pressure as functions of $[Na^+]/[H^+]$ and $[H_4SiO_4]$ [23].

theoretical evaluation of irreversible reactions in geochemical processes involving minerals and aqueous solutions).

EQUILIBRIUM MODELS APPLIED
TO LARGE BODIES OF WATER

Fresh Waters

Kramer [1,28-30] has attempted to explain the water chemistry of the Great Lakes through equilibrium models. As a good first approximation, the various chemical constituents fitted a model involving the equilibria of calcite, dolomite, apatite, kaolinite, gibbsite, Na- and K-feldspars at 5°C, 1 atm pressure (P_{CO2} = 3.5 X 10^{-4} atm) and water. Conjecture was made that equilibrium should occur since the mixing times in the lakes range from less than 10 years to more than 200 years. The constituents (variables) considered by Kramer are seen in Table VI (reference is made to the appropriate reaction in Chapter 5). It is difficult to fit Cl^- and SO_4^{2-} into an appropriate equilibrium model. In one attempt [30], the dissolution of celesite ($SrSO_{4(s)}$) (Reaction 92, Table XIII, Chapter 5) was employed for SO_4^{2-}. This, however, was unsatisfactory, and Kramer assumed two values for the (Cl^- + $2SO_4^{2-}$) concentrations: 1.2 and 13.0 X 10^{-4} eq/l. These represent the lower and upper limits for the waters of the Great Lakes. Alkalinity was assumed to be controlled by an open calcium carbonate system, with the normal atmospheric partial pressure of $CO_{2(g)}$. Dissolved silica was assumed to be controlled by the kaolinite–gibbsite Reaction 38 (Table VIII, Chapter 5). An earlier publication of Kramer [1] had reported the solubility control through the dissolution of quartz. Apparently this reaction was considered unsatisfactory since dissolved silica was undersaturated with respect to quartz in the Great Lakes.

The equilibrium concentrations for the various constituents are given in Table VI for an invariant condition provided by the Phase Rule. Kramer held the temperature (5°C), pressure (1 atm) and P_{CO2} (3.5 X 10^{-4} atm) constant, which left no degrees of freedom for the concentrations of the dissolved constituents to vary. These equilibrium concentrations were calculated by starting with an expression for electroneutrality into which the appropriate equilibria expressions were substituted [1]. Two sets of concentrations are given for each constituent except dissolved silica in Table VI. The upper value results when the sum of equivalents of Cl^- and SO_4^{2-} is 1.2 X 10^{-4} and the lower value is for the 13.0 X 10^{-4} eq/l sum.

Kramer [30] considered each constituent separately for its equilibrium fit rather than an inclusive consideration. Figure 9 shows the plots for calcium carbonate, dolomite, phosphate and silica equilibria as a function

Table VI. Equilibrium Concentrations of Constituents Considered in
Kramer's Model of the Great Lakes at 5°C [30]

Constituents	Reaction[a]	Table[a]	Solid Phases	Equilibrium Concentrations	
				log M	mg/l
Ca^{2+}	12	6	Calcite	-3.15^{b}	27.0^{b}
				-3.00^{c}	40.0^{c}
Mg^{2+}	19	6	Dolomite	-3.6	6.0
				-3.45	8.5
Na^{+}	56	10	Microcline	-4.3	1.2
			Albite	-4.2	1.4
K^{+}	53	10	Microcline	-5.4	0.15
			Kaolinite	-5.3	0.20
H_4SiO_4	39	8	Kaolinite	-4.8	0.95
			Gibbsite	-4.8	0.95
P_t	77	12	Hydroxylapatite	-7.0	$3.0\ \mu g/l$
				-7.14	$2.4\ \mu g/l$
Alkalinity	–	–	–	-2.77 eq/l	–
				-2.85 eq/l	–
pH	–	–	–	8.4	–
				8.32	–

[a]Chapter 5.
[b]$(Cl^- + 2SO_4^{2-}) = 1.2 \times 10^{-4}$ eq/l.
[c]$(Cl^- + 2SO_4^{2-}) = 13. \times 10^{-4}$ eq/l.

of temperature. These results are most interesting. The waters of Lakes
Erie and Ontario are undersaturated most of the time with respect to calcite
and aragonite. There is some tendency for the waters to reach equilibrium
or supersaturation at temperatures of 13°C and higher. Calcite undersaturation appears to be the situation whenever the water temperatures
are normally below 13°C. These observations were explained by a mixing
phenomena: "deep waters are colder and tend to be poorly mixed with
respect to the atmosphere, whereas warmer waters, representing surface
waters, do tend to be well-mixed with respect to the atmosphere." It is
doubtful that mixing alone is the reason for the poor fit to the equilibrium
model for calcite. Several other reasons may be cited and are discussed
below. The fit to the dolomite model is similar to the calcite situation:
undersaturation at the colder waters and supersaturation at the warmer
temperatures. Phosphorus appears to be undersaturated (colder temperature
data only) with respect to hydroxylapatite in the lakewaters. However,

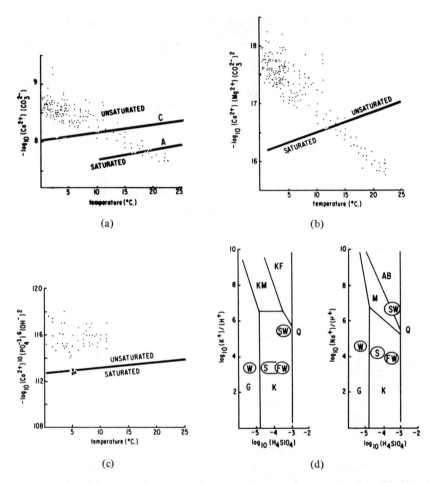

Figure 9. (a) Saturation of lake water with respect to calcite. Double line (C for calcite, A for aragonite) represents the solubility product as a function of temperature; (b) saturation of lake water with respect to dolomite; (c) saturation of lake water (winter only) with respect to hydroxylapatite as a function of temperature; and (d) silicate stability. KF, KM, G, K and Q are K+-feldspar, K+-mica, gibbsite, kaolinite and amorphous silica, respectively. M and AB are montmorillonite and albite. W, S, FW and SW represent areas of winter lake data, summer lake data, extracted freshwater sediments and extracted seawater sediments, respectively (reproduced from Kramer [30] courtesy of the American Chemical Society).

some sediment waters (x points in Figure 9c) were in equilibrium with this solid phase. Dissolved silica, Na^+, K^+ and H^+ concentrations were plotted for the winter data of Lake Ontario and for the summer data of the Great Lakes (Figure 9d). The winter waters plot in the gibbsite stability domain, whereas the summer waters fall into the kaolinite domain. This suggests an undersaturation of silica with respect to quartz, amorphous silica and kaolinite (winter data). It also suggests an incongruent dissolution of kaolinite (found in the bottom sediments of the Great Lakes) to gibbsite at the cooler water temperatures. This may be in error, however, since gibbsite has not been located in the bottom sediments. This observation may be due to the plot of water quality data collected at 5°C on 25°C stability diagrams. In summary of Kramer's effort, the water chemistry of the Great Lakes does not, in most situations, fit an equilibrium model. Many reasons may be cited for this failure, some of which are discussed below.

In another effort, several heterogeneous chemical equilibrium models were developed by Falls and Varga [31] to depict the concentrations of the principal inorganic species in Lake Keystone, Oklahoma. This lake is a riverine reservoir located on the Arkansas River in north central Oklahoma and is exposed to the geology of the Permian formation, which is composed chiefly of gypsum, halite and dolomite. It was hypothesized that the chemical composition of the Lake Keystone waters was acquired through the dissolution of these evaporites as well as the incongruent reactions of the common clay minerals. These equilibrium models were developed: an evaporite, a diluted clay-calcite, a diluted clay, a variable dilution clay-calcite, and a variable dilution clay model. In all models, 10 components were specified: H_2O, CO_2, HCl, CaO, MgO, SrO, Na_2O, SO_2, SiO_2 and Al_2O_3. Nine phases were employed to reconstruct the system: gypsum, dolomite, calcite, strontium sulfate (celestite), kaolinite, sodium and calcium montmorillonite, the solution, and air-containing variable partial pressures of $CO_{2(g)}$. This yielded a model system containing 10 components and 9 phases which, in turn, permitted three independent variables under the Phase Rule: average temperature, 15°C, a pressure of 1 atm and a [Cl⁻] of $10^{-1.5}M$ (saturation with respect to halite was not observed). A series of equations was developed by substitution of the appropriate equilibrium expression into the equation for electroneutrality. The partial pressure of $CO_{2(g)}$ was introduced as a variable since good agreement between the observed data and the equilibrium model was not observed with the normal pressure. A typical equation is as follows:

$$\frac{4}{P_{CO_2}}(10^{9.96} + 10^{9.73} + 10^{7.90})(H^+)^4 + 1.25\left(1 + \frac{10^{5.17}}{(P_{CO_2})^{0.125}}\right)(H^+)^3 =$$

$$(Cl^-)(H^+)^2 + 1.25(10^{-14.35} + 10^{-7.25}P_{CO_2}(H^+)) + 4P_{CO_2}(10^{-18.18} + 10^{-14.40}) \quad (40)$$

This equation was solved for (H^+) at several values of P_{CO_2}, whereupon the remaining constituent concentrations were computed. It was necessary to consider several modifications of the original model to provide a good fit of the observed water quality data: (1) use of a dilution factor for precipitation and runoff; (2) reduction in the number of phases with exclusion of calcite in one model; and (3) an increase in the number of variables. This led to a "variable dilution clay-calcite" model (calcite and three clay minerals were the only solid phases considered) and to a "variable dilution clay" model (three clay minerals only). Results from these modified models are seen in Figures 10 and 11. It appears that Na^+, Mg^{2+}, Sr^{2+}, SO_4^{2-} and HCO_3^- fit the "variable dilution" models reasonably well with lesser agreement for Ca^{2+} and H^+. Note the use of a P_{CO_2} of $10^{-2.9}$ atm, which is an order of magnitude higher than normal.

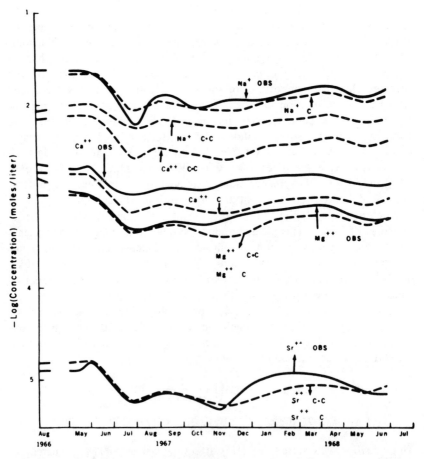

Figure 10.   Comparison of observed concentrations of major cations with the variable dilution model (reproduced from Falls and Varga [31] courtesy of the American Chemical Society). Subscripts OBS, C-C and C refer to the observed, clay-calcite model and clay model. $P_{CO_2} = 10^{-2.9}$ atm.

These results should be considered as a reasonably good approximation of naturally occurring conditions approaching equilibrium. No rational explanation may be offered for the deviations reported in this specific case study.

Truesdell and Singers [32] applied a chemical equilibrium model to an unusual water, namely, a hot water geothermal system. Assumptions were made that the fluid, 246°C and 37.6 bars abs., was produced from a single

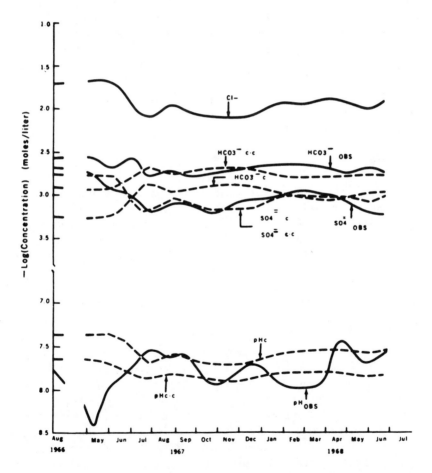

Figure 11. Comparison of observed concentrations of major anions and pH with concentrations calculated from the variable dilution equilibrium models (reproduced from Falls and Varga [31] courtesy of the American Chemical Society). Subscripts C-C, C and OBS refer to clay-calcite model, clay model and observed values. $P_{CO_2} = 10^{-2.9}$ atm.

aquifer and was homogeneous in enthalpy and chemical composition. Calculations of aquifer chemistry required many iterations with computer methods (PL-1, mentioned below). Data required included:

1. chemical analysis of the water separated from the water-stream mixture;
2. the contents of $CO_2 (g)$ and $H_2S_{(g)}$ in the separated stream;
3. the pressures of water and stream separation and the atmospheric pressure; and
4. the enthalpy of the whole fluid.

Of course the water analyses included the pH value, the temperature of the pH measurement and concentrations of all major dissolved constituents including silica. Tables VII and VIII give the appropriate reactions, thermodynamic equations, equilibrium constants and CO_2 and quartz solubilities over the temperature range 0 to 350°C. This paper represents a unique source of information for high-temperature aqueous chemical equilibrium models.

Some additional general concepts and applications of chemical equilibrium models to surface and ground waters have been reported by Jones et al. [33], Hem [34] and Stumm and Leckie [35].

Saline Waters

Sillén [36-38] was perhaps the first to propose an equilibrium model for the chemical composition of sea water. In doing so, Sillén challenged some traditional concepts of ocean chemistry. For example, the pH value of the ocean is 8.1 ± 0.2, which, presumably, has remained constant "during the last few hundred million years." It was commonly accepted that this pH value is regulated by the ocean's carbonate buffer system. Sillén took exception when he proposed "that the fine-grained silicate clay minerals have very great buffering capacity and may well determine the pH of the ocean" [37]. Needless to say, the traditionalists did not accept this proposal.

The basis for Sillén's ocean model comes from Goldschmidt's geochemical balance for formation of the sea [39]. It was suggested that, for each liter of present-day sea water, approximately 600 g of igneous rock had reacted with about 1 kg of such volatile substances of H_2O, HCl, CO_2, etc. This produced 1 liter of sea water, 600 g of sediment and 3 liters of air. Figure 12 shows the real world estimates of Goldschmidt and the equilibrium model of Sillén that is assumed to have reached "true equilibrium." Into this model Sillén fitted several solid phases, which, in turn, established the composition of the liquid and gas phases.

Sillén [37,38] actually proposed three models: (I) a simplified ocean model with five components and five phases; (II) an intermediate model with eight components and seven phases; and (III) a complete ocean model

Table VII. Solubility Constants and Dissociation Constants in KA(I) and KT(I) Arrays, Analytical Expressions [32]

Reaction	Analytical expression, if used (T in K, t in °C)	Log K 25°	Log K 250°	Note
(41) H_2CO_3 app = H^+ + HCO_3^-	Log K = -2382.3/T + 8.153 - 0.02194T	-6.38	-7.88	a
(42) H_2S = H^+ + HS^-	Log K = -3279.0/T + 11.17 - 0.02386T	-6.94	-7.6	
(43) HBO_2 + OH^- = $H_2BO_3^-$	Log K = 1573.21/T + 28.6059 + 0.012078T - 13.2258 log T	+4.76	+1.98	b
(44) H_4SiO_4 = H^+ + $H_3SiO_4^-$	See Table VIII	-9.63	-9.63	c
(45) HF = H^+ + F^-	See Table VIII	-3.18	-5.80	d
(46) HSO_4^- = H^+ + SO_4^{-2}	Log K = -557.2461/T + 5.3505 - 0.0183412T	-1.99	-5.31	
(47) H_2O = H^+ + OH^-	Log K = -4470.99/T + 6.0875 - 0.01706T	-13.995	-11.38	
(48) $NH_3(H_2O)$ = NH_4^+ + OH^-	See Table VIII	-4.75	-6.00	e
(49) HCl = H^+ + Cl^-	See Table VIII	-6.10	-0.67	
(50) NaCl = Na^+ + Cl^-	See Table VIII	+1.60	-0.25	e
(51) KCl = K^+ + Cl^-	See Table VIII	+1.59	-0.1	
(52) $MgSO_4$ = Mg^{+2} + SO_4^{-2}	See Table VIII	-2.25	-5.7	f
(53) $CaSO_4$ = Ca^{+2} + SO_4^{-2}	See Table VIII	-2.30	-4.1	c
(54) KSO_4^- = K^+ + SO_4^{-2}	See Table VIII	-0.83	-2.35	e
(55) $NaSO_4^-$ = Na^+ + SO_4^{-2}	See Table VIII	-0.83	-2.35	f
(56) $CaCO_3$ = Ca^{+2} + CO_3^{-2}	See Table VIII	-2.30	-5.90	c
(57) $MgOH^+$ = Mg^{+2} + OH^-	See Table VIII	-2.60	-4.65	c
(58) $H_3SiO_4^-$ = H^+ + $H_2SiO_4^{-2}$	Log K = -3450/T + 6.34 - 0.0216T	-11.7	-11.5	
(59) HCO_3^- = H^+ + CO_3^{-2}	Log K = -2730.7/T + 5.388 - 0.02199T	-9.12	-11.34	
(60) H_2S gas = H_2S aq	K = 357 + 15.688t - 0.038253t^2	+2.86	+3.28	
(61) CO_2 gas + H_2O = H_2CO_3 app	See Table VIII	+3.21	+3.72	a

aH_2CO_3 apparent includes H_2CO_3 and CO_2 aqueous.
bH_2BO_3 given in print out as BO_2^-.
cExtrapolated above 200°C.
dExtrapolated above 200°C.
eExtrapolated below 100°C.
fAssumed identical to K KSO_4.

Table VIII. pK Values and CO₂ and Quartz Solubilities for Table VII [32]

°C	$H_4SiO_4^0$	HF^0	$NH_3(H_2O)^0$	HCl^0	$NaCl^0$	KCl^0	$MgSO_4^0$	$CaSO_4^0$	KSO_4^-	$CaCO_3^0$	$MgOH^+$	CO_2 Solubility[a]	Quartz Solubility[b] (ppm)
0	10.2	2.96	4.87	-7.5	-1.65	-1.65	2.05	2.3	0.65	3.0	2.58	700	2.4
10	9.94	3.00	4.80	-6.8	-1.63	-1.62	2.10	2.3	0.71	3.05	2.58	1000	3.6
25	9.63	3.18	4.75	-6.1	-1.60	-1.59	2.25	2.3	0.83	3.20	2.60	1630	6.6
35	9.48	3.25	4.70	-5.7	-1.40	-1.50	2.35	2.35	0.90	3.27	2.63	2100	8.6
50	9.30	3.40	4.70	-5.0	-1.20	-1.40	2.60	2.40	1.00	3.40	2.7	2900	13.5
75	9.11	3.64	4.75	-3.8	-0.90	-1.20	2.90	2.55	1.15	3.65	2.9	4000	27
100	9.03	3.85	4.85	-2.9	-0.55	-1.00	3.20	2.7	1.30	3.90	3.1	5200	48
125	9.03	4.09	4.97	-2.0	-0.55	-0.90	3.55	2.9	1.45	4.15	3.33	6000	80
150	9.10	4.34	5.10	-1.23	-0.45	-0.75	3.90	3.1	1.60	4.50	3.6	6600	125
175	9.23	4.59	5.33	-0.60	-0.30	-0.60	4.40	3.35	1.78	4.85	3.85	6800	190
200	9.36	4.89	5.53	-0.07	-0.15	-0.40	4.80	3.6	1.93	5.20	4.1	6400	265
225	9.48	5.3	5.73	0.30	0.05	-0.20	5.25	3.8	2.10	5.55	4.35	5900	367
250	9.63	5.8	6.0	0.67	0.25	0.10	5.7	4.1	2.35	5.90	4.65	5300	490
275	9.83	6.2	6.3	0.95	0.60	0.30	6.1	4.3	2.55	6.20	4.9	4600	615
300	10.2	6.8	6.75	1.2	0.95	0.6	6.4	4.5	2.75	6.45	5.15	3900	680
325	10.5	7.1	7.25	1.6	1.35	1.0	6.7	4.75	2.9	6.65	5.45	3100	720
350	11.0	7.4	8.0	2.5	2.0	1.7	7.0	5.0	3.1	7.0	5.7	2100	(750)

[a]Henry's law constants (see p. 82).

[b]Quartz solubility in water at saturated water vapor pressure. 0 to 240°C. An incorrect 350° value is included because the lookup subroutine requires a monotonic function.

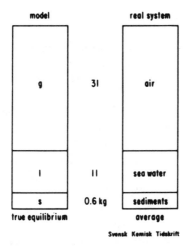

Svensk Kemisk Tidskrift

Figure 12. The true system (right) of 1 liter of sea water with the corresponding 0.6-kg sediments and 3 liters of air. On the left, the equlibrium model with the same amounts of the various components at true equilibrium (with the exception of N_2) [39].

with nine components and nine phases. Model (III) is the only one considered here since it is apparently the most successful. The nine components are: HCl, H_2O, CO_2, $NaOH$, KOH, CaO, MgO, SiO_2 and $Al(OH)_3$. The nine phases are: (1) gas phase, (2) aqueous solution, (3) quartz, (4) kaolinite, (5) illite, (6) chlorite (a Mg-rich layer mineral), (7) montmorillonite (Na-rich), (8) calcite, and (9) phillipsite (a Ca-rich clay mineral). Two degrees of freedom are calculated for this model from the Phase Rule ($F = 9 - 9 + 2 = 2$). Since Sillén chose temperature and $[Cl^-]$ for the two independent variables, composition of the aqueous and gaseous phases is established automatically.

Sillén was severely criticized for injection of the silicate clay minerals into the ocean model. The major objection was that silicates react too slowly to approach an equilibrium condition. However, there is some experimental support in the literature that suggests ion exchange reactions are complete within 24 hours and phase transformations of montmorillonite to chlorite and illite occur within "a few years." In addition, there is considerable evidence that solid phases (3) through (8) cited above exist in large amounts in oceanic sediments. The weakest part of Sillén's equilibrium model lies with the existence of phillipsite. Perhaps the strongest argument supporting the use of silicates comes from Reaction 55 of Table X, Chapter 5, (reversed below):

$$1.5 Al_2 Si_2 O_5 (OH)_{4(s)} + K^+ = H^+ + 1.5 H_2 O + KAl_3 Si_3 O_{10}(OH)_{2(s)}$$

The equilibrium constant for this reaction is defined by the $[H^+] : [K^+]$ ratio that was given a value in the $10^{-6.0}$ to $10^{-6.5}$ range for sea water temperatures by Hemley [40]. Sillén noted that the $[H^+] : [K^+]$ ratio in sea water is about $10^{-6.2}$.

Another interesting feature of Sillén's nine-component, nine-phase ocean model is the nonvariance of the partial pressure of $CO_{2(g)}$. This occurs because there are only two degrees of freedom in this model for which Sillén selected temperature and $[Cl^-]$. A constant P_{CO_2} is inconsistent with the arguments that the CO_2 content of the atmosphere has been slowly increasing in recent years due to pollution from combustion of fossil fuels.

Sillén's major contribution was the suggestion and subsequent introduction of equilibrium concepts to explain ocean chemistry. It was his feeling "that the present composition of ocean water does not result just from blind chance." Well-defined equilibria will give the composition. On the other hand, Sillén recognized that this approach is not perfect since there will be deviations from equilibrium and some of the equilibrium data are inaccurate, especially the silicate systems. Nevertheless, Sillén was very confident that with time and more accurate information equilibrium chemistry would account for the composition of the oceans.

Kramer [41] followed Sillén's example and derived an inorganic ocean from chemical equilibrium models. First, several assumptions were made concerning the rather large, oceanic body of water. A dynamic hydrologic cycle is operative in which chemical constituents are carried constantly into the oceans from land runoff, where precipitation of several solid phases occurs. Evaporation returns water from the oceans into the hydrologic cycle. The net effect of these three processes tends towards increasing concentration of all constituents in sea water. Precipitation and sedimentation processes tend to maintain an equilibrium concentration of the constituents with respect to a solid phase. One, homogeneous liquid phase was assumed through the mixing action of high-velocity currents. Also, constant temperature (average of 3.8°C) and pressure (1 atm) were assumed.

Only the major inorganic constituents were considered by Kramer [41] for his ocean model. These are seen in Table IX with the attendant solid phases. The choice of the silicate phases was based on field and laboratory studies that showed that: (1) montmorillonite and kaolinite are stable in weathering zones, (2) chlorite and illite are stable in the marine environment, (3) anorthigenic feldspars are formed in the marine environment, and (4) phillipsite is formed in the marine environment. Of course, these phases would control the cationic activity in the oceans. Carbonates are in equilibrium with calcite and/or aragonite. Dolomite was not considered

Table IX. Solid Phases in Equilibrium with Sea Water[a]

Liquid Phase	Solids Controlling Concentrations			
	Lower Limit	Most Probable	Upper Limit (a)	(b)
Na^+	Na-mont (C site)	Na-mont (E site)	Na-mont (E site)	
K^+	K-illite (C site)	K-illite (E site)	K-illite (E site)	
Cl^-	O (assumed)	0.55 (assumed)	0.55	0.77
			(assumed)	
SO_4^{2-}	O (assumed)	$SrCO_3$; $SrSO_4$	Gypsum	
Ca^{2+}	Ca-clay	Phillipsite	Phillipsite	
Mg^{2+}	Mg-clay	Chlorite	Chlorite	
PO_4^{3-}	OH-apatite	OH-apatite	OH-apatite	
CO_2	Calcite	Calcite	Aragonite	
F^-	F-CO_2-apatite	F-CO_2-apatite	F-CO_2-apatite	
H^+	Electroneutrality	Electroneutrality	Electroneutrality	

[a]Reproduced from Kramer [41] courtesy of *Geochim. Cosmochim. Acta.*

because it does not form in the oceans, according to Kramer. Control of phosphate solubility was assigned to hydroxylapatite and fluoride to a "CO_2-F-apatite" $(Na_{0.296} Ca_{9.56}) ((PO_4)_{5.35}(SO_4)_{0.30}(CO_3)_{0.33}) (F_{2.04})$. Both of these have been found in recent marine sediments. Sulfate solubility was given to gypsum as an "upper limit" with $SrSO_4$-$SrCO_3$ as "the most probable" regulator. Chloride appears to have no solubility regulation in water, whereupon Kramer argues that it was probably derived from "degassing" the earth's crust.

The results from Kramer's three models are seen in Table X. The "most probable" model fitted the ocean values reasonably well with the exceptions of Ca^{2+}, Mg^{2+}, carbonate alkalinity, CO_2 and F^-. These poor fits were blamed on inaccurate equilibrium constants and, perhaps, some erroneous assumptions about the reactions with the clay minerals. Kramer felt that his most important conclusion came from the role of Cl^- in changing "fresh" waters into "marine" waters. Ancient ocean waters may have been fresh due to no or little Cl^-. This is implied in the results of the "lower limit" model. As more and more Cl^- was added to the "fresh" waters, additional ions were added through runoff. Eventually, the present equilibria and reactions were attained. A rational explanation may be the one offered by Stumm and Morgan [17], whereby saline waters result from the reaction of a strong acid, HCl (volcanic gases), with rocks and minerals, whereas "fresh" waters form from reactions of weak acids, H_2CO_3 and rocks and from various hydrolysis reactions.

Table X. Composition of Sea Water Models (all values are in mol/l except P_{CO_2}, atm and carbonate alkalinity, eq/l)[a]

			Models		
Ion	Lower Limit	Most Probable	Upper Limit[a]	Upper Limit[b]	Present Ocean
Na^+	7.9×10^{-6}	0.45	0.70^a	0.68^b	0.47
K^+	1.4×10^{-6}	9.7×10^{-3}	1.1×10^{-2}	1.5×10^{-3}	1.0×10^{-2}
Ca^{2+}	7.9×10^{-4}	6.1×10^{-3}	1.2×10^{-2}	1.2×10^{-2}	1.0×10^{-2}
Mg^{2+}	3.5×10^{-4}	6.7×10^{-2}	0.20	0.23	5.4×10^{-2}
F^-	3.4×10^{-6}	2.4×10^{-5}	9.9×10^{-5}	2.1×10^{-5}	7.0×10^{-5}
Cl^-	0 (defined)	0.55 (defined)	0.55 (defined)	0.77 (defined)	0.55
SO_4^{2-}	0 (defined)	$3.4 \times 10^{-2}\,b$	0.29	0.21	3.8×10^{-2}
pH	8.33	7.95	7.88	7.81	7.89
CO_3^{2-} Alkalinity	2.1×10^{-3}	4.3×10^{-4}	9.0×10^{-3}	7.8×10^{-3}	2.3×10^{-3}
P_{CO_2}	4.9×10^{-4}	1.7×10^{-3}	1.7×10^{-3}	2.5×10^{-3}	4.0×10^{-4}
Total P	1.3×10^{-7}	2.7×10^{-6}	2.4×10^{-6}	2.1×10^{-6}	1.5×10^{-6}
I^c	3.6×10^{-3}	0.66	0.89	1.2	0.65

[a]Reproduced from Kramer [41] courtesy of Geochim. Cosmochim. Acta.
[b]Strontium value for this determination is 5.5×10^{-4}; ocean value is 4.0×10^{-4}.
[c]I = ionic strength.

Substantial support of Sillén's model came from Holland [42], who described his paper as "essentially a continuation and extension of Sillén's approach to the factors that control the chemistry of ocean water at present." Holland relied on the publications and data of other investigators for his supportive arguments. Only the major points are reported here. To substantiate a case for silicate minerals, the river sediments would have to undergo extensive base exchange once they are carried into the oceans. For example, the Ca-rich river sediments would be exchanged for Mg^{2+}, Na^+ and K^+ on contact with ocean water. Carroll and Starkey [43] demonstrated that base exchange of a number of clays in contact with ocean water was completed within "a few tens of hours." Another role of river sediments in ocean waters would involve the reconstitution of degraded illites and chlorites. This requires an increase in the K and Mg content of these minerals. Holland cited two studies of sediments in the James River and Rappahannock River to support the reconstitution mechanism. A third, and perhaps the most important, reaction involves the formation of new clay minerals in the oceans. Again, Holland cites a couple of studies that suggested montmorillonite may react slowly in sea water to form illite and chlorite. These two clay minerals are essential components of the ocean models of Sillén [37] and Kramer [41]. One of the strongest pieces of evidence cited by Holland comes from this reaction [44]:

$$3Na\text{-}Mont._{(s)} + H^+ + 3.5H_2O = Na^+ + 4SiO_{2(s)} + 3.5Al_2Si_2O_5(OH)_{4(s)} \qquad (62)$$

The ratio $(Na^+):(H^+)$ in a solution in equilibrium with sodium montmorillonite, kaolinite and quartz should be $10^{7.0} \pm 0.5$ at 25°C. In the present oceans, the $(Na^+):(H^+)$ ratio is $10^{7.7}$. Holland considered the agreement between the actual and the theoretical ratios to be excellent. In another important area, Holland supports Sillén's contention that the P_{CO_2} of the atmosphere is fixed by reactions occurring in the oceans. It is argued that the CO_2 pressure in the atmosphere is and has been crudely buffered by the coexistence of chlorite, dolomite, calcite and quartz. No experimental data are offered to support this argument. In summary, Holland feels that three distinct groups of constituents are encountered in the major element composition of the ocean atmosphere system. The first group, the $[Cl^-]$ of sea water and the N_2 and rare gas content of the atmosphere, was affected by degassing of the earth. The second group is controlled by mineral equilibria: the major cation ratios, pH value of sea water and the CO_2 content of the atmosphere. Biological reactions may be responsible for the third group, which is the O_2 content of the atmosphere and the $[SO_4^{2-}]$ of sea water.

Several other investigators have added important knowledge to the chemical composition of sea water. Garrels and Thompson [45] provided a chemical model for the major dissolved species at 25°C and 1 atm total pressure. Mackenzie and Garrels [46] attempted a chemical mass balance between rivers and oceans to account for the removal of excesses of dissolved constituents being carried by streams into the oceans. This model requires the synthesis of typical clay minerals from degraded aluminosilicates before burial in sediments through such a reaction as

$$\text{amorphous Al silicate} + SiO_2 + HCO_3^- + \text{cations} = \text{cation Al silicate} + CO_2 + H_2O \qquad (63)$$

This synthesis implies control of the major ion ratios in seawater and the CO_2 pressure of the atmosphere by equilibria involving aluminosilicates. Chave [47] summarizes the state-of-the-art of oceanic equilibrium models and offers some interesting observations concerning carbonate chemistry. The carbonate system(s) in sea water is generally not in equilibrium. Most surface and near-surface waters are supersaturated with respect to all stable calcium and magnesium carbonate phases. There is little evidence, however, that precipitation of any carbonates occurs to a significant extent. Deep waters, several thousand meters down, are undersaturated with respect to all $CaCO_3$ phases despite an 11% by weight average calcite content of sediments underlying these waters. Chave [47] presents some evidence that organic matter in sea water may inhibit $CaCO_3$ precipitation. Another observation is made whereby ocean water is two to three orders of magnitude supersaturated with respect to dolomite, yet this carbonate solid phase is rarely found in ocean sediments. Finally, Chave suggests that regulation of the Ca^{2+} content of sea water is a complex system involving biological and chemical processes. So it is with equilibrium models to describe ocean chemistry.

COMPUTERIZED CHEMICAL MODELS

Programs

The interpretation of naturally occurring chemical processes in aquatic systems may necessitate the use of computer programs, especially where calculations are performed in multicomponent, multiphase systems. First, however, a definition of an "aqueous chemical model" is offered [48]: "it is a theoretical construction which allows prediction of the thermodynamic properties of electrolyte solutions." Since there are many ways to construct a model, consequently there are several computer programs available. Selection of the appropriate model is made on the basis of the problem to be solved. Each model, however, has its own set of assumptions and limitations. Almost all computerized models are based on the ion association theory by which the distribution of chemical species is formulated in two distinct thermodynamic approaches: the equilibrium constant and the Gibbs free energy (see above). These approaches are subject to conditions of mass balance and chemical equilibrium. The mass balance requires the sum of the free and complexed species to be equal to the total concentration. Chemical equilibrium requires that the most stable arrangement must be found for the system under examination. This is defined by equilibrium constants for all mass action expressions of the system or by using Gibbs free energy of formation values for all components of the system. If the equilibrium approach is employed, then the mass action expressions are substituted into the mass balance conditions. This results in a set of nonlinear equations that must be solved simultaneously. The Gibbs free energy approach uses the following thermodynamic relation:

$$\Delta Gr = \Delta Gr^0 + RT \ln Keq = 0 \qquad (64)$$

Here, the total Gibbs free energy function is minimized for a particular set of species and their mole numbers are subject to the mass balance requirements. Of course it is necessary to have a data base of equilibrium and standard free energy of formation values. According to Nordstrom et al. [48], there are more reliable equilibrium constants available than the free energy values at this time. There are several numerical techniques for solution of equilibrium equations or Gibbs free energy functions. The reader is referred to the work of Nordstrom et al. [48] for details of these techniques.

Garrels and Thompson [45] were perhaps the first to use the numerical technique of successive approximation in a hand calculation of a chemical model for sea water. In turn, this sea water model influenced the development of others. Some of the early computer programs were:

1. EQUIL from Fritz and Droubi (see Nordstrom et al. [48] for original citation);
2. WATEQ from Truesdell and Jones [49];
3. SOLMNEQ from Kharaka and Barnes [50];
4. EQ3 from Wolery [48].

The last three programs were designed to accept water quality analyses with onsite values for pH, Eh and temperature. They are limited because only mass balances on cations and anions were conducted and there is no proton mass balance condition. SOLMNEQ has a data base of tabulated equilibrium constants over the range of 0 to 350°C. Similarly, EQ3's base covers the 0 to 300°C range. WATEQ uses the van't Hoff equation (see Chapter 2, p. 75) or analytical expressions for equilibrium constants over the 0 to 100°C range. WATEQ has been updated and written into FORTRAN by Plummer, et al. [51] into a program called WATEQF. In turn, WATEQF has been

expanded into WATEQ2 [52]. A shorter version of WATEQ2, called
WATSPEC, was prepared by Wigley [53] for handling routine water quality
data. There are several other programs of this general type:

1. SEAWAT [48] is designed specifically for sea water calculations.
2. MIRE [48] is designed specifically for anoxic marine pore waters.
3. IONPAIR and NOPAIR [48] and CALCITE [48] are designed for fresh
    waters in carbonate geologies.
4. KATKLE-1 [48] is designed for soil waters.

Morel and Morgan [54] produced the FORTRAN program REDEQL
for computation of multicomponent metal–ligand equilibria. It includes
a large number of metal–ligand complexes. Equilibrium constants are
employed with the Newton–Raphson iteration method for solution to a
function that compares the difference between a component's total calculated concentration and the total analytical concentration. Several modifications of the REDEQL have led to several "second generation" programs:
REDEQL2, MINEQL and GEOCHEM [48]. The major feature of these
latter three programs is the inclusion of adsorption behavior of ions on
various surfaces. Nordstrom et al. [48] have prepared an excellent summary
of the computer models.

Applications

The "state-of-the-art" in the computation of the equilibrium distribution
of various species in aquatic systems was tested by Nordstrom et al. [48].
In an effort to examine the consistency (or lack of) of the several computer
programs, two test cases (a dilute river water and an average sea water
analyses) were compiled and distributed to more than 50 researchers in
chemical modeling. The intent was to examine the results predicted by
the various models. Consequently, the differences among these models
were seen immediately in terms of actual results. Furthermore, in most
instances these differences were due to thermodynamic data bases for
the several models. Also, some differences may have resulted from the
specific purpose and design priorities of each model. Nordstrom et al. [48]
compiled and presented several tabulations of such items as: (1) general
description of the computer programs, (2) the chemical analyses of the
river and sea water, and (3) the results of computing the concentrations of
various major and minor species. Here it will suffice to present the
differences of the various models by the computed saturation indices (SI)
for selected minerals in the river water test case. That there are significant
differences in SI values from several models is seen in Table XI.

What, then, are the limitations of the chemical models that lead to
variable results? A summary of the comments by Nordstrom et al. [48] is
presented here. First, "no model will be better than the assumptions upon

Table XI. Saturation Index for Selected Minerals in River Water Test Case[a]

Mineral	Formula	EQUIL	EQ3	IONPAIR	MIRE	SOLMNEQ	WATEQF	WATEQ2	WATSPEC
Calcite	$CaCO_3$	-0.51	-0.585	-0.673	-0.461	-0.765	-0.634	-0.634	-0.63
Dolomite	$CaMg(CO_3)_2$	-0.73	-0.248	-1.340	-0.730	-1.329	-1.384	-1.386	-1.38
Siderite	$FeCO_3$	-1.37	+0.456	–	-2.329	-3.377	-7.347	-1.760	-7.13
Rhodochrosite	$MnCO_3$	-1.81	-3.225	–	-2.097	-2.136	-2.180	-2.180	–
Gypsum	$CaSO_4 \cdot 2H_2O$	-3.00	-2.962	–	-3.081	-2.942	-3.057	-2.969	-3.25
Celestite	$SrSO_4$	–	–	–	–	–	–	–	–
Hydroxylapatite	$Ca_5(PO_4)_3OH$	+2.82	–	–	+5.891	+5.046	-1.784	-1.722	–
Fluorite	CaF_2	-4.39	–	–	–	-3.338	-3.079	-3.074	–
Ferric Hydroxide (am)	$Fe(OH)_3$	–	–	–	–	-7.584	+1.304	–	–
Goethite	$FeO(OH)$	+5.59	–	–	–	-1.484	+7.810	+6.843	+5.06
Hematite	Fe_2O_3	+8.11	+18.332	–	–	-3.252	+15.144	+13.223	+13.52
Gibbsite (crypt.)	$Al(OH)_3$	-0.08	+1.948	–	–	-0.058	-0.336	-1.989	-1.19
Birnessite	MnO_2	–	–	–	–	–	-4.114	-4.114	–
Chalcedony	SiO_2	–	+0.490	–	–	+0.217	-0.142	+0.189	–
Quartz	SiO_2	+0.47	+0.776	–	–	+0.697	+0.405	+0.736	+0.74
Kaolinite	$Al_2Si_2O_5(OH)_4$	+1.83	+5.826	–	–	–	+1.638	-1.021	+2.32
Sepiolite	$Mg_2SiO_{7.5}(OH) \cdot 3H_2O$	-4.55	–	–	–	-5.734	-3.699	–	–
FeS Amorphous	FeS	–	–	–	–	–	-7.644	-2.313	–
Mackinawite	FeS	–	–	–	–	–	-6.928	-1.580	–

[a]Reproduced from Nordstrom et al. [48] courtesy of the American Chemical Society.

which it is based." For example, activity coefficients are used to describe the nonideal behavior of aquatic electrolytes. In turn, these activity coefficients are approximated by semiempirical equations, which lead to different ionic activities. The second limitation is the reliability of the equilibrium constants and/or free energy values employed in the various models (see below for discussion of this point). Third, there is another limitation imposed by assumption(s) made about the redox condition of the aquatic system under examination. Whatever redox potential is assumed or selected to dominate the chemical equilibrium will, in turn, affect the oxidation state of the ions. A fourth limitation is the total number of complexes employed by a given aquatic model. Some will have two or three times as many metal–ligand complexes as another model. Another major source of difference comes from the variable handling of the carbonate system by the models. For example, a titration of the alkalinity usually suffices to determine the inorganic carbonate(s) of natural waters. In many instances this is not the situation because there must be a correction of the total alkalinity for noncarbonate alkalinity (see Chapter 3). Lastly, there is the limitation that not all models correct for temperature. For example, the programs emanating from REDEQL contain a data base of equilibrium constants at $25°C$ and are not reliable at other temperatures. It would be extremely desirable to use equilibrium constants derived at temperatures of the natural system. In summary, Nordstrom et al. [48] indicate that "the largest single source of discrepancy is the thermodynamic data base used by each model." This may be seen by comparing the SI values in Table XI. Reasonably good agreement may be seen with the calcite and gypsum SI values. On the other hand, several minerals show supersaturation and undersaturation: hydroxylapatite, goethite, hematite and kaolinite. Needless to say, the application and interpretation of chemical models must be approached with caution.

The computerized aquatic chemical models of Truesdell and Jones [49], WATEQ and Plummer's WATEQF [51] were revised extensively by Ball et al. [52,55] into WATEQ2. This program contains an extensive thermodynamic base for which 537 chemical reactions are listed [55]. It is written in PL/1 and is compatible with the IBM 370/155 to 168 system. The complete program is given by Ball et al. [55].

WATEQ2 is an excellent computerized model for trace and major element speciation and mineral equilibria in natural waters. The revisions and expansions include: ion association and solubility equilibria for many trace metals: Ag, Cd, Cu, Mn, Ni, Pb and Zn; solubility equilibria for various metastable and/or slightly soluble solids; calculation of redox potentials for various couples; and a mass balance section for sulfide solutes. Also, the base of thermodynamic data was critically evaluated and updated.

The computer program GEOCHEM considers trace metal equilibria in contaminated soil solutions [56]. It was adapted for soil systems from the REDEQL2 program by Morel and Morgan [54]. GEOCHEM includes: data for a "few hundred" additional soluble complexes and solids; a subroutine for cation exchange processes; and another subroutine for the estimation of single ion activity coefficients at ionic strengths up to 3M. Mattigog and Sposito [56] also discuss four important theoretical problems and the lack of data for: (1) stability constants for trace metal complexes with many inorganic and mixed ligands; (2) stability constants of trace metal complexes with naturally occurring organic ligands; (3) solubility product constants for clay minerals in soils; and (4) thermodynamic exchange constants and exchange phase activity coefficients. Two applications of GEOCHEM are given for the calculation of trace metal equilibria in a mixture of irrigation water and a geothermal brine and in the aqueous phase of a soil amended with sewage sludge.

Nordstrom et al. [57] utilized the chemical model WATEQ2 to evaluate the equilibrium behavior of iron in acid mine waters (also see Chapter 4). The various aqueous complexes and solid phases of iron are given in Table XII (see Ball et al. [55] for thermodynamic data). Since an excellent correlation was found between calculated and measured Eh values, it is possible to compute the distribution of iron species based on the latter Eh values when Fe^{2+}/Fe^{3+} analyses are lacking. In addition, the authors conclude that the chemical equilibrium model approach enabled the identification of which parts of an acid mine drainage system are in equilibrium and which are not.

Table XII. Aqueous Complexes and Solid Phases of Iron Considered for Computation[a]

a) Complexes

$FeOH^+$, $Fe(OH)_2^0$, $Fe(OH)_3^-$, $FeSO_4^0$

$FeOH^{2+}$, $Fe(OH)_2^+$, $Fe(OH)_3^0$, $Fe(OH)_4^-$, $Fe_2(OH)_2^{4+}$, $Fe_3(OH)_4^{5+}$ $FeSO_4^+$, $Fe(SO_4)_2^-$

$FeCl^{2+}$, $FeCl_2^+$, $FeCl_3^0$, FeF^{2+}, FeF_2^+, FeF_3^0

b) Solids

Melanterite	$FeSO_4 \cdot 7H_2O$
Copiapite	$Fe^{2+}Fe_4^{3+}(SO_4)_6(OH)_2 \cdot 20H_2O$
Coquimbite	$Fe_2(SO_4)_3 \cdot 9H_2O$
Jarosite	$KFe_3(SO_4)_2(OH)_6$
Ferric hydroxide, amorphous	$Fe(OH)_3$

[a]Reproduced from Nordstrom et al. [57] courtesy of the American Chemical Society.

Boulegue and Michard [58] have developed a computer program to describe the chemical processes of dissolved sulfur in a reducing environment (also see Chapter 4). A simulation model of the weathering of iron pyrite ($FeS_{2(s)}$) was attempted in the pH range of 5.0 to 9.0. This model described the evolution in time and space of the composition of water percolating through a pyritic rich sandstone. In addition, there was a weathering profile of neoformed minerals from the sandstone. There was good agreement between the computer simulation of the weathering of a pyritic sandstone and an experimental study of pyrite oxidation. This work of Boulegue and Michard [58] represents an unusual application of a computer model to oxidation–reduction processes in natural waters.

LIMITATIONS

It is difficult to fit natural systems into idealized equilibrium models. Situations were described above where good fit was obtained. More than likely, however, poor fit is observed. There are inherent uncertainties in equilibrium models that contribute to the frustrations of applying chemical concepts to natural systems. These limitations are discussed below with carbonate systems as the prime example. Of course, this discussion may be extrapolated to such other systems as the silicates, the Great Lakes chemistry, oceanic chemistry, etc. (see also Nordstrom et al. [48]).

Geological

Many geological explanations have been offered to account for either the under- or oversaturation of carbonate ground waters. Back and Hanshaw [8] reported the chemical hydrogeology of the carbonate peninsulas of Florida and Yucatan that are part of a single regional geological setting. The principal artesian aquifer in Florida consists chiefly of limestone that ranges in age from Middle Eocene to Middle Miocene. These investigators observed that the "calcium bicarbonate type" of water was either under- or supersaturated with respect to calcite and dolomite. Rarely did they find a ground water in thermodynamic equilibrium with the carbonate solid phase. In general, the waters were undersaturated in the recharge areas and progressively obtained equilibrium and supersaturation downgradient. Where the waters are supersaturated, calcite and dolomite should be forming. In the practical sense, it is difficult to witness the formation of these solid phases several meters underground. For this reason the kinetic argument is usually invoked and is done so without foundation for waters that are supersaturated. The implication is that the precipitation reactions are so slow that the waters contain excessive concentrations of Ca^{2+}, Mg^{2+} and

HCO_3^- for lengthy periods of time. Back and Hanshaw [8] offer the explanation that the water may be in equilibrium with magnesium–calcite that has a higher solubility than calcite. This permits continued dissolution of calcite until an equilibrium with aragonite is obtained. This, in turn, leads to the formation of aragonite and, later, inversion to calcite. Some support is given to this explanation since the computations in Table IV indicate that the Polk City and Venus waters are supersaturated with respect to calcite, aragonite and dolomite.

It is more difficult to explain the undersaturation of carbonate waters. Thrailkill [11] attempts an answer from the related problem of cave waters in contact with calcite. It was concluded that temperature and mixing of ground waters led to conditions of undersaturation. Warmer rainwater infiltrates through cooler soils where it dissolves sufficient quantities of $CO_{2(g)}$ and, in turn, dissolves more calcite and gives undersaturated waters. This effect, however, is somewhat doubtful because the warmer rain water should equilibrate its temperature very quickly with the soils and ground waters. The second effect comes from the mixing of two solutions that are both saturated with respect to calcite but are in equilibrium with different partial pressures of $CO_{2(g)}$. This mixing will result in undersaturation. The difficulty with this explanation, however, is that one of the waters must maintain its partial pressure of $CO_{2(g)}$ for long travel paths. The partial pressures reported in Tables III and IV are all higher than 3.3×10^{-4} atm (normal for CO_2), and there is little indication of undersaturated calcitic waters. Support for this statement may be found in the work by Barnes [12], who reported the ground waters feeding Birch Creek had higher than normal partial pressures of $CO_{2(g)}$. This resulted in these waters being "slightly supersaturated to just saturated with calcite."

It is much more difficult to explain the aquatic chemistry of dolomitic waters than calcitic waters. Back and Hanshaw [8] cite that several of the Florida ground waters are undersaturated with respect to dolomite. This, however, was not confirmed as the Polk City water indicated a supersaturation. The rather weak explanation of low magnesium contents and slow kinetics of the dissolution reaction was offered. For ground waters that are supersaturated with respect to dolomite, Hanshaw et al. [59] attempted another explanation through the Mg:Ca ratio for a three-phase equilibrium of a dolomite–calcite solution system. Given sufficient time and supply of magnesium ions from an external source, dolomitization of calcitic waters should proceed spontaneously whenever the Mg:Ca activity ratio exceeds 1. Where dolomite is forming today, this ratio varies from 3 to 100. On the other hand, the Mg:Ca ratio in ocean water is 5:2 and the marine carbonates are not all dolomites. The equilibrium difficulty with dolomitic waters is summarized in Figure 13 [59]. Most of the waters

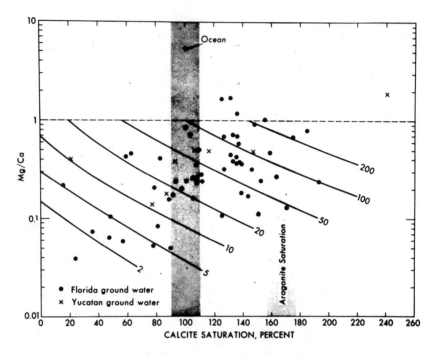

Figure 13. Relationship between the Mg:Ca ratio and equilibrium conditions for aragonite and calcite (shaded areas) and % dolomite saturation (solid lines) (reproduced from Hanshaw et al. [59] courtesy of *Econ. Geol.*).

that are undersaturated with respect to dolomite are saturated or supersaturated with respect to calcite. Apparently calcite is the "preferred" carbonate in low magnesium waters. If sufficient magnesium were introduced into these systems, dolomitization should occur but it would be a kinetically slow process. Perhaps this is the explanation for the observation that most ground waters in Tables III and IV show supersaturation with respect to dolomite.

Long and Saleem [60] and Drake and Wigley [61] present some data to support the argument of variable P_{CO_2} in ground waters. These authors found that for dolomite aquifers of the Northeastern Illinois Metropolitan area, "the major factor controlling the chemistry of the groundwater in the area is the solution of the dolomite by the *addition* of CO_2 along the groundwater flow path." Activities of Ca^{2+} for wells were plotted against their computed P_{CO_2}, and it was observed that the ground water chemistry changes are along the saturation curve. This work advances the argument that changes in P_{CO_2} rather than reaction kinetics control the $[Ca^{2+}]eq$

and $[Mg^{2+}]$eq. This is more plausible than the arguments of Holland [10] and Thrailkill [11].

Other geologic difficulties are the factors of purity and particle size. Naturally occurring calcite and dolomite are not 100% $CaCO_3$ and $CaMg(CO_3)_2$, respectively. Solubility product constants derived from laboratory studies are, of course, accomplished with the purest material possible. Miller [62] cites these problems when the solubility of three natural limestones in water was determined at various temperatures and pressures of $CO_{2(g)}$. Most of the solubility studies prior to Miller's work were achieved with finely powdered material. In nature, $CaCO_3$ occurs in rather coarsely crystalline form as massive aggregates of limestone. This would affect the type of surface presented to the aqueous phase and would affect the dissolution of Ca^{2+} and CO_3^{2-} ions. Hanshaw et al. [59] were able to determine the mineral purity of X-ray diffraction of several calcites and dolomites from several Florida aquifers. The calcite fell equally into two purity groups: 0–2 and 2–4 mol% $MgCO_3$. In wells from recharge areas, calcite was found coexisting with 20% or more of dolomite which, in turn, contained 0–2 mol% $MgCO_3$. In downgradient areas, calcites were found to coexist with dolomite containing 2–4 mol% of $MgCO_3$. In approximately one-third of the well samples, ordered dolomite was found with as much as 8 mol% excess $CaCO_3$. The dolomite texture generally was found to be massive, homogeneous, sugary and variable in grain size. It is rather apparent that difficulty is encountered when equilibrium carbonate models are applied to natural systems.

Analytical

The analytical difficulties of determining pH value and bicarbonate-carbonate alkalinity are reasonably well documented. Hem [6] cites the differences between pH values determined in the field and in the laboratory. From 165 sets of data, the average difference was about ±0.3 pH units. The difference was 1.0 pH unit for some samples. These differences are due to the loss of $CO_{2(g)}$ and temperature changes during the sampling, handling and transporting procedures. Field pH values are usually lower than laboratory values for the same sample. Back [7] reports similar analytical difficulties for limestone ground waters from Kentucky and Texas. In addition to the pH and temperature errors, bicarbonate values may be erroneous due to the occurrence of borates, phosphates and silicates in the waters. Also, loss of $CO_{2(g)}$ from the same samples will result in lower bicarbonate values as much as 10%. Barnes [63] gives an excellent account of the difficulties of measuring and reporting an accurate electrometric pH value. Barnes rejects colorimetric titrations as "unreliable or too cumbersome" for rapid field use. In turn, he employs the electrometric pH value as the endpoint in titrations of bicarbonate-carbonate alkalinity. This technique is not without error, however, as the true pH value of the endpoint is a function of temperature, ionic strength and the total carbonate concentration. Barnes reports that temperature and ionic strength have minor effects on the true endpoint, but that "the amount of bicarbonate present and the extent to which $CO_{2(g)}$ may be lost from the system during titration influence the pH at the endpoint substantially."

Thermodynamic

There are discrepancies in the thermodynamic data reported for the various soluble and solid species in the carbonate systems. Several standard free energy of formation values are seen in Table XIII for calcite, magnesite and dolomite. In turn, these values affect the equilibrium constant values cited in Table VI of Chapter 5 and also the predicted concentrations of such ionic species as Ca^{2+}, Mg^{2+}, etc. For example, Langmuir [64] cites the differences in 18 values for the solubility product constant of calcite in pure water.

Jacobsen and Langmuir [67] present perhaps the most accurate and reliable data concerning the solubility product constant of $CaCO_{3(s)}$ at $25°C$. Their study presented solubility measurements in CO_2-saturated water at 5, 15, 35 and $50°C$ and ignored $CaCO_3^0$ and $CaHCO_3^+$, which resulted in the following equation:

$$\log Kc = 13.870 - \left(\frac{3059}{T}\right) - 0.04035\ T \qquad (65)$$

Table XIII. Variations in Thermodynamic Data for Carbonate Systems (298.15° K, 1 atm)

ΔG_f^0 $CaCO_{3(s)}$ (kcal/mol)	ΔG_f^0 $MgCO_{3(s)}$ (kcal/mol)	ΔG_f^0 $CaMg(CO_3)_{2(s)}$ (kcal/mol)	$-\log K_s$ $CaCO_{3(s)}$	Reference
−269.908	−246.112	−518.734	−	4
−269.980	−	−	8.40±0.02	64
−	−	−	8.362–8.452	64
−	−241.920±0.55	−	−	65
−	−240.496±0.53	−	−	65
−	−241.416±0.37	−	−	65
−	−	−520.5	−	66

On the other hand, if $CaCO_3^0$ and $CaHCO_3^+$ are considered,

$$\log Kc = 13.543 - \left(\frac{3000}{T}\right) - 0.0401 \ T \tag{66}$$

Clearly, the solubility product constant of calcite is influenced slightly by use of the two ion pairs. Interestingly, the pKc values of Jacobsen and Langmuir differ less than those reported by Larson and Buswell [68] by 0.3 and 0.1 units, respectively, at 0 and 25°C.

The standard free energy of formation values and the solubility product constant of dolomite is perhaps more indefinite than calcite. Values for the latter constant have been reported to range from 1.5 to 3.0 $\times$ 10^{-17} [59] and from 10^{-17} to 10^{-20} [14]. Hsu [14] reported a value of 2 $\times$ 10^{-17} from an investigation of several Floridian aquifer waters that were assumed to be in equilibrium with dolomite and calcite.

Akin and Lagerwerff [69] present some appealing data to show that $CaCO_{3(s)}$ is more soluble in natural waters than in distilled water solutions. Furthermore, this increased solubility statistically correlated well with $[Mg^{2+}]$ and $[SO_4^{2-}]$, with these ions going out of solution during equilibration experiments with $CaCO_{3(s)}$. These investigators hypothesized that Mg^{2+} was substituted for Ca^{2+}, which produced a modified calcite surface. No mention was made of the role of the SO_4^{2-} ion in the enhanced solubility of $CaCO_{3(s)}$. A solubility product "enhancement factor" ($E = K_3$ (apparent)/K_s) ranged experimentally from 1.43 to 2.69 for several natural waters in which $CaCO_{3(s)}$ was placed. This observation of Akin and Lagerwerff may account for some of the apparently supersaturated calcite waters cited in Table IV.

For geochemical systems, Helgeson [27,70] argues that weathering processes are probably maintained in a state of "partial equilibrium," a state in which a system is in equilibrium with respect to at least one reaction, but out of equilibrium with respect to other processes. An example may be a system containing a single solid phase and an aqueous solution with which it is incompatible and in chemical disequilibrium. Reaction between the solution and the solid phase may lead to another solid phase that is compatible with the aqueous solution and leads to equilibrium. Helgeson claims support for this argument in the numerous mineral assemblages commonly found in metamorphic rocks; chemical and mineralogical characteristics of hydrothermal and geothermal systems; and the chemistry of ground waters involved in the weathering process. For these reasons, Helgeson [70] is critical of the oceanic equilibrium models proposed by Sillén [36] and Kramer [41]. It is argued that these models fail to account for differences in the chemical potentials of the various components from place to place in the ocean system and, by definition, no provision is made for thermodynamically irreversible reactions between sea water and the sediments contained in the ocean. Helgeson feels that a model of the world ocean should provide for different conditions of partial equilibrium in the various parts of the ocean system. These thoughts should apply to all attempts to apply equilibrium models to real world systems.

STEADY-STATE AND OPEN SYSTEMS

In the real world, most aquatic systems are open to their surroundings. Not only are they open in the thermodynamic sense, but they are subjected to the influence of many types of physical, chemical and biological reactions. Stumm and Morgan [17] and Morgan [71] provide excellent discussions of open systems and steady-state models. Only a brief summary of these discussions is given here.

For open systems, a steady-state model would be the most elementary type to consider. This brings in the concept of a time-invariant condition for concentrations of chemical constituents of a lake, ground water or ocean, for example. In an open system, external conditions control the extent to which a reaction occurs, and the Gibbs free energy is not zero in the stationary state. The condition of a steady state is established by flows of reactants and products between system and surroundings and by the kinetics of the reaction.

Morgan [71] has provided a generalized model for chemical dynamic description of natural water systems (Figure 14). A fundamental model may be written to incorporate many of the variables needed to describe the concentration of any constituent, C_i, at any point (x, y, z) in an aquatic system:

$$\frac{\partial C_i}{\partial t} + u\frac{\partial C_i}{\partial x} + v\frac{\partial C_i}{\partial y} + w\frac{\partial C_i}{\partial x} = \frac{\partial}{\partial x}\left(Dx\frac{\partial C_i}{\partial x}\right) + \frac{\partial}{\partial y}\left(Dy\frac{\partial C_i}{\partial y}\right) + \frac{\partial}{\partial z}\left(Dz\frac{\partial C_i}{\partial z}\right) + r_i'n \qquad (67)$$

where the $u(\partial C_i/\partial x)$, etc., are the advective or convective terms, $\partial/\partial x(Dx(\partial C_i/\partial x))$, etc., are the diffusion or turbulent transport terms, and $r_{i,n}$ is a kinetic term for the rate of formation or disappearance of the constituent i by process n. For the steady-state condition of i, $\partial C_i/\partial t = 0$. It should be rather obvious that these models require a rather substantial quantity of information concerning the kinetics of chemical reactions, the diffusion and mixing processes, and flows of matter and energy into and out of the system under consideration. In many cases, it is a difficult task to account for all of the terms in Equation 67 for a real world system.

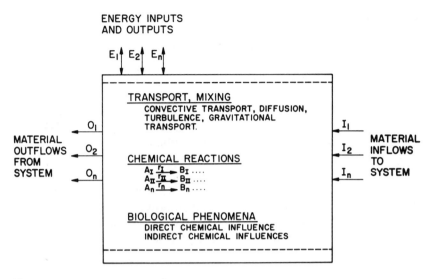

Figure 14.   Generalized model for chemical dynamic description of natural water systems (reproduced from Morgan [71] courtesy of the American Chemical Society).

Berner [72] has provided a steady-state model for the distribution of dissolved sulfate in recent sediments of the Santa Barbara Basin of southern California. In this model, "the local time rate of change of the concentration of sulfate in the interstitial water of sediments is equal to the sum of three terms, reflecting the processes of ionic diffusion, deposition plus compaction, and bacterial sulfate reduction." These three terms separately were: (1) diffusion—the one-dimensional case of Fick's second law of diffusion; (2) deposition plus compaction—a convective model; and (3) bacterial sulfate reduction—a first-order kinetic model. Combination of these three terms yielded the following:

$$\frac{\partial C}{\partial t} = D \frac{\partial^2 C}{\partial x^2} - w \frac{\partial C}{\partial t} - LkG_0 \exp\left[-\left(\frac{k}{w}\right)x\right] \qquad (68)$$

where   C = concentration
   D = diffusion coefficient
   t = time
   x = depth in the sediment measured positively downward from the sediment–water interface
   w = dx/dt, the measured "rate of deposition"
   L = stoichiometric coefficient equal to the number of sulfate ions reduced for every carbon atom oxidized (=1/2)
   G_0 = concentration of organic carbon, which can be utilized by bacteria to reduce sulfate to sulfide
   k = carbon oxidation rate constant.

Solution of Equation 68 for the steady-state condition gave the following:

$$C = (C_0 - C_\infty)\exp\left[-\left(\frac{k}{w}\right)x\right] + C_\infty \qquad (69)$$

where C_0 = concentration at zero depth
C_∞ = asymptotic concentration.

A plot of sulfate concentration of the interstitial waters versus core depth yielded the equation for the fitted curve: $C = 19 \exp(-0.015x) + 8$. Subsequently, Berner calculated k and D values from this equation that agreed reasonably well with values determined by independent methods. This agreement was considered as confirmation of Equation 69.

Other examples of steady state models are provided by Lerman and Brunskill [73] for the migration of major constituents from lake sediments into lake water and by Lerman [74] for lakes and oceans. The latter paper is an excellent presentation of the mathematical concepts underlying steady-state models.

REFERENCES

1. Kramer, J. R. University of Michigan Great Lakes Research Division, Publication #11 (1964), p. 147.
2. Langmuir, D. *Geochim. Cosmochim. Acta* 35:1023 (1971).
3. Hanshaw, B. B., et al. *Int. Assoc. Sci. Hydrol., Symposium of Dubrovnik* 601-614 (1965).
4. Robie, R. A., and D. R. Waldbaum. "Thermodynamic Properties of Minerals and Related Substances at 298.15°K. (25°C.) and One Atmosphere (1.013 Bars) Pressure and at Higher Temperatures," *Geol. Survey Bull.* 1259, Washington DC (1968).
5. White, D. E., et al. "Chemical Composition of Subsurface Waters," Chapter F. Data of Geochemistry, *U.S. Geol. Survey Prof. Paper 440-F* Washington, DC (1963).
6. Hem, J. D. "Calculation and Use of Ion Activity," Geological Survey. Water-Supply Paper 1535-C, Washington, DC (1961).
7. Back, W. "Calcium Carbonate Saturation in Ground Water," *Geol. Survey Water Supply Paper 1535-D,* Washington, DC (1961).
8. Back, W., and B. B. Hanshaw. *J. Hydrol.,* X(4):330 (1970).
9. Hostetler, P. B. *Int. Assoc. Sci. Hydrol., Commission of Subterranean Waters* 64:34 (1964).
10. Holland, H. D., et al. *J. Geol.* 72:36 (1964).
11. Thrailkill, J. *Geol. Soc. Am. Bull.* 79:19 (1968).
12. Barnes, I. *Geochim, Cosmochim. Acta* 29:85 (1965).
13. Shuster, E. T., and W. B. White. *Water Resources Res.* 8:1067 (1972).
14. Hsu, K. J. *J. Hydrol.* 1(4):288 (1963).
15. Krauskopf, K. *Introduction to Geochemistry* (New York: McGraw-Hill Book Co., 1967).
16. Garrels, R. M., and C. L. Christ. *Solution, Minerals, and Equilibria* (New York: Harper & Row, Publishers, Inc., 1965).

17. Stumm, W., and J. J. Morgan. *Aquatic Chemistry* (New York: John Wiley & Sons, Inc., 1970).
18. Siever, R. *Am. Min.* 42:821 (1957).
19. Roberson, C. E., and J. D. Hem. "Solubility of Aluminum in the Presence of Hydroxide, Fluoride, and Sulfate," *Geol. Survey Water Supply Paper 1827-C*, Washington, DC (1969).
20. Cohen, P., and C. N. Durfor. *Int. Assoc. Sci. Hydrol.*, Paris, France, Pub. No. 72 (1967), p. 193.
21. Faust, S. D., and J. Vecchioli. *J. Am. Water Works Assoc.* 66(6):371 (1974).
22. Sutherland, J. C. *Environ. Sci. Technol.* 4:826 (1970).
23. Feth, J. H., et al. "Sources of Mineral Constituents in Water from Granitic Rocks," *Geol. Survey Water Supply Paper 1535-I*, Washington, DC (1964).
24. Garrels, R. M., and F. T. Mackenzie. In: *Equilibrium Concepts in Natural Water Systems*, Advances in Chemistry Series, No. 67 (Washington, DC. American Chemical Society, 1967), p. 222.
25. Sutherland, J. C., et al. The University of Michigan Great Lakes Research Division, Publication #15, (1966), p. 439.
26. Bricker, O. P., et al. In: *Trace Inorganics in Water*, Advances in Chemistry Series, No. 73, (Washington, DC: American Chemical Society, 1968).
27. Helgeson, H. C., et al. *Geochim. Cosmochim. Acta.* 33:455 (1969).
28. Kramer, J. R. University of Michigan Great Lakes Research Division, Publication No. 7 (1961), p. 27.
29. Kramer, J. R. University of Michigan Great Lakes Research Division Publication #9 (1962), p. 21.
30. Kramer, J. R. In: *Equilibrium Concepts in Natural Water Systems*, Advances in Chemistry Series #67, (Washington, DC: American Chemical Society, 1967), p. 243.
31. Falls, C. P., and L. P. Varga. *Environ. Sci. Technol.* 7:319 (1973).
32. Truesdell, A. H., and W. Singers. *J. Res. U.S. Geol. Survey* 2(3):217 (1974).
33. Jones, B. F., et al. *Water Resources Res.* 10(4):791 (1974).
34. Hem, J. D. *Bull. Int. Assoc. Sci. Hydrol.* 19:45 (1960).
35. Stumm, W., and J. O. Leckie, *Environ. Sci. Technol.* 1(4):298 (1967).
36. Sillén, L. G. "The Physical Chemistry of Sea Water," in *Oceanography*, Mary Sears, Ed., Publication #67 (Washington, DC: American Association Advanced Science 1961), p. 549.
37. Sillén, L. G. *Science* 156(3779):1189 (1967).
38. Sillén, L. G. in: *Equilibrium Concepts in Natural Water Systems*, Advances in Chemistry Series, No. 67 (Washington, DC: American Chemical Society, 1967), p. 57.
39. Goldschmidt, V. M. *Fortschr. Mineral Krist. Petr.* 17:112 (1933).
40. Hemley, J. J. *Am. J. Sci.* 257:241 (1959).
41. Kramer, J. R. *Geochim. Cosmochim. Acta* 29:921 (1965).
42. Holland, H. D. *Proc. Nat. Acad. Sci.* 53:1173 (1965).
43. Carroll, D., and H. C. Starkey. *Proc. Nat. Conf. Clay Minerals* 5:80 (1960).
44. Hemley, J. J., et al. *U.S. Geol. Surv. Prof. Paper* 424-D, 57 (1962).
45. Garrels, R. M., and M. E. Thompson. *Am. J. Sci.* 260:57 (1962).
46. Mackenzie, F. T., and R. M. Garrels. *Am. J. Sci.* 264:507 (1966).
47. Chave, K. E. *J. Chem. Ed.* 48:148 (1971).

48. Nordstrom, D. K., et al. In: *Chemical Modeling in Aqueous Systems,* E. A. Jenne, Ed., ACS Symposium Series No. 93, Washington, DC (1979) p. 857.
49. Truesdell, A. H., and B. F. Jones. *J. Res. U.S. Geol. Survey* 2(2):233 (1974).
50. Kharaka, Y. K., and I. Barnes. "SOLMNEQ," *NTIS Tech. Report,* PB214-899, Springfield, VA (1973).
51. Plummer, L. N., et al. "WATEQF," *U.S. Geol. Survey Water Resources Invest.,* 76-13 (1976).
52. Ball, J. W., et al. "WATEQ2," In: *Chemical Modeling in Aqueous Systems,* E. A. Jenne, Ed., ACS Symposium Series No. 93, Washington, DC (1979).
53. Wigley, T. M. L. *Brit. Geomorph. Res. Group Tech. Bull.* 20 (1977).
54. Morel, F., and J. J. Morgan. *Environ. Sci. Technol.* 6:58 (1972).
55. Ball, J. W., et al. "WATEQ2," *U.S. Geol. Survey Water Resources Invest.* WRI 78-116, Menlo Park, CA (1980).
56. Mattigod, S. V., and G. Sposito. "GEOCHEM," In: *Chemical Modeling in Aqueous Systems,* E. A. Jenne, Ed., ACS Symposium Series No. 93, Washington, DC (1979).
57. Nordstrom, D. K., et al. In: *Chemical Modeling in Aqueous Systems,* E. A. Jenne, Ed., ACS Symposium Series No. 93, Washington, DC (1979).
58. Boulegue, J., and G. Michard. In: *Chemical Modeling in Aqueous Systems,* E. A. Jenne, Ed., ACS Symposium Series No. 93, Washington, DC (1979).
59. Hanshaw, B. B., et al. *Econ. Geol.* 66:710 (1971).
60. Long, D. T., and Z. A. Saleem. *Water Resources Res.* 10:1229 (1974).
61. Drake, J. J., and T. M. L. Wigley. *Water Resources Res.* 11:958 (1975).
62. Miller, J. P. *Am. J. Sci.* 250:161 (1952).
63. Barnes, I. "Field Measurement of Alkalinity and pH," Geological Survey Water-Supply Paper 1535-H, Washington, DC (1964).
64. Langmuir, D. *Geochim. Cosmochim. Acta* 32:835 (1968).
65. Langmuir, D. *J. Geol.* 73(5):730 (1965).
66. Garrels, R. M., et al. *Am. J. Sci.* 258:402 (1960).
67. Jacobson, R. L., and D. Langmuir. *Geochim. Cosmochim. Acta* 38:301 (1974).
68. Larson, T. E., and A. M. Buswell. *J. Am. Water Works Assoc.* 34:1667 (1942).
69. Akin, G. W., and J. V. Lagerwerff. *Geochim. Cosmochim. Acta* 29:252 (1965).
70. Helgeson, H. C. *Geochim. Cosmochim. Acta* 32:853 (1968).
71. Morgan, J. J., In: *Equilibrium Concepts in Natural Water Systems,* Advances in Chemistry Series No. 67, ACS, (Washington, DC: American Chemical Society, 1967).
72. Berner, R. H. *Geochim. Cosmochim. Acta* 28:1497 (1964).
73. Lerman, A., and G. J. Brunskill. *Limnol. Oceanog.* 16:880 (1971).
74. Lerman, A., In: *Nonequilibrium Systems in Natural Water Chemistry,* Advances in Chemistry Series No. 106, ACS, (Washington, DC: American Chemical Society, 1971).

CHAPTER 7

MERCURY, ARSENIC, LEAD, CADMIUM, SELENIUM AND CHROMIUM IN AQUATIC ENVIRONMENTS

INTRODUCTION

There are several heavy metals that have profound significance in man's environment. Much concern has been expressed about such "toxic" metals as Hg, Ag, Ba, Se, Cd, Pb, Cr, Ni, Zn, Cu, etc. being distributed throughout the world. This chapter details the environmental impact of five heavy metals: Hg, Pb, Cd, Se and Cr, and one semimetal, As. All of these elements have been given primary drinking water standards by the U.S. Environmental Protection Agency [1]: Hg, 0.002 mg/1; As, 0.05; Pb, 0.05; Cd, 0.01, Se, 0.01; and Cr, 0.05.

MERCURY

Background

The significance of mercury as a serious environmental hazard has been highlighted since its implication in mass poisoning in Japan. In 1953 cases of mysterious nervous diseases began to appear among the fishing population of the city of Minamata, Japan. The symptoms were due to poisoning by methylmercury, which was subsequently isolated in high concentrations from fish and shellfish caught in the bay and consumed by victims [2,3]. By the end of 1965, more than 160 cases and 52 deaths had been reported. The incident was traced to contamination of the bay by high mercury discharges from a chemical plant that used mercuric oxide in sulfuric acid as a catalyst for acetaldehyde production. The accumulation of methylmercury in the fish was suggested to be alkylation of inorganic mercury by plankton and other marine life [4]. The spent catalyst from the plant later was shown to contain 1% methylmercury and the biological methylation of mercury was thought to be insignificant.

In Sweden, mercury poisoning was responsible for a marked decrease in the wild bird population. High mercury levels were found in dead birds [5] as well as in many agricultural products [6]. This was attributed to the extensive use of methylmercury dicyanodiamide as a fungicide in seed dressing in Swedish agriculture. After the use of these compounds was banned in 1966, the mercury levels in birds and eggs were drastically reduced. However, contamination of fish and other aquatic organisms was observed even after the use of organomercury fungicides was discontinued, which then implicated all mercury discharges into Swedish waters.

Recently, mercury contamination of the fish population in Lake St. Clair—Western Lake Erie area of the Great Lakes was reported. The subsequent detection of high mercury levels in fresh water and marine fish brought the problem of mercury pollution of the aquatic environment into sharp focus in the United States and Canada. The subsequent intensive investigations documented the ubiquitous distribution of mercury in fish, the aquatic food chain and bottom sediments [7–19].

The mercury found in fish is almost entirely in the form of methylmercury [9,20-22]. Alkylmercurials are cumulative in the food chain and are more effectively concentrated than other forms of mercury by aquatic organisms. It has been reported that fish can concentrate methylmercury chloride by an overall factor of 3000 [23,24], while shellfish concentrate it by a factor of 100–100,000 [25]. Consumption of contaminated fish or other organisms by man can be disastrous because the alkyl mercury compounds are 10–100 times more toxic than the inorganic forms [26] and are almost totally absorbed by the intestinal tract, while inorganic mercury compounds are absorbed to a much lesser degree. In addition, they can cause irreversible damage to the central nervous system manifested by the loss of motor control and sensory operations. Ingestion of high concentrations of these compounds can be fatal, as was experienced in the Minamata incident.

Sources in the Environment

Natural Sources

There are several mercury-bearing minerals but only a few occur abundantly in nature. The commonest are the sulfides: cinnabar and metacinnabar, which contain 86% mercury by weight, and native mercury. Mercury is commonly present in tetrahedrite (up to 17% in the variety schwatzite), in sphalerite (up to 1%) and in wurtzite (up to 0.3%); it is present in small amounts in many other sulfides and sulfosalts.

Mercury is the only metal that occurs in liquid form in its elemental state at ordinary earth surface temperatures. Because of its high tendency to vaporize, strong ligand affinity and ease of adsorption onto surfaces, mercury is widely distributed in rocks, soils, air and water. Most igneous and sedimentary rocks contain mercury in concentrations less than 200 ppb. Shales, clays and soils, however, show considerable variation, with average contents ranging between 30 and 2300 ppb [27].

Because of the high volatility of mercury and some of its compounds, it is constantly released to the atmosphere from ore deposits, soil surfaces and volcanic emanations. Natural levels in "unpolluted air" are generally in the range of 1 to 10 ng/m^3 Hg [27]. However, concentrations as high as 20,000 ng/m^3 have been reported at ground level near mercury ore deposits [28]. Such meteorological factors as wind speed, wind direction, and seasonal and diurnal temperature variations affect the level of natural atmospheric mercury in different locations [28,29]. Most of the mercury in the air is adsorbed by atmospheric particulate matter, which is eventually removed by dry fallout or rainout [30]. Rain effectively washes mercury from the atmosphere, even near mercury ore deposits. Such scrubbing accounts for the fact that the average rain content is about 0–2 ppb [31]. Following a rainfall, the water comes in contact with soils and rocks during storm runoff or percolation into the ground, which results in a natural distribution of mercury in water. The natural content of mercury in unpolluted surface waters is generally less than 0.1 µg/l [32]; however, much higher concentrations (up to 136 µg/l) are encountered in rivers draining mercury ore deposits. Particulate matters in streams readily adsorb mercury, which results in higher concentrations in the suspended sediment and bottom muds. This suggests, therefore, that the downstream movement of mercury is greatly influenced by the amount of sediment in transport. Ground waters are liable to contain higher mercury concentrations because of the longer and more intimate contact with minerals. Oil field brines and thermal and mineral fluids contain high mercury concentrations, which are sources of pollution to surface and ground waters. Wershaw [32] reported that the oceans contain an estimated 50 million metric tons of mercury.

Anthropogenic Sources

Considerable amounts of mercury are introduced into the environment as a result of man's activity, either by deliberate release, loss or waste. In 1968 the total world production of mercury was estimated to be 8000 metric tons, of which the United States produced only 1000 metric tons. The total domestic use during that year amounted to 2500 metric tons, or 31% of the world output [32]. During the period 1930–1970, the total mercury mined in the

United States was 31,800 metric tons and 39,600 metric tons were imported. It is estimated that about 25% of this total may have leaked into the environment [32]. Electrolytic preparations of chlorine and caustic soda account for the largest industrial use of mercury and are considered the most significant source of contamination of natural waters. Before enforcement of recent regulations, the wastewater discharges from these plants contained 0.25–0.50 lb of mercury per ton of caustic soda produced. The second largest use of mercury is the manufacture of such electric apparatus as batteries, silent switches, high-intensity street lamps and fluorescent lights. Mercury compounds are also widely used by several industries as fungicide and and bactericides. For example, phenylmercury compounds were used for mildewproofing in the paint and in paper industries to prevent fungal growth in stored pulp and the growth of slimes on the machinery. Organomercurials are also used for seed dressing in agriculture. Mercury compounds are utilized to a lesser extent as catalysts in the manufacture of organic material and chemicals, as well as in pharmaceutical, cosmetic, and dental preparations. Small amounts of additive mercury in consumer goods and mercurials from laboratories, hospitals and some industries are generally discharged into wastewater treatment plants, whose outfalls act as point sources of mercury to the receiving waters. Mercury concentrations in sediment samples collected downstream from industrial and municipal sewer outfalls are several times higher than similar samples collected farther from the outfall [8].

Another major source of mercury discharge to the environment includes the burning of fossil fuels. The mercury content of coal is highly variable and shows distinct regional patterns. Recent analyses of 36 different American coals [33] revealed concentrations ranging between 70 and 33,000 $\mu g/kg$. Limited data are available for the mercury content of petroleum. Values ranging between 1.9 and 2.9 mg/kg have been reported in petroleum from oil fields in California [27]. Based on an annual world consumption of 3 X 10^9 tons of coal and 1.8 X 10^9 tons of petroleum and a conservative average concentration of 1 mg/kg of mercury in both, calculations indicate that 3000 tons and 1800 tons of mercury are released annually from burning coal and petroleum, respectively.

A potential source of environmental mercury contamination results from the smelting operation of mercury-containing ores. It is well known that ores of lead, zinc and copper are enriched in mercury, of which most probably is released to the atmosphere during smelting operations. About 10^9 tons of copper, lead and zinc ores are processed annually. Reasonable estimates of the mercury content of these ores range from 3 to 30 mg/kg, corresponding to a mercury discharge of 3000–30,000 ton/yr [34]. Most of the mercury discharged into the atmosphere reenters the terrestrial or aquatic environments by dry fallout or from rainfall.

Biological Transformations

The accumulation of methylmercury in fish and other organisms, despite the lack of inputs of organic mercury compounds into the aquatic environment, suggests that the formation of this compound is biological in origin. Jensen and Jernalöv [35] were the first to provide evidence that mono- and dimethylmercury were produced from inorganic mercury or phenylmercury by bacteria in bottom sediments. Incubation of bottom sediments from fresh water aquaria with increasing concentrations of $HgCl_2$ resulted in the formation of increasing amounts of methylmercury after seven days. At very high concentrations (>500 ppm Hg^{2+}), there was a decrease in the amount of methylmercury formed due to the toxic effects of inorganic divalent mercury on the methylating organisms. These same authors also showed that volatile dimethylmercury was formed on incubation of inorganic mercury or methylmercury, with dead fish as the biological material under anaerobic conditions. Subsequent investigations [36-43] demonstrated the widespread occurrence of bacterial species in aquatic environments, especially bottom sediments, which are capable of biological methylation of mercury even from mercuric sulfide [44], which has an extremely low solubility. This biological methylation process has prime importance in the evaluation of the mobility of mercury in the aquatic environment and its uptake and accumulation by aquatic organisms. The relative distribution of total and methylmercury between different phases in the ecosystem was shown [45] to be within the following orders of magnitude:

	Total Mercury (%)	Methylmercury (%)
Sediment	90-99	1-10
Water	1-10	<1
Biota	<1	90-99

These data indicate that methylmercury represents a small fraction of the total amount of mercury present in the ecosystem. However, mercury content in fish is the prime concern to man and, in this respect, the biological methylation of mercury has extreme environmental significance.

The conversion of mercury to methylmercury is considered to be a detoxification mechanism by bacteria. Wood [46] surveyed the methylating agents available for methyl transfer reactions in biological systems and indicated that the only known methylating agents are the methylcorrinoids, which are involved in the metabolism of methanogenic bacteria. The methylcorrinoids are methylated derivatives of vitamin B_{12} (CH_3-Co-5,6-dimethylbenzimidozolyl-cobamide). The active part of the molecule is the cobalt atom, coordinated to five nitrogen atoms from heterocyclic organic ring structures to which a methyl group has been attached. The methylation proceeds by transfer of the methyl group to Hg^{2+}. Mechanisms of the biological synthesis of methylmercury and dimethylmercury are illustrated here [46]:

$$
\begin{array}{ccc}
\text{CH}_3 \\
\text{Co}^{3+} + \text{H}_2\text{O} \\
\text{Bz} \\
\text{I}
\end{array}
\qquad
\begin{array}{cc}
\text{H} \quad \text{H} \\
\text{O} \\
\text{Co}^{3+} + \text{CH}_3^- \\
\text{Bz} \\
\text{II}
\end{array}
$$

$$
\begin{array}{cc}
\text{CH}_3 \\
\text{Co}^{3+} + \text{H}_2\text{O} + \text{Hg}^{2+} \xrightleftharpoons{K} \\
\text{Bz}
\end{array}
\qquad
\begin{array}{c}
\text{CH}_3 \\
\text{Co}^{3+} \\
\text{O} \\
\text{H} \quad \text{H} \\
\text{BzHg}^+
\end{array}
$$

Hg^{2+} — Fast reaction   Slow reaction — Hg^{2+}

$$
\begin{array}{cc}
\text{H} \quad \text{H} \\
\text{O} \\
\text{Co}^{3+} + \text{CH}_3\text{Hg}^+ \\
\text{Bz} \quad \text{III} \\
\text{II}
\end{array}
\qquad
\begin{array}{cc}
\text{H} \quad \text{H} \\
\text{O} \\
\text{CH}_3\text{Hg}^+ + \text{Co}^{3+} + \text{Hg}^{2+} \\
\text{III} \quad \text{Bz} \\
\text{II}
\end{array}
$$

$$(1)$$

Bz represents the 5,6-dimethylbenzimidazole in a methyl-B_{12} (vitamin B_{12}) derivative (I) that is converted to aquo-B_{12} (II). The synthesis of methylmercury (III) depends on the equilibrium constant, K. The slow reaction is about 1000 times slower than the fast reaction. Dimethylmercury is synthesized by an identical mechanism, except that CH_3Hg^+ is the reacting species instead of Hg^{2+}. However, the synthesis of dimethylmercury is about 6000 times slower than that of methylmercury. The optimum pH value for synthesis of methylmercury, either under laboratory conditions or in natural sediments, is 4.5. Dimethylmercury is volatile and, once in the atmosphere, is probably photolyzed by ultraviolet light to give Hg^0 plus methane and ethane.

Several investigators have shown that the methylating activity of a natural bacterial community in the sediments is counterbalanced by the mineralizing activity of other organisms. Spangler et al. [41] have demonstrated the biological degradation of methylmercury in lake sediments. The bacterial species isolated in pure culture caused conversion of methylmercury into volatile elemental mercury Hg^0 and methane. Billen et al. [47] showed that the mineralizing activity of bacterial community in the sediment can be increased in response to the increased concentration of methylmercury.

Another important biological transformation of mercury is the conversion of Hg^{2+} to Hg^0, which is regarded as a detoxification mechanism by bacteria. Aerobes can solubilize Hg^{2+} from HgS (solubility product constant $\sim 10^{-53}$) by oxidizing the sulfide to sulfate [48]. Once Hg^{2+} has been solubilized, it can be reduced to Hg^0 by an enzyme that is present in a number of bacteria. This enzyme has been shown to require reduced nicotinamide adenine dinucleotide (NADH) as a coenzyme for catalysis [46]:

$$Hg^{2+} + NADH + H^+ \rightleftharpoons Hg^0 + 2H^+ + NAD^+ \qquad (2)$$

The Hg^0 has sufficient vapor pressure to be lost from the aqueous environment to the atmosphere.

The above examples indicate there are biological cycles for the synthesis and degradation of mercury compounds in aquatic environments. The biological cycle for mercury [46] is summarized in Figure 1. The possibility of interconversions of mercury compounds establishes a dynamic system of reversible reactions, which lead to steady-state concentrations of various metallic and methylated forms in sediments. Disturbances produced in steady-state concentrations caused by introduction of mercury into the natural ecosystem will affect the natural equilibrium, which will, in turn, affect the concentration of the toxic intermediates. The steady-state concentration of methylmercury does not necessarily reflect the rate of its formation. Wood [49] indicated that the rate of synthesis of methylmercury does not have to be rapid in sediments for fish to accumulate dangerous levels. When the rate at which methylmercury is produced and taken up by fish exceeds the rate of its metabolism, then accumulation will occur. Therefore, in assessing the fate of mercury in the aquatic environment, the kinetics of all the processes that allow significant levels of toxic intermediates to accumulate is important.

Chemistry in Aqueous Systems

Equilibrium Reactions

Mercury in aqueous systems can exist in one of three oxidation states: as the free metal, Hg^0; as the mercurous ion, Hg_2^{2+}; or as the mercuric ion, Hg^{2+}. The relative stability of the different forms under a variety of conditions is obtained from available thermodynamic data of the most significant species in water. The standard free energies of formation values are given below for these species. Table I shows the equilibrium constants for formation of some of these mercury compounds.

Mineral Name	Formula[a]	ΔG_f°[b] (kcal/mol)
Metallic Mercury	$Hg_{(l)}^{\circ}$	0.0
Dissolved Mercury	Hg°	9.4
Mercurous Ion	Hg_2^{2+}	36.7
Mercuric Ion	Hg^{2+}	39.3
Calomel	$Hg_2Cl_{2(s)}$	−50.35
Mercuric Chloride	$HgCl_{2(s)}$	−42.7
Red Oxide	$HgO_{(s)}$	−13.995
Yellow Oxide	$HgO_{(s)}$	−13.964
Cinnabar	$HgS_{(s)}$	−12.1
Metacinnabar	$HgS_{(s)}$	−11.4
	$Hg_2SO_4^{\circ}$	−140.6
	$Hg_2SO_{4(s)}$	−149.589
	$Hg_2CO_{3(s)}$	−105.8
	$HgCl_2^{\circ}$	−41.4
	$HgCl_4^{2-}$	−107.7

[a] All species are aqueous unless noted otherwise.
[b] From Pourbaix [56] and the National Bureau of Standards [57].

Mercuric [58] and organomercuric perchlorates, nitrates and sulfates are strong electrolytes and are thus completely dissociated in aqueous solution [51-55,59]:

$$HgX_2 \rightleftharpoons Hg^{+2} + 2X^- \qquad (3)$$

$$RHgX \rightleftharpoons RHg^+ + X^- \qquad (4)$$

where   X = any electron withdrawing ligand that forms an ionic bond with mercury, and
R = an organic group such as methyl or phenyl.

The dissociation tendencies of mercuric compounds (HgX_2 or RHgX) depend on the nature of the ligand X^-:

$$F^- > OCOCH_3^- > HPO_4^{2-} \simeq Cl^- > Br^- > NH_3 > OH^- > SR^- > S^{2-}$$

The thiol and sulfide compounds are particularly stable and, therefore, have low tendencies to dissociate.

Mercuric and organomercuric ions undergo rapid hydrolysis to form the corresponding hydroxides, whose equilibrium constants are shown in

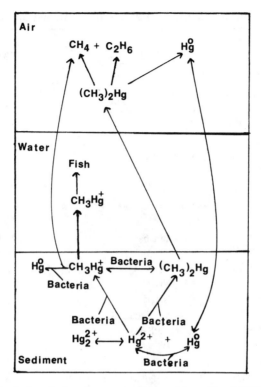

Figure 1. Biological cycle for mercury (reproduced from Wood [46] courtesy of the American Association for the Advancement of Science).

Table II. The hydrolysis reactions are pH dependent and will not occur to a significant extent at pH values less than 3-4; however, within the pH ranges usually encountered in natural waters (pH 5-9), the ions are almost completely hydrolyzed, as shown in Figures 2, 3 and 4 [60].

Mercuric and organomercuric compounds can undergo "ligand exchange reactions" with the variety of chemical species that exist in natural waters. These reactions involve the interaction of mercuric or organomercuric complexes with some chemical species, Y, to form a new complex:

$$HgX_2 + Y \rightleftharpoons HgXY + X \tag{56}$$

$$HgXY + Y \rightleftharpoons HgY_2 + X \tag{57}$$

$$RHgX + Y \rightleftharpoons RHgY + X \tag{58}$$

Table I. Equilibrium Constants for Formation of Mercury Species at 25°C

Reaction Number	Reaction	log K_{eq}	Reference
(5)	$Hg_2^{2+} = Hg_{(l)}^0 + Hg^{2+}$	−2.22	50
(6)	$Hg_{(aq)} = Hg_{(l)}^0$	6.89	50
(7)	$Hg_2^{2+} + 2Cl^- = Hg_2Cl_{2(s)}$	17.96	50
(8)	$Hg^{2+} + 2Cl^- = HgCl_2^0$	13.25	50
(9)	$Hg^{2+} + 3Cl^- = HgCl_3^-$	15.35	50
(10)	$Hg^{2+} + 2Br^- = HgBr_2^0$	16.80	51
(11)	$Hg^{2+} + 2I^- = HgI_2^0$	23.8	51
(12)	$Hg^{2+} + 2OH^- = Hg(OH)_2$	21.7	51
(13)	$Hg^{2+} + SO_4^{2-} = HgSO_{4(aq)}$	1.42	50
(14)	$Hg^{2+} + S^{2-} = HgS_{(cinnabar)}$	52.37	50
(15)	$Hg^{2+} + S^{2-} = HgS_{(metacinnabar)}$	53.68	50
(16)	$HgS_2^{2-} = HgS_{(s)} + S^{2-}$	−4.57	50
(17)	$Hg^{2+} + 2HS^- = Hg(HS)_2^0$	37.73	50
(18)	$Hg^{2+} + 2NH_{3(aq)} = Hg(NH_3)_2$	17.40	51
(19)	$Hg^{2+} + 4NH_{3(aq)} = Hg(NH_3)_2^{2+}$	19.28	50
(20)	$Hg^{2+} + 2NH_2R_{(histidine)} = Hg(NH_2R)_2$	21.20	51
(21)	$Hg^{2+} + 2SR_{(cysteine)}^- = Hg(SR)_2$	41.00	51
(22)	$Hg^{2+} + 2SCN^- = Hg(SCN)_2$	17.4	51
(23)	$Hg^{2+} + 2CN^- = Hg(CN)_2$	34.7	51
(24)	$Hg^{2+} + 2CH_3COOH_{(aq)} = Hg(CH_3CO_2)_{2(s)} + 2H^+$	3.11	50
(25)	$Hg^{2+} + 2CH_{4(aq)} = Hg(CH_3)_{2(l)} + 2H^+$	−7.8	50
(26)	$Hg_{(l)} + CH_3OH_{(aq)} + CH_{4(aq)} = Hg(CH_3)_{2(l)} + H_2O$	−19.74	50
(27)	$CH_3Hg^+ + F^- = CH_3HgF$	9.37 (9.5)[a]	52[b]
(28)	$CH_3Hg^+ + Cl^- = CH_3HgCl$	5.25 (5.45)[a]	52[b]
(29)	$CH_3HgCl_{(aq)} = CH_3HgCl_{(l)}$	1.70	53
(30)	$CH_3Hg^+ + Br^- = CH_3HgBr$	6.62 (6.70)[a]	52[b]
(31)	$CH_3Hg^+ + I^- = CH_3HgI$	8.60 (8.70)[a]	52[b]
(32)	$CH_3Hg^+ + OH^- = CH_3HgOH$	9.37 (9.5)[a]	52[b]
(33)	$CH_3Hg^+ + OC_6H_5^- = CH_3HgOC_6H_5$	~6.50	51
(34)	$CH_3Hg^+ + OCOCH_3^- = CH_3HgOCOCH_3$	~3.60	51
(35)	$CH_3Hg^+ + HPO_4^{2-} = CH_3HgHPO_4^-$	5.03	52[b]
(36)	$CH_3Hg^+ + HPO_3^{2-} = CH_3HgHPO_3^-$	4.67	52[b]
(37)	$CH_3Hg^+ + S^{2-} = CH_3HgS^-$	21.20	52[b]
(38)	$CH_3Hg^+ + SCH_2CH_2OH^- = CH_3HgSCH_2CH_2OH$	16.12	52[b]
(39)	$CH_3Hg^+ + SR_{(cysteine)}^- = CH_3HgSR$	15.70	59

Table 1, continued

Reaction Number	Reaction	log K_{eq}	Reference
(40)	$CH_3Hg^+ + SO_3^{2-} = CH_3HgSO_3^-$	8.11	52[b]
(41)	$CH_3Hg^+ + S_2O_3^{2-} = CH_3HgS_2O_3^-$	10.90	52[b]
(42)	$CH_3Hg^+ + SCN^- = CH_3HgSCN$	6.10	51
(43)	$CH_3Hg^+ + NH_{3(aq)} = CH_3Hg(NH_3)$	7.60 (8.4)[a]	52[b]
(44)	$CH_3Hg^+ + NH_2CH_2CH_2NH_2 = CH_3HgNH_2CH_2CH_2NH_2$	8.25	52[b]
(45)	$CH_3Hg^+ + CN^- = CH_3HgCN$	14.2	52[b]
(46)	$C_6H_5Hg^+ + OH^- = C_6H_5HgOH$	9.89	54
(47)	$C_6H_5Hg^+ + OCOCH_3^- = C_6H_5HgOCOCH_3$	4.82	54
(48)	$C_6H_5Hg^+ + OCOCH_2CH_3^- = C_6H_5HgOCOCH_2CH_3$	4.51	54
(49)	$C_6H_5Hg^+ + SC_6H_5^- = C_6H_5HgSC_6H_5$	>16.0	55

[a]Value in parentheses taken from Simpson [51]: I = 0.5, at 25°C.
[b]I = 0.1, at 25°C.

Table II. Equilibrium Constants for Hydrolysis of Mercuric, Alkylmercuric and Phenylmercuric Ions at 25°C

Reaction No.	Reaction	pK_a[a]	Reference
(50)	$Hg^{2+} + H_2O = HgOH^+ + H^+$	3.70±0.70	59
(51)	$HgOH^+ + H_2O = Hg(OH)_2 + H^+$	2.60±0.09	59
(52)	$Hg^{2+} + 2H_2O = Hg(OH)_2 + 2H^+$	6.30±0.05	59
(53)	$CH_3Hg^+ + H_2O = CH_3HgOH + H^+$	4.50[b]	52
(54)	$C_2H_5Hg^+ + H_2O = C_2H_5HgOH + H^+$	4.90[b]	53
(55)	$C_6H_5Hg^+ + H_2O = C_6H_5HgOH + H^+$	4.11[b]	60

[a]Ionic strength was 0.5 for all constants listed.
[b]Standard deviations were not given.

A number of kinetic studies of the ligand exchange reactions of protein sulfhydryl groups (–SH) were prompted by the interest in the function of these groups in enzyme catalysis. Early studies by Boyer [61] showed that kinetically different types of –SH groups in proteins could be distinguished by interaction with p-chloromercuribenzoate. Second-order rate constants ranged from about 10^2 liter/mol·sec for "masked" –SH groups to 10^5 liter/mol·sec for free –SH groups, indicating that unhindered thiol-mercury complexes would exchange much more rapidly.

Simpson [62] determined the rate constants for ligand exchange reactions of methylmercury compounds. These reactions were frequently close to diffusion controlled, with the slowest rate constants about 10 liter/mol·sec. These kinetic data indicate that in natural waters that contain multiple chemical species and mercury compounds, the establishment of ligand exchange equilibria is quite rapid, even at very low concentrations.

Equilibrium Diagrams of Aqueous Inorganic Mercury Species

Figure 2 illustrates a stability-field (Eh-pH) diagram constructed by Hem [50] for the solid and liquid forms of mercury in a model system

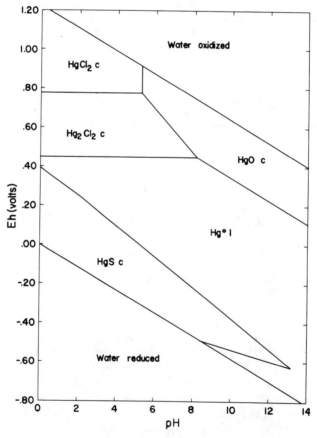

Figure 2.   Stability fields for solid (c) and liquid (l) mercury species at 25°C and 1 atm [50].

containing chlorine and sulfur species at a concentration of $10^{-3}M$ of each in water at 25°C.

Elemental mercury (Hg^0) and cinnabar ($HgS_{(s)}$) are the principal species most likely to enter into equilibria affecting the solubility of mercury in natural waters where the pH values range from 5 to 9 and Eh values are less than 0.5. Organic mercury compounds, which play an important role in the fate of mercury in natural water systems, are thermodynamically unstable in this system.

Figure 3 illustrates the areas of dominance of the solute species in equilibrium under the same conditions specified in Figure 2. These two diagrams present the main features of aqueous inorganic chemistry of mercury.

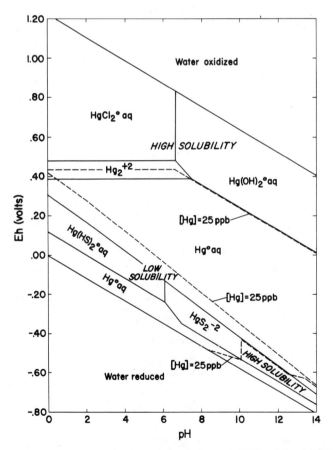

Figure 3.   Stability fields for aqueous mercury at 25°C and 1 atm [50].

Above a pH value of 5.0, elemental mercury is the predominant species in solution over most of the area of moderately oxidizing conditions. The solubility of metallic mercury is about 25 $\mu g/l$ at 25°C and is nearly constant over the whole area where the metal is stable. In acidic, oxygenated water with a high chloride ion concentration, the solubility of mercury is greatly increased due to the formation of the uncharged $HgCl_2$ or such anionic complexes as $HgCl_4^{2-}$. As the pH value is increased to the neutral or alkaline range, $Hg(OH)_{2(aq)}$, becomes the predominant species and the solubility of mercury is also increased.

At low redox potentials, mercury is precipitated as the sulfide cinnabar, which has extremely low solubility. The equilibrium solubility of mercury, near neutral pH, is only 0.002 $\mu g/l$ in the field of dominance of $Hg(HS)_{2(aq)}$ and HgS_2^{2-}. At pH values above 9.0, if much reduced sulfur is present mercuric sulfide anions become very soluble. The complexing of mercury by sulfide ions seems to be one of the major mechanisms of immobilization of mercury in reducing sediments. Under very strongly reducing conditions, the solubility of mercury may be increased due to formation of the free metal.

In the previous discussion, the relative stability of the different species of mercury under a variety of conditions was obtained from available thermodynamic data of the most significant species in water. However, frequent departure of natural systems from equilibrium is known, and often species that are not in equilibrium can be important components of natural waters. The following aspects of the chemistry of mercury are important sources of departure from theoretical predictions:

1. Elemental mercury has high volatility and can be vaporized at the air–water interface.
2. Inorganic mercury species have a strong tendency to form organic complexes through ligand exchange reactions. In addition, microbial transformation of inorganic mercury to organic forms in natural systems is well documented. These compounds are thermodynamically unstable with respect to water. However, kinetic barriers rather than thermodynamic ones are responsible for their stability in aquatic environments.
3. Elemental mercury [63] and most mercury species [50] exhibit higher solubilities in organic solvents than in water. This suggests that these forms are preferentially soluble in the lipid-rich membranes of living cells and, therefore, can be removed from water by living organisms [62].
4. Mercury tends to participate in dismutation reactions of the following type:

$$Hg_2^{2+} = Hg_{(1)}^0 + Hg^{2+} \tag{5}$$

This and similar reactions provide means by which mercury could be converted to the liquid form and escape as vapor [62].

Equilibrium Diagrams of Organic Mercury Species

The equilibrium concentrations of methylmercuric complexes were calculated by Baughman et al. [60] for an aqueous system containing hydroxides, chloride, hydrogen sulfide and its dissociated forms, thiols (RSH), amines (RNH$_2$), phenols (humic acid), ammonia, and orthophosphate at concentrations normally found in natural waters. Figures 4 and 5 indicate that methylmercuric sulfide complexes account for 95% of the methylmercury complexes in natural waters containing reduced sulfur species. Complexation of the methylmercury group by reduced sulfur species (H$_2$S, SH$^-$, and S^{2-}) results in the formation of two species: methylmercuric sulfide ion (CH$_3$HgS$^-$) and *bis* (methylmercury) sulfide ((CH$_3$Hg)$_2$S). These two species are in equilibrium:

$$2CH_3HgS^- + H^+ \rightleftharpoons (CH_3Hg)_2S + SH^- \tag{59}$$

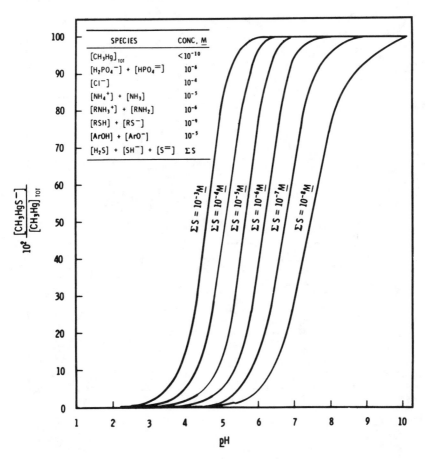

Figure 4.   Relative concentrations of CH$_3$HgS$^-$ in systems with many chemical species [60].

The relative concentrations of CH_3HgS^- and $(CH_3Hg)_2S$ depend on the pH value and total concentration of the reduced sulfur species. Methylmercuric sulfide ion, CH_3HgS^-, accounts for almost all of the complexed methylmercury in the pH range of 5 to 9 in the presence of high concentrations of reduced sulfur species ($10^{-3}-10^{-4}M$) (Figure 4). When the reduced sulfur concentration is decreased, CH_3HgS^- remains the predominant complex in basic waters. However, $(CH_3Hg)_2S$ becomes the predominant species in acidic waters (Figure 5).

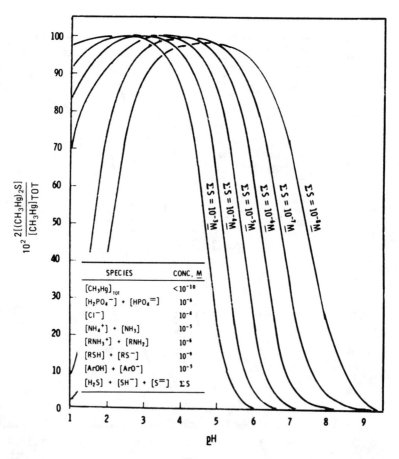

Figure 5. Relative concentrations of $(CH_3Hg)_2S$ in systems with many chemical species [60].

Chemical and Physical Reactions of Organomercury Compounds in Water

Acidolysis Reactions

Cleavage of the carbon–mercury bond in dialkyl- or diarylmercury compounds and organomercuric salts by protic acids is considered acidolysis [64,65]:

$$R_2Hg + HX \rightarrow RHgX + RH \tag{60}$$

$$RHgX + HX \rightarrow HgX_2 + RH \tag{61}$$

Several investigators [21,66] have suggested this reaction to be a pathway for the chemical transformation of dimethylmercury to methylmercury derivatives in the aquatic environment:

$$CH_3HgCH_3 + HX \xrightarrow{H_2O} CH_3HgX + CH_4 \tag{62}$$

where $X = Cl^-, Br^-, I^-, ClO_4^-$ or NO_3^- .

Studies [60] of the acidolysis of dimethylmercury by HCl, HBr and $HClO_4$ in water at 65-85°C indicated that the reaction followed second-order kinetics. Rate constants were independent of the nature of the acid, indicating no anion dependency. The extrapolated rate constant at 25°C was 7.33×10^{-5} liter/mol·sec.

The kinetics of acidolysis of diphenylmercury by $HClO_4$ was also studied [60] in an aqueous–ethanol mixture and in water. Aqueous ethanol was used because the solubility of diphenylmercury ($10^{-6}M$) was too low for convenient determination of the rate constants. The reaction was shown to follow second-order kinetics. Rate constants were increased as the ethanol concentration was reduced due to an increase in acidity. Extrapolation of the rate data obtained at higher temperatures to 25°C gave a rate constant of 9.67×10^{-3} liter/mol·sec for water. Product studies indicated that the reaction proceeded as follows:

$$C_6H_5HgC_6H_5 + H^+ + ClO_4^- \xrightarrow{H_2O} C_6H_5Hg^+ + ClO_4^- + C_6H_6 \tag{63}$$

The acidolysis half-lives at a pH value of 5 were calculated to be 33 years for dimethylmercury and 0.25 years for diphenylmercury at 25°C. Earlier studies on the acidolysis of organomercuric salts revealed that these reactions were also slow under natural conditions. The extrapolated second-order rate constant for the acidolysis of methylmercuric iodide by $1M$ H_2SO_4 was reported [67] to be 3×10^{-9} liter/mol·sec and a half-life time of 3×10^3 days at 25°C. Acidolysis of phenylmercuric chloride [68] in water containing 10% ethyl alcohol gave an extrapolated second-order rate constant of 1.7×10^{-6} liter/mol·sec at 25°C. These kinetic studies indicate that acidolysis reactions would be extremely slow under normal conditions in the natural aquatic environment.

Demercuration Reactions

Demercuration reactions constitute another possible route for chemical degradation of organomercuric salts:

$$RHgX \xrightarrow{\text{H}_2\text{O}} ROH + Hg^o + X^-$$ (64)

This reaction was shown to be kinetically first order, and the reaction rate is very rapid for such branched alkylmercuric salts as *t*-butylmercuric and cyclohexylmercuric halides [69-72]. However, demercuration of methylmercuric salts is very slow and has a first-order rate constant, at 25°C, of 8×10^{-13} sec^{-1}, or a half-life of about 3×10^4 years.

Another demercuration reaction important in the analytical chemistry of organomercurials is the halodemercuration reaction. Talmi and Mesmer [73] studied the stoichiometry and kinetics of the interaction of halogens with methylmercuric compounds in water. The reaction of I_2 with methylmercury chloride gave the following stoichiometry:

$$CH_3HgCl_{(aq)} + I_2 \rightleftharpoons CH_3I + \frac{1}{2} HgI_{2(s)} + \frac{1}{2} HgCl_2$$ (65)

where enough chloride is present to complex the CH_3Hg^+ ion. This reaction does not proceed in the dark but has to be initiated in the light. The reaction rates showed strong $[H_3^+O]$ dependence, being high at pH values below 6, and were decreased sharply with increasing $[OH^-]$ to 8 or 9. Bromine reacted similarly but the reaction rates were about 33 times slower than the I_2 reaction. No reaction was observed in Cl_2-saturated water.

Evaporative Loss of Organomercurials

Many organomercury compounds have high vapor pressures, and evaporative loss from the aqueous phase may constitute one of the important factors affecting the fate of mercury in the aquatic environment.

Dimethylmercury is a volatile liquid at ordinary temperatures (bp 93-96°C), with a solubility in water of $10^{-6}M$ (0-0.23 mg/l). The distribution coefficient for partition of $(CH_3)_2Hg$ between air and water is 0.31 at 25°C and 0.15 at 0°C, which leads to a heat of dissolution of $(CH_3)_2Hg$ gas of −4.5 kcal/mol [73]:

$$(CH_3)_2Hg_{(g)} \rightleftharpoons (CH_3)_2Hg_{(aq)} \qquad (66)$$

The solubility of methylmercuric chloride was estimated [73] to be $0.0236M$ (5930 ± 115 mg/l) in water and $0.0230M$ (5780 ± 180 mg/l) in $1M$ NaCl at 25°C. The distribution coefficient for CH_3HgCl between water and air at 22°C is 2.8×10^{-5}, which is extremely low compared to $(CH_3)_2Hg$. Table III shows the vapor pressure of CH_3HgX above the solids $CH_3HgCl_{(s)}$ and $CH_3Hg(OH)_{(s)}$ and $0.004M$ aqueous solution of CH_3HgCl. The volatility of methylmercuric chloride from water shows a strong pH dependence [73]. Figure 6 shows the effect of $[H_3O^+]$ on the vapor composition above a solution of $0.004M$ CH_3HgCl at 22°C. At pH values less than 5, CH_3HgCl accounts for almost all of the mercury species in the vapor phase because of its predominance in the aqueous phase. As the pH value rises, the concentration of CH_3HgX drops sharply in the vapor phase due to formation of the less volatile CH_3HgOH. It seems, therefore, that under normal environmental conditions, evaporative loss to the atmosphere of dissolved ionic mercurials is insignificant.

An estimate of the evaporative loss can be obtained by using Tsivoglou's [74] experimental data, which indicated that for a moderately turbulent river, dimethylmercury would have an evaporative half-life of about 12 hours. These same data predict that elemental mercury would be lost from the river at a rate 2.3 times faster than dimethylmercury. Bisogni and Lawrence [43] studied the methylation of inorganic mercury under aerobic and anaerobic conditions in a series of laboratory-scale reactors. A mass balance performed on all the mercury species in the reactors under steady-state conditions

Table III. Vapor Pressures of CH_3HgX [73]

Sample	N^a	Temperature (°C)	CH_3HgX Concentration in Vapor Phase (ng/ml)	Vapor Pressure (μm Hg)
$CH_3HgCl_{(s)}$	15	22	154 ±11.4	11.1 ±0.8
$CH_3HgCl_{(s)}$	3	0	12.1± 2.3	0.80±0.2
$CH_3HgOH_{(s)}$	2	22	77.4± 7.6	5.6 ±0.6
1000 ppm aq CH_3HgCl	12	22	27.8± 4.1	2.0 ±0.3

aNumber of determinations.

indicated that a large percentage of the inorganic mercury was transformed to the metallic form. It was stripped from the aqueous phase (Table IV) when such a carrier gas as air, methane or carbon dioxide was forced through these mercuric–mercury aqueous systems: aerobic and anaerobic microbial reactors, and abiotic systems. The reduction of mercuric salts to the free metal can be readily explained by comparing Figure 3 for the predominance of mercury species to the conditions of the test. Dimethylmercury was formed in small amounts, all of which was found in the effluent gas from the reactors. Monomethylmercury was the predominant methylation product at a neutral pH value and all was recovered in the aqueous phase. Although this study illustrated the evaporative loss of the volatile mercury species under conditions approximating a biological treatment process, the importance of this process under natural environmental conditions should not be overlooked.

ARSENIC

Background

Arsenic is one of the most widely distributed elements in the earth's crust and in the biosphere. The therapeutic properties of inorganic arsenic compounds have been recognized for centuries. Demonstrations of the therapeutic properties of organic arsenicals in the treatment of syphilis in the early 1900s had laid the foundation for modern chemotherapy. Organic arsenicals have since been used as spirochetocides, amebicides, trypanocides and inhibitors of many parasites. In addition, arsenic compounds have been used as stimulants and tonics. In modern medicine, arsenicals are used only for the treatment of trypanosomiasis and amebiasis and in the therapy of canine filariasis [75]. They are also used as growth stimulants for poultry, swine and calves and are administered as dietary supplements at concentrations of 0.005–0.01% of food [76].

Arsenic has also been recognized through the years for the toxic properties of some of its compounds, so that the word "arsenic" has become synonymous with "poison." It is these toxic properties of arsenic and the possible adverse health effects in drinking water supplies that are of major environmental concern.

The toxicity of arsenic compounds is not universal. Generally, trivalent arsenic is much more toxic than the pentavalent form [75,77]. Trivalent arsenicals exert their effects by reacting with the sulfhydryl groups in enzyme systems essential for metabolism, which inhibits their action [75, 77,78]. Arsenates, on the other hand, are not inhibitory to enzyme systems and can substitute for phosphates in some phorphorylases [78]. Toxicity

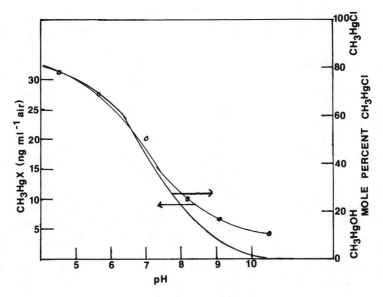

Figure 6. Effect of pH on the vapor pressure and the distribution of mercury species in the aqueous phase at 22°C (reproduced from Talmi and Mesmer [73] courtesy of *Water Res.*).

Table IV. Steady-State Distribution of Mercury Forms for Aerobic and Anaerobic Reactors [43]^a

Daily Input of Hg/l of Reactor (mg/day)	Weight percentage transformed to				
	Hg^0 (effluent gas) (%)	$(CH_3)_2Hg$ (effluent gas) (%)	CH_3Hg^+ (reactor) (%)	Remaining in Inorganic Form (reactor) (%)	Recovery of All Forms (%)
Aerobic Reactors					
0.0167	71.1	0.1	7.6	1.1	79.9
0.167	79.3	<0.1	1.3	2.1	82.8
1.67	88.5	<0.1	0.3	1.3	90.1
16.7	78.6	<0.1	0.1	0.4	79.1
Anaerobic Reactors					
0.0167	71.9	0.3	5.9	14.3	92.4
0.167	40.9	0.1	1.0	49.3	91.3
1.67	42.5	<0.1	0.2	52.0	94.6
16.7	48.0	<0.1	0.1	44.9	92.9

^{a}Net specific growth rate $\frac{1}{6}$/day at 30°C.

of arsenate to humans has been reported [77] to be several times less than that of arsenites. Arsine, however, appears to be the most toxic compound to man of all arsenicals.

Several studies have been reported on the toxicity of arsenical compounds to various species of fish and aquatic life [79,80]. Generally, pentavalent arsenates were found to be much less toxic than trivalent arsenites. These findings led McKee and Wolf [81] to recommend a limit of 1.0 mg/l As for marine fish and aquatic life.

Arsenic has been implicated as a carcinogenic agent, but there is considerable disagreement among researchers on this point. In experimental animals it has not been possible to induce carcinoma. There is, however, strong epidemiological evidence from Taiwan that relates the incidence of skin cancer to the arsenic content of drinking water [82,83]. A general survey was made on 40,421 inhabitants in 37 villages in the affected area. Of this population, 18.54% showed skin manifestations of chronic arsenicism. Furthermore, the highest incidence of the disease appeared among those inhabitants who used artesian wells for their water supply. The researchers Chen and Wu [83] speculated that arsenic was strongly suspected as the main cause of the disease. No arsenic contents of the artesian wells were reported.

Tseng et al. [84] reported a detailed epidemiological study of the relationship between the arsenic contents of artesian well waters and the occurrence of skin cancer, hyperpigmentation, keratosis and Blackfoot disease in Taiwan. In the villages surveyed, the arsenic contents of 114 well waters ranged from 0.01–1.82 ppm. Most of the well waters in the endemic area had an arsenic content around 0.4–0.6 ppm. Tseng et al. [84] concluded that "as was the case for skin cancer, the prevalence rates for hyperpigmentation, keratosis, and Blackfoot disease revealed clean-cut relationships between these conditions and the arsenic content of the water in the artesian wells. The greater the arsenic content, the higher was the prevalence."

Another report [85] relates several instances of arsenical cancer caused by drinking water. An interesting example occurred in the town of Reichenstein in the county of Glatz, Silesia, where for many centuries gold was produced from gold-containing arsenical ores. The smelting of large quantities of these ores led to the escape of arsenical fumes, which precipitated with the rain. In one of the brooks, 11.6 ppm of arsenic was found in the water and 475 ppm was found in a mud sample. Several serious diseases (called locally "Reichensteiner Krankheit" or Disease of Reichensteiner) and, in some cases, cancer, plagued this town for centuries. A change in the methods of smelting and a different water supply (whose arsenic content was 0.014 ppm) led to the disappearance of the "Reichenstein disease." In another area of the world, many cancers of the skin of peasants were observed in the town of Bellville, Cardoba province, Argentina. The wells of Bellville had arsenic contents of 2.66 and 4.3 ppm. This same report by Neubauer [85] mentioned (and unrelated to the above) that the Report of the Royal Commission on Arsenical Poisoning in 1903 established a limit of 0.14 ppm as As_2O_3 for arsenic in liquid foods and water.

Baroni and co-workers [86] reinvestigated the possible carcinogenic action of arsenic trioxide (As_2O_3) and sodium arsenate (Na_2HAsO_4) by using skin application and feeding tests in Swiss mice. They concluded that all tests failed to show any carcinogenic, initiating or promoting activity of the two arsenicals under the experimental conditions used. In view of available evidence that certain exposures to arsenic represent a human cancer hazard, further studies on the role of arsenic in carcinogenesis were considered necessary.

To date, the role of arsenic and arsenical compounds in producing cancer remains uncertain, and no definite proof has been reported.

Sources in the Environment

Natural Sources

Arsenic exists in the native state as the sulfide ores: orpiment (As_2S_3) and realgar (AsS), or as the arsenides or sulfarsenides of heavy metals, especially arsenopyrite (FeAsS). The latter mineral is the most common. Arsenic sulfides often occur wherever metallic sulfides are found, and small amounts regularly appear in conjunction with silver, lead, copper, nickel, antimony, cobalt and iron.

Arsenic is found at 1.0 ppm in siliceous sediments (cherts), and limestones have a mean concentration of 1.0 ppm, compared to 2 ppm in igneous rocks and 10 ppm in shale [87]. Fumarolic volcanic gases in Japan have shown that volatile arsenic and detritus around volcanoes may show high concentrations combined with metals or as sulfides. Soils contain arsenic in varying concentrations. The average content of soils from different countries was reported to be 5 ppm and of Russian soils, 3.6 ppm, whereas soils in regions of recent volcanism contain about 20 ppm [88]. In such areas as Buns, Switzerland, and in the Waiotapu Valley, New Zealand, arsenic in soil may reach 10,000 ppm [89]. In the presence of O_2 and Fe_2O_3, trivalent arsenic in soil is oxidized to the pentavalent form [84].

Weathering of rocks constitutes the major pathway of transport of arsenic from the continents to the oceans. The weathered arsenic may be in solution or part of the sediment load of rivers. Arsenic content in rivers ranges between 0 and 10 μg/l (Table V) [90]. Spring waters high in bicarbonate from California, Rumania, Kamachatka, USSR and New Zealand had 400–1300 μg/l [87]. Arsenic has also been detected in thermal waters, but the largest amounts have been found in a water of high salinity, Searles Lake, California, which showed 198–243 mg/l [77]. For an average value of 1 μg/l in rivers free from arsenic pollution and a volume of runoff of 3.3×10^{16} liter/yr, the soluble arsenic in rivers has been estimated [90] to be 33×10^3 ton/yr. For an average concentration of 2 μg/l in the oceans (Table V), the total arsenic content is estimated to be 2.8×10^9 metric tons.

Anthropogenic Sources

Man's activities have resulted in the introduction of large quantities of arsenic into the environment, whether by direct release, loss or waste. The annual world production of arsenic has been increased during the past 60 years [90]. At present, there is a steady production of about 50,000 metric tons/yr, with U.S. production about 15% and U.S. consumption about 50% of the total. Most of this arsenic is introduced eventually into the environment and reaches rivers, lakes and oceans. Arsenic is not mined as such, but is produced in the form of white arsenic, As_2O_3, as a by-product of smelting of arsenic-containing ores of copper, iron, zinc, lead, gold, silver, manganese and tin. Therefore, the quantity produced is directly related to the output of these metals. Arsenic evolves as dust with the flue gases and is trapped by filters or electrostatic precipitators for further processing. Most of the arsenic produced is used in the manufacture of such pesticides as lead arsenate, calcium arsenate and Cu acetate meta-arsenate (Paris Green), as well as in wood preservatives, such as Wolman salts, which contain 25% sodium arsenate. Because of replacement of arsenicals by organic pesticides, consumption of lead and calcium arsenates in the U.S. has declined considerably,

Table V. Arsenic Content of Natural Waters—Selected Reported Occurrence [90]

Fresh Waters	Concentrations (μg/l)	Reference
Lakes and Rivers		
Lakes in Greece	1.1-54.5	91
Lakes in Japan	0.16-1.9	92
Lakes in Wisconsin, U.S.A.	2-56	91
Rivers and lakes, U.S.A.	10-1100	93
Rivers in Sweden	0.2-0.4	90
Rivers in Japan	0.25-7.7 (weighted avg. 1.7)	92
Elbe River, Germany	20-25	92
Columbia River, U.S.A.	avg. 1.6	92
Oceans		
English Channel	2-4	92
Pacific coastal water	3-6	92
Northwest Pacific	0.15-2.5 (avg. 1.2)	92
Indian Ocean	1.3-2.2 (avg. 1.6)	92
South West Indian Ocean	1.4-5.0 (avg. 3)	92

while increasing abroad [77]. Smaller amounts of arsenic were employed in the manufacture of pharmaceuticals, glass, cattle and sheep dips, hide preservatives and poisoned bait weed killers. Arsenicals have also been used in the manufacture of dyestuffs and chemical warfare gases. Elemental arsenic is used in the production of various alloys, e.g., a small amount is added to lead in the manufacture of shot and produces hardened and more perfect spheres. The element increases the resistance of copper to corrosion, increases its machinability; and raises the annealing temperature. Arsenicals are also used as debarking agents in the paper industry, and arsanilic acid in small amounts (25–45 ppm) is added to swine and poultry feeds to promote growth [76].

Another significant source of arsenic release to the environment is the smelting or roasting of sulifde-containing ores and the combustion of fossil fuels. Smelting operations result in the release of large quantities of As_2O_3 to the atmosphere; however, the magnitude of this source has never been evaluated quantitatively. Crude petroleum oils contain a low concentration of As estimated [94] less than 1 mg/kg. Therefore, As release from the combustion of petroleum is considered insignificant. However, coal contains from 3–45 mg/kg, with an average value of 5 mg/kg [92]. Ferguson and Gavis [90] estimated that 2.5 g arsenic would be released to the atmosphere for every ton of coal consumed that assumes one-half of the arsenic is volatilized during combustion. Based on a total consumption of 117×10^9 tons of coal and lignite between 1900 and 1971, the quantity of arsenic released was calculated to be 29×10^4 tons in 71 years. Arsenic released to the atmosphere from fuel combustion and smelting operation is in the particulate form. Therefore, it is removed rapidly by dry fallout or rainout and eventually reaches the natural water systems.

Accumulation in the Food Chain

Aquatic organisms can concentrate and accumulate arsenic from the traces found in most natural waters. Marine organisms apparently accumulate higher concentrations than fresh water ones [92]. Oysters, prawns, shrimp and bass collected off the British Isles contained arsenic at concentrations of 3, 174, 42 and 40 mg/kg, respectively [95]. Vinogradov [88] reports that there is roughly ten times as much arsenic in marine water, silt, plants and animals as in their fresh water counterparts. All of the marine organisms examined [88] contained arsenic: sponges (8–24 mg/kg), molluscs (1–68 mg/kg), crustacea (10–79 mg/kg, wet weight); and other marine animals (2–50 mg/kg). The concentration in different varieties of fish ranged from 2 to 15 mg/kg by wet weight or 2 to 25 mg/kg by dry weight.

There have been few studies on the accumulation of arsenic in the biological food chain and on the accumulation in fresh-water fish. Ellis et al. [96]

reported on the arsenic content of 681 fresh-water fish representing 15 species taken from 15 inland waters in Florida, Georgia, Alabama and Texas. For these 681 fish, the maximum value of As as As_2O_3 was 2.78 ppm (total wet weight): average was 0.71 ppm and the minimum value was 0.02 ppm. Some typical average values (in ppm) for individual species were: shad, 0.86; German carp, 0.68; golden shiner, 1.30; spotted sucker, 0.28; black bass, 0.83; and top minnows, 0.55. No trout was reported. The body distribution of arsenic in 14 large-mouthed black bass was (all values in ppm wet weight): total fish, 0.66; eviscerated fish, 0.50; total viscera, 2.61; viscera without liver, 2.15; liver only, 6.06; and liver oil fraction, 40.51. It appears that arsenic is concentrated in the oily fraction of the liver of these fishes, which must be regarded as potential sources of arsenic in foods and other commercial products utilizing freshwater fish material. The arsenic contents of fresh-water amphipods were 3.25; isopods, 4.16; and crayfish, 5.46 (ppm, total dry weight). Apparently, these fresh-water crustaceans supply the arsenic to the fresh-water fish.

It should be indicated that rats were fed a diet of shrimp containing 17.9 ppm arsenic for 12 months with no ill effects. There was no evidence of toxicity in their growth, physical appearance and activity, or by histological examination of the liver, spleen and kidney [97].

Ullmann et al. [98] reported on the arsenic accumulation by fish in lakes treated with sodium arsenite. For Calico bass, the average arsenic content of the viscera was 0.33 ppm (wet weight) and of the fillet, 0.38 ppm, before the lakes were treated. After sodium arsenite treatment, the average arsenic content of the viscera was 0.56 ppm and of the fillet, 0.28 ppm. In two other lakes, however, the arsenic contents of the viscera and fillet were less than 0.10 ppm before and after treatment. The authors [98] felt that 21 days was not enough time for a significant accumulation of arsenic to occur in these fish.

Biological Transformations

Arsenic, like mercury, undergoes transformation in the environment through the metabolic activities of microorganisms, especially bacteria and and fungi. McBride and Wolf [99] demonstrated that microorganisms can reduce pentavalent arsenic (arsenate) to the trivalent form (arsenite). On the other hand, several investigators [100,101] showed that microorganisms are able to oxidize arsenite to less toxic arsenate:

$$HO - \overset{\overset{\displaystyle OH}{|}}{\underset{\underset{\displaystyle O}{\|}}{As}} - OH \xrightarrow{2e} \overset{}{\underset{\underset{\displaystyle O}{\|}}{As}} - OH \qquad (67)$$

Inorganic arsenic can also be transformed to organic forms by biological methylation. McBride and Wolfe [99], using cell-free extracts and whole cells of *Methanobacterium*, demonstrated that microorganisms can methylate inorganic arsenic to dimethylarsine under anaerobic conditions.

The biochemical pathway leading to formation of methylated arsenic compounds proceeds in this manner [99]: Arsenate is first reduced to arsenite, which is then methylated to form methylarsonic acid. Reductive methylation of the latter results in the formation of dimethylarsinic acid (cacodylic acid). This acid is then reduced to form dimethylarsine. Methylarsonic acid is not reduced in the absence of a methyl donor, which suggests a second methylation must occur before the compound can be reduced to arsine. Dimethylarsonic acid, on the other hand, is reduced rapidly to alkylarsine, even in the absence of a methyl donor. The analytical methods used to identify the alkylarsine could not distinguish dimethylarsine from equimolar quantities of mono- or trimethylarsine derivatives. However, the ratio of methyl groups to arsenic in the arsine product indicated that dimethylarsine was the principal product.

It has long been recognized that several species of fungi are capable of biosynthesis of alkylarsenes [102,103] from inorganic arsenic. Cox and Alexander [104] have shown that *Candida humicola* is capable of converting arsenate, arsenite, monomethylarsonic or dimethylarsonic acid, which are pesticides or are formed from pesticides, to trimethylarsine. The identity of the biologically produced trimethylarsine was verified by mass spectrometry. High phosphate concentrations in the culture media inhibited the formation of trimethylarsine from arsenate, arsenite and monomethylarsenate, but not from dimethylarsenate, indicating that the latter compound is the principal precursor for formation of the alkylarsine.

Dimethylarsinic acid (cacodylic acid), which is used as a herbicide and defoliant, undergoes biological degradation in soils [105] and aquatic systems [106]. Degradation proceeds under aerobic and anaerobic conditions by reduction of the acid to a volatile alkylarsine, presumably dimethylarsine, which is lost to the atmosphere. In addition, cleavage of the C-As bond occurs under aerobic conditions to yield carbon dioxide and arsenate [105,106].

The above discussion indicates that there is a biological cycle for arsenic, just as there is one for mercury, in the aquatic environment (Figure 7). Unlike alkylmercury compounds, the alkylarsines do not accumulate to a significant extent in fish and other organisms in the upper levels of the food chain [106]. Alkylarsenicals were shown [107] to be widely distributed in nature. Analyses of a wide range of natural waters in and around Tampa, Florida (Table VI) revealed the presence of both methylarsonic acid and dimethylarsinic acid in almost all samples [107]. Dimethylarsinic acid was found to be a major form of arsenic in the environment. Methylarsonic acid

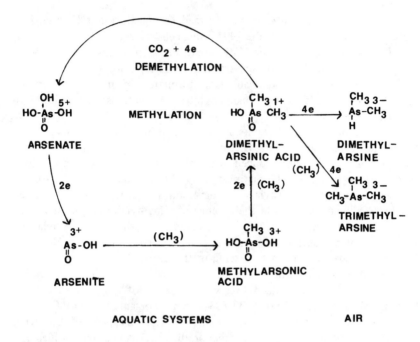

Figure 7.   Biological cycle for arsenic [46].

was present in lower concentrations, which is a probable consequence of being an intermediate in the arsenic methylation sequence.

The methylation of arsenic is considered, similar to mercury, as a detoxification mechanism by microorganisms. However, these biological transformations are of extreme importance in determining the mobility and fate of the element in the environment.

Chemistry in Aqueous Systems

Equilibrium Reactions

General. Arsenic is a metalloid that exists in four oxidation states in aqueous systems: -3, 0, $+3$, $+5$. The reduced state, -3, is represented by gaseous arsine, $AsH_{3(g)}$, the elemental state, As^0, is of rare occurrence, whereas the two oxidation states are represented by arsenites, $+3$, and arsenates, $+5$. The standard free energy of formation for some arsenic species is shown in Table VII. Arsenic forms several such solid phases as arsenious anhydride, As_2O_3, arsenic anhydride, As_2O_5, arsenious sulfide, As_2S_3, arsenic

Table VI. Analysis of Environmental Samples (The Results Are Given in ppb as Arsenic; the Precision is ±10% Relative or ±0.01 ppb [107])

Sample	As(III) (ppb)	As(III) (%)	As(V) (ppb)	As(V) (%)	Methylarsonic acid (ppb)	Methylarsonic acid (%)	Dimethylarsenic acid (ppb)	Dimethylarsenic acid (%)	Total (ppb)
			Fresh water samples						
Hillsborough River	<0.02	<10	0.25	100	<0.02	<10	<0.02	<10	0.25
Withlacoochee River	<0.02	<5	0.16	30.7	0.06	11.5	0.30	57.7	0.52
Wellwater Near Withlacoochee River	<0.02	<3	0.27	46.6	0.11	19	0.20	34.5	0.58
Remote Pond, Withlacoochee Forest	<0.02	<2	0.32	30	0.12	11	0.62	58.5	1.06
University Research Pond, USF	0.79	40.5	0.96	49	0.05	2.6	0.15	7.7	1.95
Lake Echols, Tampa	2.74	76.5	0.41	11.4	0.11	3.1	0.32	8.9	3.58
Lake Magdalene, Tampa	0.89	51	0.49	28	0.22	12.6	0.15	8.6	1.75
			Saline waters						
Bay, Causeway	0.12	6.8	1.45	81.9	<0.02	<1	0.20	11.3	1.77
Tidal Flat	0.62	27	1.29	56.6	0.08	3.5	0.29	12.7	2.28
McKay Bay	0.06	4	0.35	23.6	0.07	4.7	1.00	68	1.48

Table VII. Standard Free Energies of Formation of Some Arsenic
Species at 25°C [56,108]

Mineral Name	Formula[a]	ΔG_f^0 (kcal/mol)
Arsenic	$As_{(s)}^0$	0.0
Arsenolite	$As_2O_{3(s)}$	−137.731
Arsenic Anhydride	$As_2O_{5(s)}$	−184.60
Arsenyl Ion	AsO^+	−39.1
m-Arsenious Acid	$HAsO_2$	−96.25
m-Arsenite Ion	AsO_2^-	−83.7
o-Arsenious Acid	H_3AsO_3	−152.94
Mono-o-arsenite Ion	$H_2AsO_3^-$	−140.40
	$HAsO_3^{-2}$	−125.3
o-Arsenic Acid	H_3AsO_4	−183.80
Mono-o-Arsenate Ion	$H_2AsO_4^-$	−178.90
Di-o-Arsenate Ion	$HAsO_4^{-2}$	−169.00
Tri-o-Arsenate Ion	AsO_4^{-3}	−152.00
Arsine	$AsH_{3(g)}$	16.50
	AsH_3	23.8
	$HAsS_2$	−11.6
	AsS_2^-	−6.56
Realgar	$AsS_{(s)}$	−16.806
Orpiment	$As_2S_{3(s)}$	−40.25
Magnesium Arsenate	$Mg_3(AsO_4)_{2(s)}$	−679.3
Calcium Arsenate	$Ca_3(AsO_4)_{2(s)}$	−725.32

[a]All species are aqueous unless otherwise noted.

sulfide, As_2S_5, calcium arsenate, $Ca_3(AsO_4)_2$, and magnesium arsenate, $Mg_3(AsO_4)_2$. The solubility equilibria for some of these solid phases are shown in Table VIII.

Acid–Base Equilibria. Arsenic enters into several acid–base equilibria in water through the formation of two acids: arsenious (+3) and arsenic (+5). These reactions and their equilibrium constants are seen in Table IX. Arsenious acid, $HAsO_2$, is formed from the dissolution of arsenious anhydride, As_2O_3, which has a solubility in water of 21 g/l (0.106M) at 25°C. Arsenious acid is an amphoteric compound and forms arsenyl ions, AsO^+, at pH values below −0.9 and arsenite ions, AsO_2^-, at pH values above 9.2. Figure 8 shows the acid–base equilibria for arsenious acid. It can be seen that

Table VIII. Solubility Equilibria of Arsenious and Arsenic Solid Phases at 25°C

Reaction No.	Reaction	log K_{eq}
(68)	$2AsO^+ + H_2O = As_2O_{3(s)} + 2H^+$	-2.04[a]
(69)	$As_2O_{3(s)} + H_2O = 2HAsO_2$	-1.36[a]
(70)	$As_2O_{3(s)} + H_2O = 2AsO_2^- + 2H^+$	-19.78[a]
(71)	$As_2O_{5(s)} + 3H_2O = 2H_3AsO_4$	9.48[a]
(72)	$As_2O_{5(s)} + 3H_3O = 2H_2AsO_4^-$	2.30[a]
(73)	$As_2O_{5(s)} + 3H_2O = 2HAsO_4^{2-} + 4H^+$	-12.24[a]
(74)	$As_2O_{5(s)} + 3H_2O = 2AsO_4^{3-} + 6H^+$	-37.18[a]
(75)	$Ca_3(AsO_4)_{2(s)} = 3Ca^{2+} + 2AsO_4^{3-}$	-18.17[b]
(76)	$Mg_3(AsO_4)_{2(s)} = 3Mg^{2+} + 2AsO_4^{3-}$	-19.68[b]

[a]From Pourbaix [56].
[b]From Bjerrum [108], at 20°C.

Table IX. Acid–Base Equilibria of Arsenious and Arsenic Acids at 25°C [56]

Reaction No.	Reaction			log K_a
	+3			
(77)	AsO^+	$+ H_2O = HAsO_2 + H^+$		0.34
(78)	$HAsO_2$	$= AsO_2^- + H^+$		-9.21
	+5			
(79)	H_3AsO_4	$= H_2AsO_4^- + H^+$		-3.60
(80)	$H_2AsO_4^-$	$= HAsO_4^{2-} + H^+$		-7.26
(81)	$HAsO_4^{2-}$	$= AsO_4^{3-} + H^+$		-12.47

$HAsO_2$, a very weak acid, is the predominant form between the pH values of -0.9 and 9.2 when the arsenic content lies below 21 g/l.

Arsenic acid is formed from the dissolution of arsenic anhydride, As_2O_5, which is extremely soluble in water, 658 g/l at 20°C ($2.86M$). Arsenic acid is a triprotic acid and has three protolysis constants, as seen in Table IX. Consequently, three anionic species are formed in the water phase: $H_2AsO_4^-$, $HAsO_4^{2-}$, and AsO_4^{3-}. The predominant species will depend, of course, on the pH value of the water phase. Figure 9 shows the acid–base equilibria for arsenic acid.

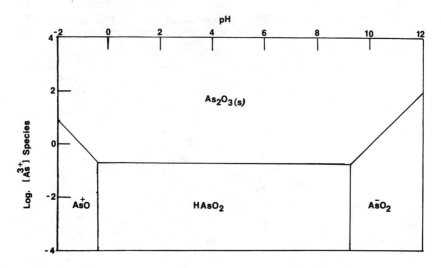

Figure 8.   Acid–base equilibria for arsenious compounds.

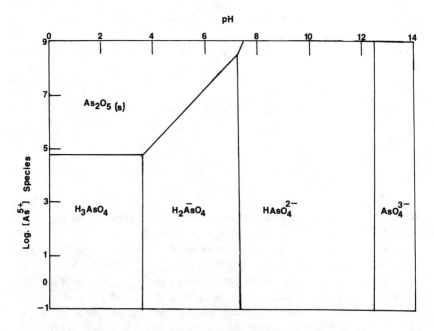

Figure 9.   Acid–base equilibria for arsenic compounds.

Oxidation–Reduction Equilibria. Figure 10 shows an Eh-pH diagram for arsenic (total As concentration $10^{-5}M$) in an aqueous system containing sulfur (total S concentration $10^{-3}M$) [90] that was constructed from the available thermodynamic data in Tables VII–X. This diagram shows the predominant soluble species and the solids whose solubilities are low enough to occur in this system. The cross-hatched area indicates regions of solubility less than $10^{-5.3}M$, and the solid species are enclosed in parentheses.

Arsenic acid species (H_3AsO_4, $H_2AsO_4^-$, $HAsO_4^{2-}$ and AsO_4^{3-}) are stable at the high Eh values encountered in oxygenated waters. Under mildly reducing conditions or low Eh values, arsenious acid species (H_3AsO_3, $H_2AsO_3^-$ and $HAsO_3^{2-}$) become stable. The arsenic oxides, As_2O_5 and As_2O_3, do not appear on the diagram because of their high solubility. At a pH value below 5.5 and Eh value about 0.0 V, realgar (AsS) and orpiment (As_2S_3), which have low solubilities, occur as stable solids. The predominant species at low pH values in the presence of sulfides is $HAsS_{2(aq)}$, which has a maximum solubility of $10^{-6.5}M$ (0.025 mg/l As), whereas AsS_2^- predominates at pH values higher than 3.7. The solubility of AsS_2^- at pH values above 5.5 is $10^{-5}M$. Metallic As is thermodynamically stable at lower Eh values. Arsine, AsH_3, may be formed at extremely low Eh values. The solubility of arsine in water

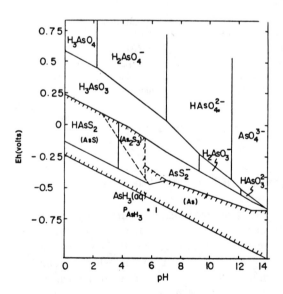

Figure 10.   The Eh–pH diagram for As at 25°C and 1 atm with $[As]_t = 10^{-5}M$ and $[S]_t = 10^{-3}M$. Solid species in ( ) in cross-hatched area have solubilities less than $10^{-5.3}M$ (reproduced from Ferguson and Gavis [90] courtesy of *Water Res.*).

Table X. Reduction Reaction of Arsenious and Arsenic Compounds at 25°C [56]

Reaction No.	Reaction	$E_h^{o\ a}$ (V)
(82)	$H_3AsO_4 + 3H^+ + 2e = AsO^+ + 3H_2O$	0.550
(83)	$H_3AsO_4 + 2H^+ + 2e = HAsO_2 + 2H_2O$	0.560
(84)	$H_2AsO_4 + 3H^+ + 2e = HAsO_2 + 2H_2O$	0.666
(85)	$HAsO_4^{2-} + 4H^+ + 2e = HAsO_2 + 2H_2O$	0.881
(86)	$HAsO_4^{2-} + 3H^+ + 2e = AsO_2^- + 2H_2O$	0.609
(87)	$AsO_4^{3-} + 4H^+ + 2c = AsO_2^- + 2H_2O$	0.977
(88)	$As_2O_{3(s)} + 6H^+ + 6e = 2As + 3H_2O$	0.234
(89)	$As_2O_{5(s)} + 10H^+ + 10e = 2As + 5H_2O$	0.429
(90)	$As_2O_{5(s)} + 4H^+ + 4e = As_2O_{3(s)} + 2H_2O$	0.721
(91)	$AsO^+ + 2H^+ + 3e = As + H_2O$	0.254
(92)	$HAsO_2 + 3H^+ + 3e = As + 2H_2O$	0.248
(93)	$AsO_2^- + 4H^+ + 3e = As + 2H_2O$	0.429
(94)	$AsO_4^{3-} + 8H^+ + 5e = As + 4H_2O$	0.648
(95)	$2H_3AsO_4 + 4H^+ + 4e = As_2O_{3(s)} + 5H_2O$	0.580
(96)	$2H_2AsO_4^- + 6H^+ + 4e = As_2O_{3(s)} + 5H_2O$	0.687
(97)	$2HAsO_4^{-2} + 8H^+ + 4e = As_2O_{3(s)} + 5H_2O$	0.901
(98)	$2AsO_4^{3-} + 10H^+ + 4e = As_2O_{3(s)} + 5H_2O$	1.270
(99)	$As + 3H^+ + 3e = AsH_{3(g)}$	−0.608
(100)	$O_2 + 4H^+ + 4e = 2H_2O$	1.229
(101)	$2H_2O + 2e = H_2 + 2OH^-$	−0.826
(102)	$As + 3H_2O + 3e = AsH_{3(g)} + 3OH^-$	−1.21

[a]Signs of the half-cells potentials are in accord with the IUPAC convention (see Chapter 4).

is very low, at a partial pressure of about 1 atm the calculated solubility is $10^{-5.3}M$.

No other ions encountered in natural systems, except iron, show significant reactions with arsenic to change the areas of predominance of the soluble species shown in the diagram (Figure 10). In the presence of iron, a small region of stability will occur at a pH value below 2.3 and an Eh value above +0.74 V. Organic arsenicals do not appear in the diagram because they are thermodynamically unstable in the system.

The stability relationship discussed in the previous section should be considered as an attempt to understand the possible oxidation–reduction patterns in natural waters. As discussed previously, natural systems are highly dynamic, and the kinetics of the different equilibrium reactions, which are not known for arsenic species, should be evaluated before a precise model can be established. In addition, biological transformations result in the production or organic intermediates that have major environmental significance. These cannot be predicted by thermodynamic models but they exist in the environment because of kinetic barriers.

LEAD

Background

Lead is widely distributed in the earth's crust and is also found in the atmosphere and hydrosphere. Because of its low melting point (327°C), malleability, ductility and resistance to corrosion, it has been used in the manufacture of metal products for thousands of years. The extensive, long-term mining, smelting and commercial uses have resulted in a considerable amount of information regarding the harmful effects of industrial exposure to man, animals and plants. Thus, it has long been recognized that lead could be absorbed by inhalation and ingestion and that lead absorption was responsible for loss of movement in printers' fingers exposed to heated lead type and for "dry grippes" in pottery and glass workers.

Lead is not considered an essential element to the nutrition of animals and humans [109]. Being foreign to the human body, it is a cumulative poison that tends to be deposited in bone [110] but that is also found in brain, liver, kidney, aorta and muscles. Absorbed lead is transported to these tissues via the blood system, and some portion is removed from the body through the alimentary tract and urinary system. Lead poisoning in its advanced forms causes permanent damage to the body or may cause death. The poisoning usually results from the cumulative toxic effects of lead after continuous, long-term consumption, rather than from occasional small doses [111].

The major sources of lead input to man are food, air and tobacco smoke, as well as water and other beverages. The daily average intake of lead by adults in North America is about 0.33 mg. Of this quantity, 0.01–0.03 mg/day are derived from water used for cooking and drinking [112,113]. A total intake of lead in excess of 0.6 mg/day may result in the accumulation of a dangerous quantity during a lifetime. Drinking water containing lead in concentrations varying from 0.042 to 1.0 mg/l or more has caused lead poisoning among human beings [81]. On the other hand, concentrations ranging from 0.01 to 0.16 mg/l apparently have been not poisonous over long periods of time [114]. The maximum acceptable concentration of lead in water used as drinking water is 0.05 mg/l. This limit was established to provide protection based on the combined exposure from both food and water [1].

Sources in the Environment

Natural Sources

Lead exists in nature mainly as the sulfide (galena). Lead carbonate (cerussite), lead sulfate (anglesite) and lead chlorophosphate (pyromorphite, $Pb_5Cl(PO_4)_3$) are also naturally occurring forms. The low solubility of lead in the aqueous phase of natural systems and the formation of stable complexes with organic matter are manifested in the low uptake by some plants and animals. There are extremely low concentrations of lead in natural bodies of water in proportion to the concentration in the beds of lakes and streams. The net effect of these sluggish dynamics is a high degree of accumulation with prolonged exposure.

The natural concentration of lead in soils is in the range of 2 to 200 ppm, exclusive of areas near deposits of lead ores [115,116]. Soils in rural areas of the U.S. have lead concentrations usually similar to the average content in the earth's crust, 10-15 $\mu g/g$ [117]. Higher concentrations in soils of lead in specific areas are attributed to such manmade sources as mining, fallout, industrial operations and motor vehicles. The lead contents of soils along high-traffic density highways have been studied by several investigators. Substantial increase was observed in the top layer of soils adjacent to highways (150-522 $\mu g/g$), which decreased with distance and with depth [118]. In addition to these sources, soils receive, on the average, 1 $\mu g/cm^2/yr$ from precipitation, and 0.2 $\mu g/cm^2/yr$ may be deposited in dustfall.

The contribution of natural sources to present day concentrations of lead in the atmosphere apparently is insignificant. Natural concentrations have been estimated to be about 0.0005 $\mu g/m^3$ of air [119] and result from airborne dust containing on the average 10–15 ppm of lead [117] and from gases diffusing from earth's crust [120].

Natural concentrations of lead determined in the annual ice layers from the interior of northern Greenland [121] show that lead concentrations were increased from less than 0.0005 $\mu g/kg$ of ice at 800 B.C. to more than 0.2 $\mu g/kg$ in 1965. The sharpest rise occurred after 1940 and today's lead concentrations in Greenland are about 400 times the natural levels. The increases in lead content in other ecosystems have been documented. For example, it is estimated that the preindustrial lead content in marine water was about 0.02-0.04 $\mu g/kg$ [122]. At the present time, surface waters in some areas of the Mediterranean Sea and Pacific Ocean contain as much as 0.20 and 0.35 $\mu g/kg$, respectively, and only deep waters, below 1000 meters, appear uncontaminated [122]. The lead content of fresh waters also has increased in recent times. The mean global natural lead content of lakes and rivers is estimated to be 1-10 $\mu g/l$ [123].

Anthropogenic Sources

The current world production of lead is about 2.5 million ton/yr, of which 40% is in the United States. The yearly consumption of primary and second-ary (recycled) lead in the U.S. has nearly doubled in the last 20 years, reaching about 1.3 million tons [124]. The largest single user is the electric storage battery industry, which accounted for 38.7% of the lead consumed in 1968. The metallic form is used to make the grids and lugs, and the lead oxides (litharge, PbO_2, red lead, Pb_3O_4; and black oxide, Pb_2O) constitute the active material pasted on the plates. The petroleum industry is the second largest consumer of lead and accounts for 20% of the total, used mostly for gasoline additives (lead alkyls). The paint industry also uses several lead compounds in the manufacture of different paints. White lead (basic lead carbonate, $2PbCO_3 \cdot Pb(OH)_2$) and red lead are used in exterior house paints and other outdoor uses because of their superior weathering characteristics. Lead chromates provide yellow, green and red pigments used in traffic marking paints and printing inks. Lead is also used in the ceramics industry; lead oxides and silicates are principal components in glazes for china and structural clays. Other uses of lead include the manufacture of such insecti-cides as lead arsenate (decreasing since 1946) and in electric cable insulation, hose, pipe, sheet and floor coverings, as well as a stabilizer in vinyl plastics. Wastewaters from these industrial and manufacturing operations contain varying concentrations of lead, which may reach the natural waters in the effluent discharges.

Lead in the Atmosphere

Considerable amounts of lead are emitted to the atmosphere in the form of particulate matter from a variety of sources. An inventory of the sources of lead emissions in the U.S. [124] for 1968 is shown in Table XI. These data indicate that the combustion of leaded gasoline constitutes approximately 98% of the total emission of lead from all sources. The burning of coal has not been a serious source of atmospheric lead in comparison with the burning of gasoline. Analysis of 827 samples of domestic coal provided a weighted average of 7 ppm [125]. This concentration would produce 3528 tons of lead of which 747 either remained in the clinker or were collected with fly ash. In modern furnaces burning pulverized coal, about 95–99% of the fly ash is removed by electrostatic precipitators and only a small portion of the lead is emitted as aerosols [126]. Improved industrial controls have also resulted in a considerable reduction of the amounts of lead emitted to the atmosphere from smelting of lead ores. Currently, the con-tribution of this source to atmospheric lead content is considered small.

Table XI. Lead Emission in the United States, 1968 [124]

Emission Source	Lead Emitted (ton/yr)
Gasoline Combustion	181,000
Coal Combustion	920
Fuel Oil Combustion	24
Lead Alkyl Manufacturing	810
Primary Lead Smelting	174
Secondary Lead Smelting	811
Brass Manufacturing	521
Lead Oxide Manufacturing	20
Gasoline Transfer	36
Total	184,316

The concentration of lead in ambient air has been closely correlated with the density of vehicular traffic [127,128]. It is highest in what might be termed "vehicular microclimates" in large cities. For example, a comparison of estimated emissions of lead from several sources in the Cincinnati area indicates that the principal contributor is the combustion of leaded gasoline [129], each gallon of which contains about 2.4 g of lead as an antiknock additive. The concentration decreases gradually as one moves from urban areas into the suburbs or smaller towns and finally into rural areas. Results of the National Air Surveillance Networks indicated that the average concentration of lead in the atmosphere of most cities in 1953-1966 was 1–3 $\mu g/m^3$, but suburban and nonurban stations averaged 0.1–0.5 $\mu g/m^3$. Many concentrations at some of the very rural stations were less than 0.05 $\mu g/m^3$ [130,131]. Studies have indicated that atmospheric fallout of lead-bearing particulates can be a significant source of lead input into surface waters, either directly or by transport of deposited particles on land surfaces in runoff [118,132-134].

Lead aerosols that are largely airborne for long distances from the emission sources because of their small size are removed from the atmosphere by precipitation. Lazrus et al. [135] determined the lead concentration in precipitation at 32 U.S. stations and found a correlation between the gallons of gasoline used and the concentration of lead in rainfall in each area. The average lead concentration was 34 $\mu g/l$ in precipitation, with a median of about 10 $\mu g/l$. The highest concentration was about 300 $\mu g/l$. The deposition rate by rainfall was estimated as high as 138 g/ha/month, and rates over 25 g/ha/month were observed at most stations in the northeastern part of the U.S. Hem and Durum [133] determined the concentration of lead in rain water at Menlo Park, California during 1971. The concentrations varied from a few $\mu g/l$ to more than 100 $\mu g/l$. The total lead concentration varied widely from storm to storm and within individual storms, but the total rainout of lead was small compared to the relatively constant dry fallout rate at this location. However, in semirural areas an average of 1 $\mu g/cm^2/yr$ of lead is precipitated from the atmosphere [136] corresponding to a rate of 9 g/ha/ month. The residence time of lead in the atmosphere was estimated from an analysis of Pb^{210} in rain water to range from 7 to 30 days, indicating the efficiency with which aerosols are removed by precipitation [137,138].

Lead in Natural Waters

Atmospheric fallout and rainout of particulate lead are considered the most significant sources of lead input into natural surface waters, especially in urban areas. Storm runoff can bring significant amounts of lead into solution, in addition to the transport of leadbearing sediments into receiving waters. The runoff rates, however, will determine the total lead concentration in transport. Generally, areas with high runoff rates are less likely to have high lead fallout rates, but runoff originating in urban areas will tend to be high in lead concentration.

The chemical composition of runoff waters is also an important factor in determining the concentration of dissolved lead. Hem and Durum [133] suggested that runoff waters with a pH value near 6.5 and an alkalinity below 30 mg/l (as HCO_3^-) are expected to contain high concentrations of soluble lead ranging from 40 $\mu g/l$ up to several hundred $\mu g/l$. Such waters may occur in high runoff areas with significant lead fallout and rainout. However, waters with alkalinities exceeding 60 mg/l and a pH value near 8.0 are expected to contain low concentrations of dissolved lead, generally below 10 $\mu g/l$. Durum et al. [93] reported the dissolved concentration of several minor elements, including lead, in 726 sites on rivers and lakes in the U.S. and Puerto Rico. The limit of detection for lead in that study was 1 $\mu g/l$. Detectable concentrations were found in 63% of the samples, with only three samples containing more than 50 $\mu g/l$. However, there was a distinct regional pattern in the data, as shown in Table XII. This regional distribution of lead in the stream waters is in agreement with the suggestion that the water composition in the eastern states is commonly favorable for the dissolution of lead [133]. Also, rainfall data [135] show that lead is supplied to this region in larger amounts than the nationwide average.

Most of the available information on the lead content of rivers and lakes is concerned with dissolved lead. However, a significant portion of lead input into natural waters is in the form of particulate lead that is transported in the runoff from rain storms, especially in urban areas. Samples of runoff originating from paved streets and other impervious surfaces in the city of

Table XII. Regional Summary of Lead in Surface Water of the U.S. [93]

Region	Maximum (μg/l)	Minimum (μg/l)	Median (μg/l)	<1 μg/l (%)	>1 μg/l (%)	>10 μg/l (%)
New England and Northeastern	890	<1	6	8	92	16
Southeastern	44	<1	4	27	73	22
Central	84	<1	<1	51	49	8
Southwestern	34	<1	<1	61	39	16
Northwestern	23	<1	<1	62	38	2

Palo Alto, California and discharging into the nearby Matadero Creek showed a total lead content of 93 μg/l, of which 91 μg/l could be removed by filtration. The lead content of the sediment fraction was about 0.11%, which is about the same percentage observed in the dry fallout in that area [133].

Samples of flood runoff collected in November, 1971 from two small streams draining urbanized areas around San Francisco Bay contained high concentrations of suspended lead ranging from 200 to 680 μg/l; more than 90% of the total lead was associated with the sediment fraction [134].

Studies on the distribution of lead in Lake Michigan [139] indicated that the concentration in the upper few centimeters of water near the air–water interface ranged from 2 μg/l offshore at Chicago to approximately 0.4 μg/l ten miles west of West Haven, Michigan. Substantially higher concentrations of lead were found in fine-grained sediments near the sediment–water interface, which decreased in underlying sediments of southern Lake Michigan. The mean concentration of lead in surficial sediments was 88 μg/g, compared to 20 μg/g in sediments at depths ranging from 15 to 100 cm. The mean concentration in suspended matter was 56 μg/g [139].

Effects of Lead on Aquatic Life

The toxicity of lead to aquatic organisms has long been recognized in England. The disappearance of fish from British streams receiving effluents from lead mines and the deaths of waterfowl and other animals in the vicinity of streams were reported as early as 1874 [140]. The continued absence of fish in rivers passing through old mining areas led Carpenter [141] to conduct the first definitive experiments on lead poisoning in fish. These studies led to the formulation of the "coagulation film anoxia" theory [140]. When fish are placed in lead-contaminated water, a film of coagulated mucous is formed over the entire body and is particularly prominent over the gills. The insoluble material interferes with the respiratory function of the gills, resulting in acute respiratory distress and death by suffocation. This effect, however, could also be produced by toxic concentrations of other heavy metal ions, including zinc, copper, cadmium and mercury [140].

The toxicity of lead to fresh-water fish is influenced by the physicochemical characteristics of the water: temperature, dissolved oxygen concentration, calcium content and, most importantly, the water hardness, which determines the lead solubility. The effect of hardness on the toxicity of lead was demonstrated by acute toxicity tests on several species of fish in waters of varying hardness. In soft waters (20–45 mg/l $CaCO_3$), the 96-hour LC_{50} values were: 1.0 mg/l for rainbow trout, 4.0–5.0 mg/l for brook trout and 5.0–7.0 mg/l for fathead minnows [80]. In hard waters, 96-hour LC_{50} values were 442 mg/l for brook trout and 482 mg/l for fathead minnows. Detrimental effects to rainbow and brook trout were observed at a concentration of 0.1 mg/l in soft waters [80].

There is neither any available information in the literature concerning the ability of fresh-water fish to concentrate lead from the surrounding medium, nor any studies in which the lead concentration was measured simultaneously in fish and water. Concentrations as high as 12 ppm in liver, 5.7 ppm in gills and 1.4 ppm in muscle of fish taken from a lake near a rich lead mine were reported [142]. However, the concentration of lead in the lake water was not reported.

Marine organisms, on the other hand, have the ability to concentrate lead up to 40,000 times [81]. The normal concentration of lead in sea water is 0.03 μg/l [143] in the bulk of the water in deep ocean basins. However, the concentration can be as high as 0.3 μg/l near the surface and shallow water near continents [144]. Although the lead content reported in seafood is relatively low, it indicates a considerable ability to be concentrated by these organisms. The reported lead concentration range in seafood is 0.17–2.5 ppm, with an average of 0.5 ppm [145]. Average wet weight concentrations in eastern oysters and soft-shell clams were reported to be 0.47 and 0.7 ppm, respectively [145]. The remarkable ability of the eastern oyster to concentrate lead was demonstrated by exposing oysters to flowing sea water containing lead concentrations of 0.025, 0.05, 0.1 and 0.2 mg/l. The total accumulation of lead amounted to 17, 35, 75 and 200 ppm (wet weight) after 49 days, with the highest concentrations in the liver and the lowest in muscle tissue [146]. The higher lead concentrations in the sea water (0.1 and 0.2 mg/l), however, had deleterious effects on the oysters. The recommended maximum lead concentration in estuarine waters is 0.05 mg/l [80].

Chemistry in Aqueous Systems

Equilibrium solubilities of the inorganic species of lead that are most likely to be present in natural waters can be computed from the available thermodynamic properties of lead and its compounds. The data in Table XIII [133]

Table XIII. Standard Free Energies of Formation of Some Lead Species at 25°C [133]

Mineral Name	Formula[a]	ΔG^0_f (kcal/mol)
Lead	$Pb^0_{(s)}$	0.0
Plumbous Ion	Pb^{2+}	−5.3
	$PbOH^+$	−51.4 (20°C)
	$Pb(OH)^0_2$	−95.8
	$Pb(OH)^-_3$	−137.6
	$PbCl^+$	−39.39
	$PbCl^0_2$	−71.03
Massicot (yellow)	$PbO_{(s)}$	−44.91
Litharge (red)	$PbO_{(s)}$	−45.16
	$Pb(OH)_{2(s)}$	−108.1
Galena	$PbS_{(s)}$	−22.96
Cerussite	$PbCO_{3(s)}$	−150.3
Hydrocerussite	$Pb_3(OH)_2(CO_3)_{2(s)}$	−409.1
Anglesite	$PbSO_{4(s)}$	−194.36
Plumbic Ion	Pb^{4+}	72.3
Dioxide	$PbO_{2(s)}$	−51.95

[a]All species are aqueous unless otherwise noted.

indicate that the least soluble common forms of lead in oxidizing systems are probably the carbonate, $PbCO_3$ (cerussite), the hydroxide, $Pb(OH)_2$, and the hydroxycarbonate, $Pb_3(OH)_2(CO_3)_2$ (hydrocerussite). In the presence of sulfur under reducing conditions, lead sulfide, PbS (galena), would be the stable solid.

In absence of other ligands, the ionic species below pH 7.0 is Pb^{2+}; under alkaline oxidizing conditions, the stable form is $PbO_{2(s)}$. However, in the presence of dissolved CO_2, lead carbonate precipitates. Bilinski and Stumm [147] reported that the following species of lead may occur in natural waters: Pb^{2+}, $PbCO_3$, $Pb(CO_3)^{2-}_2$, $PbOH^+$ and $Pb(OH)_2$.

Figure 11 shows an Eh-pH diagram for lead constructed by Hem and Durum [133]. This shows the stability fields for the important species in the presence of different amounts of all dissolved forms of CO_2 at a total concentration of $10^{-3}M$, equivalent to 61 mg/l of HCO^-_3 and a sulfur content of $10^{-3}M$, equivalent to 96 mg/l as SO^{2-}_4. The solubility of lead sulfate at the specified sulfate level in the system is much higher than the maximum shown in Figure 14. Therefore, lead sulfate would not be a stable solid. The solid lines represent the equilibrium concentrations of dissolved lead for the minimum considered, 1.0 μg/l, and for three higher values. The shaded areas

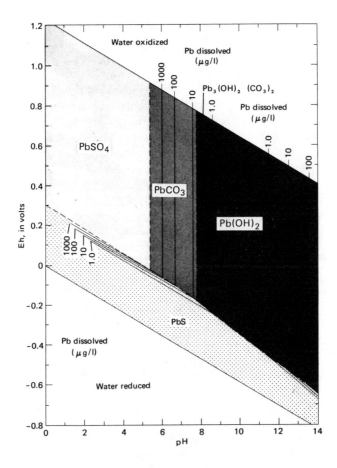

Figure 11.   Stability fields for solids and solubility of lead in the system Pb-S-CO$_2$-H$_2$O at 25°C and 1 atm (I. = 0.005) [133].

bounded by dashed lines represent the stability fields for the five solid forms of lead that predominate in this system. Lead sulfide (PbS$_{(s)}$), which has very low solubility, is stable only under reducing conditions but is converted to lead hydroxide or carbonate under oxidizing conditions. The concentration of the dissolved CO$_2$ species determines the positions of the boundaries between the solid phases Pb(OH)$_{2(s)}$, Pb$_3$(OH)$_2$(CO$_3$)$_{2(s)}$ and PbCO$_{3(s)}$. The carbonate field would be absent at total CO_2 concentration less than $10^{-4.87}M$ (about 0.8 mg/l HCO_3^-) or at a partial pressure of CO_2 less than $10^{-3.44}$ atm in a system open to the atmosphere.

Using the equilibrium constants in Table XIV, Hem and Durum [133] also calculated the equilibrium solubility of lead for systems devoid of carbon dioxide and in the presence of CO_2 at concentrations of $10^{-3}M$ (alkalinity 61 mg/l HCO_3^-) and $10^{-2}M$ (alkalinity 610 mg/l) under normal oxidizing conditions. Figures 12, 13 and 14 show the results of these solubility calculations expressed as total dissolved lead concentration versus pH. The two curves in each figure are for ionic strengths of 0–0.1 covering the range of total dissolved solids concentrations from 0–5000 mg/l. This includes levels present in natural waters used as drinking water supplies. All the figures are for systems at 25°C, 1 atm pressure, and in the absence of sulfur species.

These results indicate the important influence of $[H_3^+O]$ on the solubility of lead in aqueous systems. The solubility of lead is below 10 μg/l above pH 8.0, regardless of the alkalinity. However, in slightly acidic waters (near pH 6.5) and with a low alkalinity, the solubility of lead could increase to

Table XIV. Chemical Equilibria Constants Used in Calculating Total Lead Solubility [133][a]

Equation No.	Equation	Log K_{eq}
(103)	$[Pb^{2+}][H^+]^{-2}$	8.15
(104)	$[PbOH^+][H^+][Pb^{2+}]^{-1}$	−8.12
(105)	$[Pb(OH)_2^0][H^+]^2[Pb^{2+}]^{-1}$	−17.16
(106)	$[Pb(OH)_3^-][H^+]^3[Pb^{2+}]^{-1}$	−28.08
(107)	$[Pb^{2+}][HCO_3^-][H^+]^{-1}$	−3.09
(108)	$[H_2CO_3]P_{CO_2}{}^{b}$	−1.43
(109)	$[HCO_3^-][H^+][H_2CO_3]^{-1}$	−6.35
(110)	$[CO_3^{2-}][H^+][HCO_3^-]^{-1}$	−10.33

[a]Total dissolved lead at equilibrium is

$$\Sigma Pb_{diss.} = \frac{[Pb^{2+}]}{\gamma Pb^{2+}} + \frac{[PbOH^+]}{\gamma PbOH^+} + \frac{[Pb(OH)_2^0]}{1} + \frac{[Pb(OH)_3^-]}{\gamma Pb(OH)_3^-}$$

and

$$\Sigma CO_{2diss.} = \frac{[H_2CO_3]}{1} + \frac{[HCO_3^-]}{\gamma HCO_3^-} + \frac{[CO_3^{2-}]}{\gamma CO_3^{2-}}$$

where γ is the thermodynamic activity coefficient.
[b]P is the partial pressure in atmospheres.

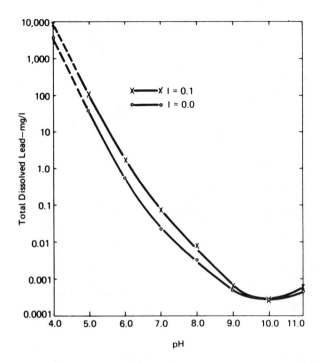

Figure 12.   Equilibrium solubility of lead at 25°C and 1 atm with dissolved $[CO_2]_t =$ $10^{-3} M$ and I. = 0.00 and 0.10 (reproduced from Hem and Durum [133] courtesy of the American Water Works Association).

values greater than 100 μg/l, which exceeds the 50-μg/l maximum allowable level in drinking waters.

In the above discussion, the solubility data are based on equilibrium considerations that frequently are not achieved in natural water systems. In addition, the mobility of lead, like other heavy metals, may be influenced by complexation with organic matter or sorption on hydrous oxides and clay minerals. Therefore, it is important to recognize the limitations of the equilibrium model in predicting the distribution of different forms of lead in the natural aquatic environment.

CADMIUM

Background

Cadmium is recognized as one of the serious environmental contaminants because of its high toxicity, accumulation and retention in the human body.

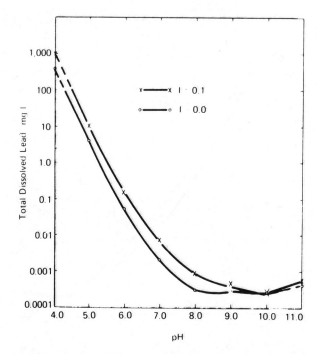

Figure 13.   Equilibrium solubility of lead at 25°C and 1 atm with dissolved $[CO_2]_t = 10^{-2}M$ and I. = 0.00 and 0.10 (reproduced from Hem and Durum [133] courtesy of the American Water Works Association).

This element is distinguished by having a long biological half-life in man in the range of 10 to 30 years [148]. Cadmium accumulates mainly in the kidneys, liver, pancreas and thyroid of humans and other animals [80]. Although many plant and animal tissues contain about 1 mg of Cd/kg of tissue, there is no evidence that Cd is biologically essential or beneficial. Consumption of cadmium salts causes cramps, nausea, vomiting and diarrhea. Cadmium-contaminated ice cubes in cold drinks have caused acute gastritis symptoms within one hour [81]. A boy was reported to have died within 1½ hours after consuming a dose of about 8.9 g of cadmium chloride [81]. Itai-itai disease syndrome, a severe endemic illness, has been associated with the ingestion of as little as 600 μg/day of Cd [80].

The significance of cadmium as a potential contaminant of the aquatic environment was realized after a ground water pollution incident was reported in Long Island, New York [149]. Industrial waste discharges from electroplating industries resulted in contamination of the ground water where

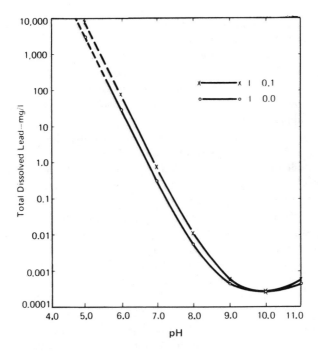

Figure 14.   Equilibrium solubility of lead at 25°C and 1 atm with dissolved $[CO_2]_t$ less than $10^{-4.87}M$ and I. = 0.00 and 0.10 (reproduced from Hem and Durum [133] courtesy of the American Water Works Association).

the [Cd] in one of the well waters reached 3.2 mg/l. Kobayashi [150] reported an incident of cadmium poisoning in Japan in which mining wastes were released for more than 20 years into a stream used for irrigation of rice and as a water supply. There was a substantial increase of the cadmium content in the soil and crops. High concentrations of cadmium have also been reported in Missouri mine waters: one spring there had a concentration of 1000 mg/l of Cd [81].

Because cadmium is a cumulative toxicant causing progressive chronic poisoning in man, the maximum permissible level in drinking water is 0.01 mg/l [1].

Sources in the Environment

Cadmium occurs in nature largely as the sulfide, greenockite, or as a cadmium blend, often as an impurity in zinc, copper and lead ores. Zinc is the most attractive economical ore. The abundance ratio of zinc to cadimum in the earth's outer crust is approximately 1000:1 [151]. The cadmium content of soils was reported in the range of 0.01 to 0.7 ppm, with an average of 0.05 ppm [152]. The average content in sandstone was 0.05 ppm; in shale it was 0.3 ppm [152]. Cadmium may occur naturally in crude oil and gasoline. Soil and vegetation adjacent to heavily traveled highways showed an increased cadmium content [118].

Cadmium appears in some ratio to zinc as an impurity in any use of this metal. The cadmium impurity is usually higher if it is not recovered during the processing of the zinc ore. Cadmium is recovered by heating the ore until it is volatilized. The cadmium fume is recycled and the dust, either cadmium oxide or sulfide, is collected from the stack gases in electrostatic precipitators. Some manufacturers process both zinc and cadmium, whereas others process the flue dust from zinc, copper and lead smelters. The cadmium fumes are recycled several times until the content is high enough to collect the dust. Emissions to the atmosphere occur at this stage of purification. Based on estimates of the cadmium content of extracted ores, imported flue dusts and atmospheric emissions, 478 tons of Cd were lost to the environment as impurities in zinc metal and in tailings from ore processing [153]. Kobayashi [154] reported that substantial amounts of zinc and cadmium were added to the environment in stack gases from a Japanese zinc refinery.

The total world production of cadmium in 1968 was estimated to be 17,000 tons, of which the United States used about 6700 tons [154]. The principal use of cadmium is in the electroplating industry, which accounts for 45% of the total consumption. The pigments and plastics industries use 21% and 15%, respectively. Cadmium pigments are used in paint, printing ink and plastics. Cadmium compounds are used as stabilizers in the manufacture of polyvinyl plastics. Cadmium salts are also used in fluorescent tubes, television tubes, some types of batteries, and as insecticides and helminthcides. Elemental cadmium is used to alloy with copper, lead, silver and aluminum. All of these industrial activities are expected to result in the release of Cd into the environment.

Occurrence in Natural Waters

The most significant study on the occurrence of cadmium in natural waters was conducted by the U.S. Geological Survey [93]. Water samples were collected at 726 locations, mostly on rivers throughout the United States, and the concentrations of some heavy metals, including cadmium, were determined. The samples were filtered through a 0.45-μ porosity membrane filter prior to analysis by atomic absorption spectroscopy (detection limit for cadmium was 1 μg/l). These results indicate the dissolved species levels. About 46% of the samples contained 1 μg/l or more of cadmium and the median value was less than 1 $\mu g/l$. However, cadmium showed a strongly defined geographical distribution in the U.S. rivers. Cadmium was present above the detection limit in 64% of the samples from New England and the northeastern states: in 45% of the samples from the southeastern and central states: and in only 35% and 22%, respectively, of the samples from the southwestern and northwestern states. This pattern led Hem [151] to suggest that pollution sources and atmospheric rainout may be the most important contributors of cadmium to river waters. There is no available information, however, on the cadmium content of rain water, but a distribution pattern similar to that of zinc may be expected [151] since cadmium usually accompanies zinc in industrial discharges. More information on cadmium rainout and fallout is needed for a satisfactory understanding of the circulation patterns of cadmium in the hydrosphere [151].

There is a lack of data also on the cadmium content of particulate matter and bottom sediments in lakes and streams, but the levels are expected to be higher there than in the overlying waters. Emerson et al. [155] showed that marine sediments collected near an industrial wastewater outfall in the Los Angeles area contained cadmium concentrations ranging from 3.3 to 3.9 ppm. The concentration levels in the sediments were reduced with greater distances from the outfall. The natural level of cadmium in unpolluted coastal regions is reported to be less than 0.1 $\mu g/l$ [156,157].

Occurrence in Aquatic Life

Cadmium is extremely toxic to aquatic organisms causing progressive chronic poisoning in mammals, fish and probably other animals because this metal is not excreted [81]. The eggs and larvae of fish are apparently more sensitive than adult fish to poisoning by cadmium. The safe levels of cadmium for fathead minnows and bluegills in hard waters are between 0.06 and 0.03 mg/l, and safe levels for coho salmon fry have been reported as 0.004–0.001 mg/l in soft waters [80]. A concentration of 0.005 mg/l was observed to reduce reproduction of *Daphnia* in a one-generation exposure lasting three weeks [80].

Few studies have been reported on the accumulation of cadmium in fresh water organisms. Mount [158] found accumulations in living bluegills as high as 100 $\mu g/g$ (dry weight) and in the gills of dead catfish up to 1000 $\mu g/g$. Marine organisms, on the other hand, are known to concentrate cadmium to extremely high levels. Concentration factors of 1000 in fish muscle [159], 3000 in marine plants, and up to 29,600 in certain marine animals [151] have been reported. Pringle et al. [146] reported that soft-shell clams accumulated cadmium at a rate of 0.1 $\mu g/g/day$ wet weight from a level in sea water as low as 0.05 mg/l. Kerfoot and Jacobs [160] studied the accumulation of cadmium in two types of marine phytoplanktonic algae and two species of shellfish using a prototype wastewater treatment aquaculture system. The algae showed a rapid increase in cadmium concentration in the cells proportional to the initial concentration in solution (range 0.01 to 0.12 mg/l) until an equilibrium was reached. Green algae accumulated more cadmium than diatoms. On a dry weight basis, green algae accumulated cadmium to about 6700 times the concentration initially in solution. The algae exposed to cadmium retained their original content after resuspension in uncontaminated sea water for a long period of time. Both oysters and clams also responded to cadmium enrichment of the sea water with a linear accumulation and retention of the metal in their tissues by direct absorption from solution. The rates of accumulation were proportional to the concentration of cadmium in the sea water. Shellfish also assimilated up to 10% of the cadmium presented to them in an enriched algal diet. This study illustrates the potential hazards of cadimum contamination of the aquatic environment. The concentration, accumulation and retention of cadmium in the aquatic food chain eventually can affect man through ingestion of contaminated seafood.

Cadmium acts synergistically with other metals to increase toxicity. A cadmium concentration of 0.03 mg/l, in combination with 0.15 mg/l of zinc, will kill chinook salmon fry [161]. Copper concentrations at 1 mg/l or more substantially increase the toxicity of cadmium [162]. Therefore, marine or estuarine waters containing copper or zinc in excess of 1 mg/l should be protected from cadmium contamination.

Chemistry in Aqueous Systems

Figure 15 shows an Eh-pH diagram for cadmium that was constructed by Hem [151] in the system $Cd-S-CO_2-H_2O$ at one atm and $25°C$. The stability fields for the more important species are shown in the presence of $10^{-3}M$ of dissolved carbon dioxide and sulfur species, and $10^{-7.05}M$ dissolved Cd, which is equivalent to 10 $\mu g/l$. The diagram is constructed from thermodynamic data compiled by Hem [151], which are shown in Table XV. The stability constants represent systems at zero ionic strength.

The solid species that have stability fields in this system are the carbonate (otavite) and sulfide (greenockite). The latter is formed over a wide pH range under reducing conditions in the presence of low concentrations of sulfide ions. Precipitation of the sulfide, which has extremely low solubility, may be an important factor in the control of cadmium in ground- and sea-water systems. Bottom muds in lakes and reservoirs constitute a sink for the deposited sulfide, which is highly resistant to carbonation and weathering. The dissolved species in the system include Cd^{2+} and $Cd(OH)_3^-$.

Cadmium solubility is below 10 $\mu g/l$ (the maximum permissible limit for drinking waters) at high pH values (between pH 9.9 and 10.7) or in reduced systems under anaerobic conditions. The solubilities of cadmium carbonate

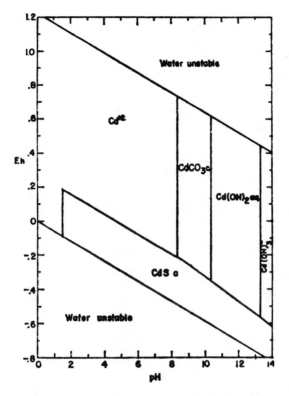

Figure 15.   Stability of solids and predominant dissolved cadmium species in the
system Cd-CO$_2$-S-H$_2$O at 25°C and 1 atm in relation to Eh and pH. Dissolved [Cd]$_t$ =
10$^{-7.05}$M. Dissolved [CO$_2$] and [S] = 10^{-3}M (reproduced from Hem [151] courtesy
of the American Geophysical Union).

and hydroxide are relatively high at or below a pH value of 7.0. To make an
exact evaluation of solubility, Hem [151] made a detailed set of solubility
calculations using the data in Table XV and the following equations:

$$\frac{[Cd^{2+}]}{[H^+]^2} = 10^{13.61} \tag{111}$$

$$\frac{[Cd^{2+}] \, [HCO_3^-]}{[H^+]} = 10^{-1.74} \tag{112}$$

$$\frac{[CdOH^+] \, [H^+]}{[Cd^{2+}]} = 10^{-9.39} \tag{113}$$

Table XV. Standard Free Energies of Formation for Cadmium Species at 25°C [151]

Mineral Name	Formula[a]	ΔG_f^0 (kcal/mol)
Cadmium, grey	$Cd^0_{(s)}$	0.0
Cadmium Oxide, inactive	$CdO_{(s)}$	−56.44
Cadmium Hydroxide, inactive	$Cd(OH)_{2(s)}$	−113.35
Cadmous Ion	Cd^{2+}	−18.54
	$CdOH^+$	−62.4
	$Cd(OH)_2^0$	−105.9
	$Cd(OH)_3^-$	−144.41
	$CdSiO_{2(s)}$	−264.20
	$CdCl^+$	−52.64
	$CdOHCl$	−95.49
	$CdSO_4^0$	−199.64
Greenockite	$CdS_{(s)}$	−34.8
Otavite	$CdCO_{3(s)}$	−159.964

[a]All species are aqueous unless otherwise noted.

$$\frac{[Cd(OH)_2][H^+]^2}{[Cd^{2+}]} = 10^{-19.08} \tag{114}$$

$$\frac{[Cd(OH)_3^-][H^+]^3}{[Cd^{2+}]} = 10^{-32.4} \tag{115}$$

$$C_t = \frac{[Cd^{2+}]}{\gamma Cd^{2+}} + \frac{[CdOH^+]}{\gamma CdOH^+} + \frac{[Cd(OH)_2]}{\gamma Cd(OH)_2} + \frac{[Cd(OH)_3^-]}{\gamma Cd(OH)_3^-} \tag{116}$$

The soluble species considered in these computations are: Cd^{2+}, $CdOH^+$, $Cd(OH)_{2(aq)}$ and $Cd(OH)_3^-$. The cadmium solubility as a function of $[H_3^+O]$ in the presence of a total concentration of $10^{-3}M$ of dissolved carbon dioxide species is shown in Figure 16. Figure 17 shows a system in which the total dissolved carbon dioxide species is $10^{-2}M$. There is a wide range of values reported for the solubility of $Cd(OH)_{2(s)}$; however, it is evident that its solubility is relatively high and is unlikely to control cadmium solubility in natural systems. In the systems presented in Figures 16 and 17, the least soluble species is the carbonate in the pH range of 8.9 to 10.0. However, concentrations of cadmium higher than 10 μg/l may be stable in river and ground waters having low alkalinity values. Figure 17 indicates that an increase in the total dissolved carbon dioxide species by one log unit results

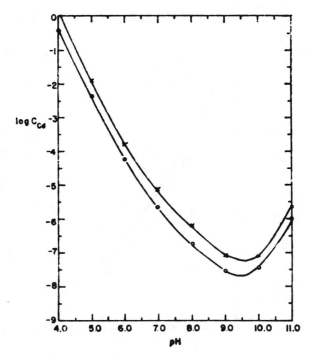

Figure 16.   Solubility of cadmium as a function of pH. Dissolved $[CO_2]_t = 10^{-3}M$,
I. = 0.00 (○) and 0.10 (×) (reproduced from Hem [151] courtesy of the American
Geophysical Union).

in a decrease of the solubility of cadmium by an equal amount. This suggests
that a single graph can be used for any system having the same solid and
solute species. For example, at any given pH value in a system containing
$10^{-2.7}M$ of dissolved carbon dioxide species, the solubility of cadmium is
0.3 log unit less than that shown in Figure 16 or 0.7 log unit greater than
that shown in Figure 17.

The solubility of cadmium in a system containing dissolved carbon dioxide
species less than $10^{-3.59}M$, which is equivalent to 15 mg/l HCO_3^- at pH 8.3, is
shown in Figure 18. The solid in this system is $Cd(OH)_2$, and the solubility
does not get below 1 mg/l, even at pH 10.0, and is close to 400 μg/l at pH
11.0. Few natural waters have this low $[CO_{2(g)}]$, but these levels may be
encountered. It must be noted that all of the solubility calculations for
cadmium are for systems under oxidizing conditions at 25°C and in the
absence of ions not considered in the calculations. Such conditions may
be encountered in many rivers, lakes and underground waters. The effect of

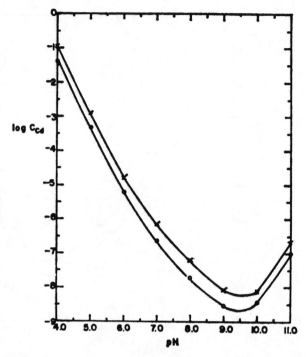

Figure 17.   Solubility of cadmium as a function of pH. Dissolved $[CO_2]_t = 10^{-2}M$, I. = 0.00 (○) and 0.10 (×) (reproduced from Hem [151] courtesy of the American Geophysical Union).

temperature on the solubility of cadmium carbonate cannot be evaluated because of the lack of any reported data.

Strong evidence for the formation of a 1:1 soluble cadmium carbonate complex has been reported [163]. The stability constant of the carbonate complex

$$K_{eq} = \frac{[CdCO_3^0]}{[Cd^{2+}][CO_3^{2-}]}$$ (117)

is $10^{4.02}$ at 20°C, I = $10^{-2}M$. At high carbonate contents, the complex $Cd(CO_3)_3^{4-}$ is known to occur [164] ($\beta_3 = 1.7 \times 10^6 M^{-3}$).

The solubility calculations for the systems discussed above considered only certain dissolved cadmium species [151]. However, there are other systems in which other ionic species may be important. For example, in solutions

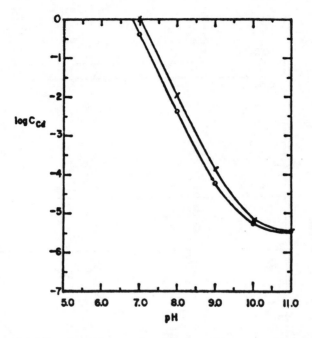

Figure 18.   Solubility of cadmium as a function of pH. CO_2 species absent. I. = 0.00 (○) and 0.10 (×) (reproduced from Hem [151] courtesy of the American Geophysical Union).

having a $0.1M$ or higher chloride concentration, cadmium forms complex ions, e.g.,

$$Cd^{2+} + Cl^- = CdCl^+ \ (K_{eq} = 10^2) \tag{118}$$

and a mixed chloride-hydroxide complex, e.g.,

$$CdCl^+ + H_2O = CdClOH^0 + H^+ \ (K_{eq} = 10^{-10.15}) \tag{119}$$

Such complexation can be important in seawater and will influence the degree of speciation of cadmium ions. Figure 19 shows the distribution of species of cadmium in sea water at 25°C as a function of pH [165]. The following complexing anions were considered: Cl^- = 0.554 molal, SO_4^{2-} = $0.0284M$, HCO_3^- = $1.62 \times 10^{-3}M$, and CO_3^{2-} = $2.6 \times 10^{-6}M$ at a pH value of 7.0. Cadmium is strongly associated with chloride ions in sea water over the entire pH range shown in Figure 19, and the distribution is only negligibly

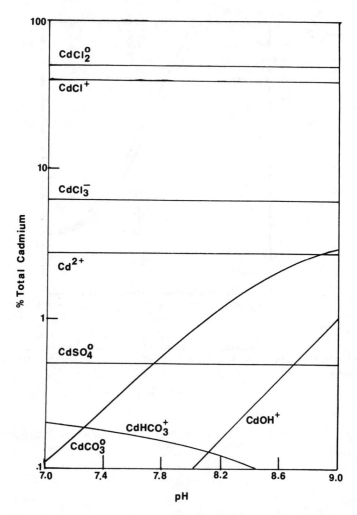

Figure 19. Calculated distribution of the chemical species of cadmium in sea water at 25°C and 1 atm as a function of pH (reproduced from Zirino and Yamamoto [165] courtesy of the American Society of Limnology and Oceanography).

affected by $[H_3O]$. The predominant species are [165]: $CdCl_2^0$ (51%), $CdCl^+$ (39%) and $CdCl_3^-$ (6%). Uncomplexed Cd^{2+} constitutes about 2.5% of the total. Although some $CdCO_3^0$ is formed at higher pH values, its percentage remains below 3%. At higher pH values, substitution of Cl^- by OH^- could be expected with formation of CdOHCl.

General Distribution in Nature

The equilibrium calculations presented in the above discussion should be considered as an approach to understanding, in a general way, the anticipated behavior of cadmium in aqueous systems. In natural water systems, however, the occurrence and mobility of cadmium can be influenced by other factors, such as uptake by aquatic biota, sorption on particulate matter and bottom sediments, and complexation with organic ligands.

The uptake of cadmium by unicellular marine algae was shown to be extremely rapid with an increase in the content of cells proportional to an increase in concentration of cadmium in seawater [160]. The accumulated cadmium was tightly bound to the cells and could not be released back to solution, probably due to complexation with cell constituents. When green algae were exposed to a concentration of 100 μg/l cadmium, about 69 μg remained in solution and 31 μg of cadmium were associated with the algal cells. On dry weight basis, the algae accumulated cadmium to about 6700 times the concentration initially in solution. Oysters and clams also have absorbed and accumulated cadmium at rates proportional to the initial solution concentration and were able to assimilate some of the cadmium associated with their algal diets. Apart from the fact that accumulation of the metal in the higher members of the food chain may reach toxic levels to man, the ability of aquatic organisms to remove, accumulate and retain cadmium illustrates the significant role they can play in affecting the mobility and fate of the metal in aquatic environments. Fresh water organisms are expected to behave similarly, especially in highly productive lakes. The death and decay of these organisms may result in the release of cadmium back into the aquatic system, or it may be fixed as the sulfide in the anaerobic bottom sediments.

Cadmium undergoes complex reactions with humic substances [163]. The complexed fraction was found to be slightly dependent on pH value, and the ratio of complexed to uncomplexed cadmium ion did not increase linearly with humic acid concentration (Table XVI):

$$\frac{[\text{Cd complex}]}{[\text{Cd}^{2+}]} \alpha \frac{[\text{Ligand}]^{0.64}}{[\text{H}^+]^{0.2}} \tag{120}$$

The ligand dependence shows that some of the ligand molecules complex more than one cadmium ion, which is consistent with a polymeric ligand structure and perhaps with polymerization of the complex. Whether the interaction is considered as a complex formation or adsorption will depend on the size of the humic molecule. Other common potential ligands that are likely to give complexation with cadmium in natural systems were also evaluated [163]. The ratio of complexed to uncomplexed cadmium depends

Table XVI. Dependence of the Degree of Complexation of
Cadmium by Humic Substances on pH Value [163]

pH Value		6.0	6.5	7.0	7.5	8.0	8.5
				[Cd complex]/[Cd^{2+}]			
Humic	86	1.00	1.26	1.59	2.00	2.50	3.20
Concentrations, mg/l	20	0.40	0.50	0.62	0.77	1.00	1.26
	6.6	0.20	0.25	0.32	0.40	0.50	0.62

only on the stability constant and the concentration of ligand and not on the total cadmium concentration, provided the ligand remains in excess. For a ligand to be effective at a low concentration $\approx 10^{-8}M$, the stability constant of its complex with cadmium would have to be $10^8 M^{-1}$ or greater. This excludes all naturally occurring ligands except any free cysteine that may be present and perhaps certain highly active humic fractions. With such synthetic chelating agents as NTA and EDTA, the latter may be present in sewage effluents and could be important in this respect.

The sorption of cadmium on particulate matter and bottom sediments is considered a major factor affecting its concentration in natural waters [166]. The sorption of cadmium on bottom muds obtained from four British rivers is shown in Table XVII. The results are expressed in terms of a concentration factor, which is empirical and denotes a distribution coefficient of the solute between the solid and aqueous phases. The concentration factors for the bottom muds varied from 6,000–40,000 indicating the variation of the composition of the particular sample and the surface area of the particles. The concentration factor for a particular system decreased in the presence of added EDTA, which is a complexing agent for cadmium. River muds are

Table XVII. Measurement of Concentration Factors for Cadmium on Four Samples
of River Mud at an Added Cadmium Concentration of 2.1 μg/l [166]

River or stream	Concentration Factors[a]	
	K after 2 hr	K after 24 hr
River Ivel	17,000	40,000
River Hiz	3,000	6,000
Pix Brook	14,000	26,000
Stevenage Brook	6,000-25,000	

[a]K = C_s/C_t, where C_s is the concentration of adsorbed species on the solid and C_t is the concentration of the solute remaining in solution after adsorption has occurred.

Table XVIII. Concentration Factors for the Adsorption of Cadmium from Solution on to Components of River Mud, and on Other Solids Expected to be Present in Rivers, Expressed Relative to the Dry Weight of Each Material [166]

Solid	Added Cadmium Concentration (μg/l)	Concentration Factors[a]	
		K after 2 hr	K after 24 hr
Silica	7.3	430 (±40)	1,000 (±100)
Kaolin	7.3	250 (±30)	380 (±50)
Humic acid	2.1-4.3	14,000 (±2,000)	18,000 (±3,000)
Fish fecal matter and other material accumulating in fish tanks	2.1	100-300	200-1,000
Plant material (watercress, *Nasturtium officinale*)	2.1	800	1,000

[a]See Table XVII for explanation.

composed mainly of clay, silica and humic material and the ratio of these components depends on the origin of the sample. The extent of adsorption of cadmium on these solids and other materials expected to be present in rivers was measured [166] and these concentration factors are shown in Table XVIII. The concentration factors for humic acids were much closer to those of river muds; therefore, it is probable that these constituents of mud are responsible for its adsorptive properties. The concentration factors for silica and kaolin were relatively low, probably due to their low cation exchange capacities. Fish fecal matter and live plant material gave concentration factors that were relatively low compared with humic acid. The uptake of cadmium by growing plants is again shown to be an important factor in the transport and removal of cadmium from aquatic systems.

SELENIUM

Background

Selenium is widely distributed in nature but does not occur evenly in the lithosphere. Selenium compounds are toxic to animals feeding on highly seleniferous plants, causing the so-called alkali disease [167]. Chronic toxicity results from the ingestion of foods containing 5–40 ppm of the element [168]. Large doses result in acute toxicity and sudden death.

Symptoms of chronic toxicity in animals include lameness, hoof malformation, loss of hair and emaciation. Soluble selenium compounds are completely absorbed in the intestines and the ingested element becomes widely distributed in the internal organs, with the largest amounts occurring in the liver and kidneys. The main excretory pathway for selenium is the urine as the trimethylselenium ion $(CH_3)_3Se^+$ [169]. Under condition of selenium toxicity, the volatile dimethyl selenide is excreted via the lungs [170].

Although selenium in high concentrations is toxic to animals, the element is an essential nutrient in low concentrations. Selenium deficiency became an agricultural problem after World War II because of changes in animal nutrition, which resulted in serious losses of livestock in the U.S. This prompted the Food and Drug Administration to approve the use of selenium as selenite or selenate in concentrations up to 0.2 ppm as an additive in animal feed. The element was recently shown to be a part of glutathione peroxidase enzyme [171] and plays a role in the electron transport into the cytochrome system [172]. The average selenium content in human blood in the U.S. is 0.2 ppm [173]. Inorganic selenium compounds act as detoxifying agents for subacute doses of cadmium and mercury [174] and in some circumstances act antagonistically to arsenic.

Most selenium intake is normally from food such as cereals and seafood. The average human diet in the U.S. contains 1.8 mg of selenium per month [175]. The minimum nutritional requirements have been estimated to be 1 mg/month. Most evidence indicates that there is a greater overall potential for selenium deficiency than for toxicity at current levels of selenium intake. The maximum contaminant level in drinking waters is 10 μg/l, a level that just barely provides a minimum nutritional amount of selenium with a consumption of 2 liter/day [1].

Sources in the Environment

Natural Sources

Sulfides or native sulfur deposits often contain selenium in significant amounts because of the chemical similarity between sulfur and selenium. Thus, sulfides of bismuth, iron, mercury, silver, copper, lead and zinc have been found to contain selenium, sometimes at levels over 20% [176]. Low levels have been found in the sulfate minerals jarosite and barite. Crude sulfur also often contains selenium at levels over 0.1%.

Selenium is not evenly distributed in geologic materials. Igneous rocks have been estimated to contain selenium in the range of 0.03 to 0.8 ppm, with an average of 0.09 ppm [177]. This value was accepted as the average crustal abundance of the element, but it has been recently revised downward to an estimated abundance of 0.05 ppm [178]. Selenium was found in volcanic gases, tuffs and sulfur [179], which suggested that volcanic activity played an important role in the development of seleniferous geologic beds. In addition to volcanism, weathering of igneous rocks and leaching by rain water are considered to be important sources of the element in sedimentary rocks [178]. Sandstone and limestone contain variable amounts of selenium, usually below 1 ppm [167]; however, higher concentrations have been reported in certain locations [180]. Shales are consistently higher in selenium content than sandstone and limestone.

Selenium occurs in soils in concentrations ranging from 0.1 ppm [181] in selenium-deficient areas to 1200 ppm in organic-rich soils in Ireland [182]. In soils, its concentration depends on the selenium content of the parent materials and the intensity of weathering and leaching. Selenium apparently occurs in unweathered rocks as the free element or as selenides, which are readily oxidized during the weathering of the parent material to soils. In acid soils, the element is probably present as the selenite, which is firmly bound in iron oxide colloids and is unavailable to plants. Soils in Hawaii that contain selenium in concentrations ranging from 6 to 15 ppm do not produce seleniferous vegetation [179]. In alkaline soils, however, the selenium is further oxidized to selenate ions, which are readily available to plants. The soils of South Dakota and Kansas contain less than 1 ppm selenium but produce seleniferous (toxic) vegetation [183].

Selenium also occurs in coal deposits, probably due to the presence of selenium-enriched biomaterials, which decomposed during coal formation. The selenium content of 138 samples of coal from U.S. deposits in 22 states ranged from 0.46 to 10.65 ppm and averaged 2.8 ppm [184], which is over 50 times the crustal abundance of the element.

Fuel oils apparently contain less selenium than do coals. Five samples of raw petroleum in Japan contained an average of 0.82 ppm selenium (0.5–9.5), and nine samples of heavy petroleum averaged 0.99 ppm (0.5–1.65) [185]. An average of 0.17 ppm selenium (0.06–0.35 ppm) was reported in 42 U.S. crude oil samples [184]. Analysis of 47 samples of crude and fuel oils from various parts of the world gave values of <0.006–2.2 ppm, the average being about 0.6 ppm [186].

Anthropogenic Sources

The free world production of selenium from 1964 through 1973 averaged 2.3 million pounds annually (Table XIX). The output has been increased from a low of 1.7 million pounds in 1965 to a high of 2.9 million pounds in 1970, with the United States being the leading producer for most of these years, followed by Canada, Japan and Sweden [186].

The known deposits of selenium do not contain sufficient quantities to permit their being mined for selenium alone. Almost all primary selenium is

Table XIX. Selenium: Free World Refinery Production, by Country (1000 lb) [186]

Country	1964	1965	1966	1967	1968	1969	1970	1971	1972	1973
Australia	4	5	4	4	4	4	7	7	7	8
Belgium–Luxembourg	87	93	91	90	54	46	68	120	147	106
Canada	466	512	575	52	636	820	854	886	655	598
Finland	15	13	12	15	16	14	15	14	16	12
Japan	326	348	421	422	399	435	467	524	738	789
Mexico	7	18	4	–	2	65	278	115	97	86
Peru	17	19	13	11	13	15	15	16	18	18
Sweden	181	176	154	158	168	168	139	134	140	120
United States	899	510	590	568	603	1217	975	627	739	627
Yugoslavia	8	17	21	10	21	20	35	54	55	94
Total	2010	1711	1885	2030	1916	2804	2853	2497	2612	2458

produced from copper refinery slimes. The present production technology consists of processes designed primarily for effective recovery of precious metals, and selenium recoveries have secondary importance, which is reflected in low recovery efficiencies. Selenium is recovered from slimes by volatilization during roasting or furnacing as the oxide, which is scrubbed from the exhaust gas, or by leaching of roasted calcine or furnace slag. All processes use sulfur dioxide to precipitate selenium metal from solutions of sodium selenite and selenious acid.

The principal commercial uses of selenium are listed in Table XX [186]. The annual consumption in the United States increased 50% from 1964 to 1973. The most significant increase has been in its use in the electronic industry, including use in rectifiers, xerographic copying machines and photoelectric cells, which accounted for 45% of 1973 demand. The use of selenium in manufacturing glass and ceramics, probably its oldest application, increased 67% during the 1970s because of the increasing quantities of selenium-containing tinted glass used in the construction and transportation industries. The chemical industry accounts for an estimated one-eighth of the selenium consumed. Most of this is used in pigment manufacture to color plastics, paints, enamels, inks and rubber. Selenium compounds are also used as components of plating solutions and as chemical agents in the preparation of many products. The chief use of the element in agriculture is in prevention of selenium deficiency in livestock and poultry.

Occurrence in the Atmosphere

Selenium is introduced into the atmosphere by several natural sources. It has been suggested that volcanoes may be major contributors of selenium to the air [179]. Volatile selenium compounds are released by accumulator

Table XX. Uses of Some Inorganic Selenium Compounds [186]

Aluminum Selenide, Al_2Se_3	In preparation of hydrogen selenide; in semiconductor research.
Ammonium Selenite, $(NH_4)_2SeO_3$	In manufacture of red glass; as reagent for alkaloids.
Arsenic Hemiselenide, As_2Se	In manufacture of glass.
Bismuth Selenide, Bi_2Se_3	In semiconductor research.
Cadmium Selenide, CdSe	In photoconductors, semiconductors, photoelectric cells and rectifiers; in phosphors.
Calcium Selenide, CaSe	In electron emitters.
Cupric Selenate, $CuSeO_4$	In coloring Cu or Cu alloys black.
Cupric Selenide, CuSe	As catalyst in Kjeldahl digestions; in semiconductors.
Indium Selenide, InSe	In semiconductor research.
Potassium Selenate, K_2SeO_4	As reagent.
Selenium Disulfide, SeS_2	In remedies for eczemas and fungus infections in dogs and cats; as antidandruff agent in shampoos for human use; usually employed as a mixture with the monosulfide.
Selenium Hexafluoride, SeF_6	As gaseous electric insulator.
Selenium Monosulfide, SeS	Topically against eczemas, fungus infections, demodectic mange, fleabites in small animals; usually employed as a mixture with the disulfide.
Selenium Dioxide, SeO_2	In the manufacture of other selenium compounds; as a reagent for alkaloids.
Sodium Selenate, Na_2SeO_4	As veterinary therapeutic agent.
Sodium Selenite, Na_2SeO_3	In removing green color from glass during its manufacture; as veterinary therapeutic agent.

plants mainly as dimethyl selenide [187] and, to a lesser extent, as dimethyl diselenide [188]. Dimethyl selenide is also given off by certain microorganisms [189] and is exhaled by animals fed seleniferous diets [170]. There are no accurate estimates, however, of the quantities contributed to the air by each of the above sources.

Potential industrial sources of atmospheric selenium are summarized in Table XXI [186]. The total industrial emissions of selenium for 1970 were estimated to be 2.43 million pounds. Burning of coal accounted for 62% of the total emissions from all sources. Losses in nonferrous mining, smelting and refining operations accounted for 26% of the total. The remainder was equally contributed by precious metal refinery operations, where all primary selenium is now a by-product; the loss of volatilized metal in glass manufacturing; and the burning of fuel oil [186].

The estimated industrial emissions of selenium seems to be a small quantity to cause a serious pollution problem in comparison with the amounts of other industrial pollutants emitted. Most of the selenium emissions probably occur as finely divided particulates, either as the element or selenium dioxide, which are readily removed from the atmosphere by dry fallout or in rain and

Table XXI. Estimates for Selenium Emission Factors [186]

Mining and Milling	
Copper	0.015 lb/1000 tons ore mined
Lead	0.047 lb/1000 tons ore mined
Zinc	0.032 lb/1000 tons ore mined
Phosphate (western)	0.350 lb/1000 tons ore mined
Uranium	0.350 lb/1000 tons ore mined
Smelting and Refining	
Copper	0.25 lb/ton copper produced
Lead	0.05 lb/ton lead produced
Zinc	0.04 lb/ton zinc produced
Selenium Refining	
Primary (from copper by-product)	277 lb/ton selenium recovered
Secondary	100 lb/ton selenium recovered
End Product Manufacturing	
Glass and ceramics	700 lb/ton selenium consumed
Electronics and electrical	2 lb/ton selenium consumed
Duplicating	2 lb/ton selenium consumed
Pigments	15 lb/ton selenium consumed
Iron and steel alloys	1000 lb/ton selenium consumed
Other	10 lb/ton selenium consumed
Other Emission Sources	
Coal	2.90 lb/1000 tons coal burned
Oil	0.21 lb/1000 barrels oil burned
Incineration	0.02 lb/1000 tons of refuse burned

snow. The selenium concentration in the air in the vicinity of two electrolytic copper plants in Russia [190] was found to be 0.5 $\mu g/m^3$ at the first plant; that was reduced to 0.07 $\mu g/m^3$ 2 km away. At the second plant, the concentration was 0.39 $\mu g/m^3$ and none was found 2 km away. Atmospheric dust collected from air-conditioning filters in 10 U.S. cities contained 0.05–10 ppm of selenium [191]. Seven air samples collected during the spring of 1965 in Cambridge, Massachusetts [192] contained an average of 0.001 $\mu g/m^3$. Analyses of selenium in ambient air of 21 metropolitan areas was less than 0.04 $\mu g/m^3$ of air in all but two cities [193]. Slightly higher concentrations were reported from Los Angeles and Denver. In general, it appears that selenium continuously enters and is removed from the atmosphere and that its average concentration in air is very low, probably well below 0.01 $\mu g/m^3$ [186].

Occurrence in Natural Waters

The occurrence of soluble selenium compounds in some soils and other geologic materials suggests that natural waters would be effective in leaching

Table XXII. Selenium Content of Various Streams [194]

River	Sampled At	Industrial Area	Se (µg/l)
Mississippi	Minneapolis, Minnesota	Yes?	0.114
Susquehanna	Marietta, Pennsylvania	Yes	0.325
Mad	Blue Lake, California	No	0.348
Klamath	Klamath Glenn, California	No	0.122
Russian	California Hwy 116, California	No	0.142
Eel	California Hwy 101, California	No	0.237
Brazos	U.S. Hwy 59, Texas	No	0.177
Rhone	Avignon, France	Yes?	0.153
Amazon	Santarem, Brazil	No	0.21
		Average of all rivers	0.2

and transporting the element from drainage areas to floodplains and other large water bodies. The levels encountered in surface waters, however, are generally "low." The average selenium content of nine rivers was 0.2 µg/l (Table XXII) [194]. This average level was not affected by the presence of industrial activity in the drainage basins. For example, the Susquehanna River, situated in a coal mining industrial area, contained 0.325 µg/l of selenium, which was not appreciably different from the level found in the Mad River in a nonindustrial area. The Amazon River, which drains a non-industrialized area, contained 0.21 µg/l of selenium, as much as the average. However, rivers receiving drainage from seleniferous soils may contain high concentrations of selenium under the proper conditions of pH (Table XXIII) [195]. Tributaries of the Colorado River receiving such drainage contained up to 400 µg/l. The selenium content was below 1 µg/l in acidic waters, probably due to the formation of insoluble iron complexes with selenites. In alkaline waters, selenites are oxidized to the soluble selenates and higher concentrations of selenium were encountered.

Ground waters from deep wells seem to contain only a few µg/l of selenium, but higher concentrations may be encountered in shallow wells [186]. Samples from 22 Australian villages contained less than 1 µg/l [196] and tap and mineral waters from Stuttgart, Germany have been reported to contain 1.6 and 5.3 µg/l, respectively [197]. Analyses of 535 samples of waters from the major watersheds of the United States over a four-year period showed only two samples containing more than 10 µg/l of selenium, with the higher of the two being 14 µg/l [198].

Although most natural waters contain low levels of selenium (few µg/l), it is estimated that river flows deposit 8000 ton/yr of selenium in the ocean. The average selenium content in ocean waters examined by Schutz and

Table XXIII. Selenium Content of Colorado Surface Waters [195]

Stream	County	pH	Se (μg/l)
	pH 6.1–6.9		
Animas River	San Juan	6.1	1
Mineral Creek		6.1	1
		6.4	1
Animas River		6.5	1
	LaPlata	6.9	1
Los Pinos River		6.7	1
Animas River		6.8	1
Vallecito Creek		6.8	1
San Juan River	Archuleta	6.9	1
		6.7	1
East Fork, San Juan River		6.7	1
	pH 7.8–8.2		
East Fork, San Juan River	Archuleta	7.8	10
Rio Blanco		8.0	50
Navajo River		7.9	270
San Juan River		8.1	20
Hermosa Creek	LaPlata	7.9	60
Florida River		8.0	1
		8.2	400
Spring Creek		7.9	30
Animas River		8.0	40

Turekian [199] was 0.09 μg/l and the value was relatively constant (Tables XXIV and XXV). Low values were also reported in the English Channel and the Irish Sea, which contained 0.5 and 0.34 μg/l, respectively. These low levels in sea waters have been attributed to the precipitation of selenite with oxides of metals such as iron and manganese [187].

Accumulation in the Food Chain

Selenium enters the food chain almost entirely via plants. The level and form of selenium in soils determines the concentration of the element in plants. Selenate is the major form absorbed by plants, while elemental selenium or selenites are not readily available. In seleniferous soils, certain plants, referred to as selenium accumulators or indicators, can concentrate the element to levels over 1000 ppm without being injured. The weed *Astragulus* was reported to contain as much as 4500 mg/kg of selenium [200]. Other plants, referred to as secondary selenium absorbers, rarely contain more than a few hundred ppm. Most crop plants, grains and grasses rarely

Table XXIV. Selenium Content of Sea Water [199]

Source	Number of Samples	Se, µg/l Range	Se, µg/l Average
Caribbean	2	0.10–0.12	0.11
Western North Atlantic	5	0.084–0.13	0.096
Eastern North Atlantic	6	0.076–0.11	0.088
Western South Atlantic	2	0.070–0.08	0.075
Eastern Pacific	6	0.061–0.12	0.104
Antarctic	1	0.052	0.052
Long Island Sound	5	0.10–0.13	0.11

Table XXV. Variations of Selenium Content of Sea Water
with Depth of Sample [199]

Source	Depth (meters)	Se (μg/l)
Western North Atlantic, lat. 39° 15' north, long. 63° 09' west	5	0.11
	500	0.10
Eastern North Atlantic, lat. 21° 21' north, long. 24° 03' west	8	0.076
	600	0.088
Eastern North Atlantic, lat. 9° 20' north, long. 18° 36' west	500	0.083
	800	0.088
Western South Atlantic, lat. 21° 49' south, long. 35° 43' west	10	0.080
	100	0.070
Eastern Pacific, lat. 3° 32' north, long. 81° 11' west	14	0.080
	200	0.055
	400	0.12

contain more than 30 ppm. The selenium content of various species of algae collected from waters in industrial and nonindustrial areas ranged between 0.05 to 0.24 ppm. The variation was due more to the species than the source of the algae [178].

Various species of dehydrated fish [201] and fish meal [202] from different sources showed an average selenium content in the range of 1.47 to 2.45 ppm, averaging 2.0 ppm. The selenium content showed more variation between the species of fish from the same water than between multiple samples of a single species from different waters. The selenium is probably passed through the food chain to fish, which further accumulate it to relatively higher levels.

Recent data for a cross section of the American diet [203] indicated an average selenium content of 0.01–0.38 ppm with seafoods having slightly higher content, averaging 0.5 ppm. The present food processing methods, food habits and transportation capabilities seem to preclude the possibility of serious excess or deficiency in the human diet in the United States.

Biological Transformations

Selenium, like mercury and arsenic, undergoes biological transformations in the environment. Several species of microorganisms are able to reduce selenite. The reduction of selenite by extracts of *Micrococcus* sp. consisted of two steps [204]: a rapid reduction to elemental selenium followed by slower reduction of the colloidal selenium to selenide. Chemically prepared suspensions of colloidal selenium were also found to be reduced to selenide. Selenate, on the other hand, was not reduced. Other species of microorganisms included: *Candida albicans* [205], *Salmonella* [206] and *Streptococcus* [207], which reduced selenite to elemental selenium rather than to the selenide. The selenite appeared to be bound to cell proteins through vicinal thiol groups. The following pathway of electron flow for reduction of selenite was suggested [205]:

Since the reduction of selenite can be inhibited by a variety of sulfhydryl blocking agents, these groups are present at the active site of the selenium-reducing enzyme, selenoreductase.

Certain microorganisms, e.g., *E. coli*, developed a permanent adaptation to selenate [208]. It was also shown that microorganisms taken from geologic zones high in selenium were less susceptible to the toxic effects of the element than were microorganisms taken from soils low in selenium. This increased resistance in adapted strains was attributed to higher levels of selenoreductase [209].

There is relatively little evidence of existence of oxidative pathways of selenium metabolism in microorganisms, by contrast to the well-documented ability to reduce soluble selenium compounds to the insoluble and nontoxic elemental state. This fact illustrates the important role of microorganisms in controlling the relative abundance of selenium and could have important consequences for the environmental cycling of the element.

Inorganic selenium, like arsenic and mercury, undergoes biological methylation by various organisms to give volatile compounds with a characteristic garlic-like odor [210]. For example, fungi of several genera produce dimethyl selenide from inorganic selenium compounds. The mechanism suggested for the conversion is as follows [210]:

$$H_2SeO_3 \rightarrow H^+ + :\overset{O-}{\underset{O}{Se}}-OH \xrightarrow{CH_3^+} CH_3\overset{O}{\underset{O}{Se}}-OH \xrightarrow{\text{ionization}} CH_3\overset{O-}{\underset{O}{Se}}:$$

| ION | METHANE-SELENONIC ACID | ION OF METHANE-SELENINIC ACID |

$$\xrightarrow{CH_3^+} (CH_3)_2\overset{O}{\underset{O}{Se}} \xrightarrow{\text{reduction}} {}^.(CH_3)_2Se:$$

| | DIMETHYL SELENONE | DIMETHYL SELENIDE | (121) |

Methionine was reported to be the donor of the methyl group in this reaction.

Fleming and Alexander [211] isolated a strain of *Penicillium* from raw sewage, which produced dimethylselenide from inorganic selenium compounds. Raw sewage has been reported to contain as much as 280 μg/l of selenium and secondary effluent contained 50 μg/l. The methylation of selenium was also shown to occur in soils enriched with sodium selenite [189].

Biological methylation of selenium was shown to occur in the aquatic environment. Microorganisms in lake sediment were shown to convert inorganic selenium from sodium selenite or sodium selenate and organic selenium from selenocystine, selenourea and seleno-DL-methionine into methylated compounds [212]. The main products were the volatile dimethyl selenide $(CH_3)_2Se$, dimethyl diselenide $(CH_3)_2Se_2$ and an unknown compound (Table XXVI). The production of volatile selenium was temperature dependent. About 79% as much $(CH_3)_2Se$ was produced with incubation at 10°C as at 20°C. At 4°C the production was lowered to about 10%. The temperature effect was even more drastic for the conversion of selenate. The production at 10°C was about 15% of that at 20°C; no production was

Table XXVI. Methylation of Selenium Compounds in Sediment Samples from 12 Lakes in the Sudbury Area. In Experiments Where Selenium Compounds were Added, the Concentration of Added Selenium was 5 mg/l. Volatile Selenium was Measured in Nanograms. Abbreviation: U, unknown compound [212]

Lake	Se in Sediment (μg/g, dry wt)	No Addition			Sodium Selenite			Sodium Selenate		
		$(CH_3)_2$Se (ng)	$(CH_3)_2$Se$_2$ (ng)	U (ng)	$(CH_3)_2$Se (ng)	$(CH_3)_2$Se$_2$ (ng)	U (ng)	$(CH_3)_2$Se (ng)	$(CH_3)_2$Se$_2$ (ng)	U (ng)
Elbow	0.48	34	0	3.3	33.3	0	0	34	0	3.3
Ramsey	1.64	0	0	0	0	0	0	33.6	0	0
Kelley	20.48	2.7	3.3	2.3	14	0	0	20	3.3	5.3
Long	0.90	1.7	0	0	27.3	0	0	24.7	0	0
Simons	16.28	0	0	0	18.7	0	18	20	4.7	19.7
Vermillion	0.52	0	0	0	20.3	0	0	9.7	8	5.3
Windy	0.65	0	0	0	12.3	0	0	0	0	0
Moose	0.67	0	0	0	5	0	0	10.5	0	2
Kukagami	0.44	0	0	0	33.3	0	0	23.4	0	0
Nepewassi	0.53	0	0	0	25.3	0	0	28.7	0	0
Johnnie	0.55	6.3	0	3.7	43.3	16.5	0	94.6	19.4	54.8
George	0.67	0	0	0	56.8	0	8.7	94.7	11	7.3

observed at 4°C. The conversion proceeded under both aerobic and anaerobic conditions. Fine-structural examination of bacteria, after growth in the presence of selenite, revealed electron-opaque deposits of irregular shapes within the cytoplasm. These deposits were shown by energy-dispersive X-ray microanalysis to contain selenium.

The above discussion indicates that there is a biological cycle for selenium similar to the ones for arsenic and mercury in the aquatic environment (Figure 20). The methylation of selenium is considered to be a highly effective biological detoxification mechanism. Dimethyl selenide was shown to be about one five-hundredth as toxic as selenite [170], and there is no evidence for its accumulation in fish or other levels of the food chain. It seems probable, however, that the biological methylation of selenium does not pose the same ecological threat as the methylation of inorganic mercury compounds. However, the biological transformations of the different species of selenium should be considered as an important factor in evaluating its fate and cycling in the aquatic environment.

Chemistry in Aqueous Systems

Selenium exists in four oxidation states in aqueous systems: −2, 0, +4, +6. The reduced state, −2, is represented by the selenide species and the elemental state, Se^0, exists in different allotropic forms, whereas the two higher oxidation states are represented by selenites, +4, and selenate, +6. The

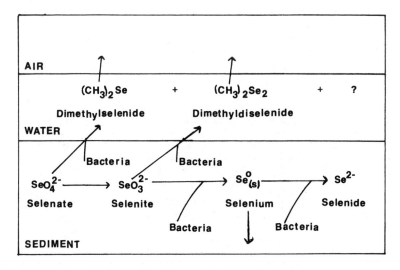

Figure 20.   Biological cycle of selenium.

standard free energies of formation for some selenium species are shown in Table XXVII. The relative stability of the different species of selenium can be obtained from the thermodynamic data presented in Tables XXVIII and XXIX. Figure 21 shows the stability (Eh-pH) diagram for the different species of selenium in aqueous systems [186].

A large portion of the stability domain of elemental selenium, Se^0, covers that of water. Therefore, it is stable in water and aqueous solutions at all pH values free of oxidizing and reducing agents. The stability of elemental

Table XXVII. Standard Free Energies of Formation of Some Selenium Species at 25°C [57]

Mineral Name	Formula[a]	ΔG_f^0 (kcal/mol)
Selenium	$Se_{(s)}^0$	0.0
Hydrogen Selenide	H_2Se	5.3
Hydrogen Selenide Ion	HSe^-	10.5
Selenide Ion	Se^{2-}	30.9
Selenious Acid	H_2SeO_3	−101.87
Acid Selenite Ion	$HSeO_3^-$	−98.36
Selenite Ion	SeO_3^{2-}	−88.40
Selenic Acid	H_2SeO_4	−105.42
Acid Selenate Ion	$HSeO_4^-$	−108.10
Selenate Ion	SeO_4^{2-}	−105.50
Selenium Dioxide	$SeO_{2(s)}$	−41.5

[a]All species are aqueous unless otherwise stated.

Table XXVIII. Equilibrium Constants for Selenium Species at 25°C [213]

Reaction No.	Reaction	$\log K_a$
(122)	$H_2Se = HSe^- + H^+$	−3.74
(123)	$HSe^- = Se^{2-} + H^+$	−14.01
(124)	$H_2SeO_3 = HSeO_3^- + H^+$	−2.57
(125)	$H_2SeO_3 = SeO_3^{2-} + H^+$	−6.58
(126)	$H_2SeO_4 = HSeO_4^- + H^+$	2.05
(127)	$H_2SeO_4 = SeO_4^{2-} + H^+$	−2.05

Table XXIX. Oxidation Potentials of Selenium Species at 25°C [213]

Reaction No.	Reaction	$E_0{}^a$ (V)
(128)	$H_2Se + 3H_2O = H_2SeO_3 + 6H^+ + 6e$	+0.36
(129)	$H_2Se + 3H_2O = HSO_3^- + 7H^+ + 6e$	+0.386
(130)	$HSe^- + 3H_2O = H_2SeO_3^- + 6H^+ + 6e$	+0.349
(131)	$HSe^- + 3H_2O = SeO_3^{2-} + 7H^+ + 6e$	+0.414
(132)	$Se^{2-} + 3H_2O = SeO_3^{2-} + 6H^+ + 6e$	+0.276
(133)	$H_2SeO_3 + H_2O = HSeO_4^- + 3H^+ + 2e$	+1.090
(134)	$H_2SeO_3 + H_2O = SeO_4^{2-} + 4H^+ + 2e$	+1.151
(135)	$HSeO_3^- + H_2O = SeO_4^{2-} + 3H^+ + 2e$	+1.075
(136)	$SeO_3^{2-} + H_2O = SeO_3^{2-} + 2H^+ + 2e$	+0.880
(137)	$H_2Se = Se_{(s)} + 2H^+ + 2e$	−0.399
(138)	$HSe^- = Se_{(s)} + H^+ + 2e$	−0.510
(139)	$Se^{2-} = Se_{(s)} + 2e$	−0.924
(140)	$Se_{(s)} + 3H_2O = H_2SeO_3 + 4H^+ + 4e$	+0.741
(141)	$Se_{(s)} + 3H_2O = HSeO_3^- + 5H^+ + 4e$	+0.778
(142)	$Se_{(s)} + 3H_2O = SeO_3^{2-} + 6H^+ + 4e$	+0.875

[a]Signs of the half-cell potentials are in accord with the IUPAC convention (see Chapter 4).

selenium is demonstrated by its occurrence in sandstone in dry alkaline environments [186]. Selenium exists in three allotropic forms: gray rhombic, which is the most stable, red-monoclinic, and the black amorphous forms. The red and black amorphous allotropes are the most likely forms of Se^0 in soils and the aquatic environment [214]. The red allotrope is found when selenium is precipitated from aqueous solution. At temperatures higher than 30°C, the red forms gradually revert to the black amorphous Se^0 [215]. This latter form appears to be an indefinite transition stage between the red and the gray allotrope. Elemental selenium burns in air to form selenium dioxide, SeO_2, which is a solid substance (mp 340°C). During the combustion of fossil fuels SeO_2 is formed, but it is readily reduced to elemental selenium by sulfur dioxide, which is also formed during the combustion of these materials.

The stability of elemental selenium indicates that it is probably a major inert "sink" when it is introduced into the environment in various ways. Contamination of water by elemental selenium would probably pose a minimal toxicity hazard.

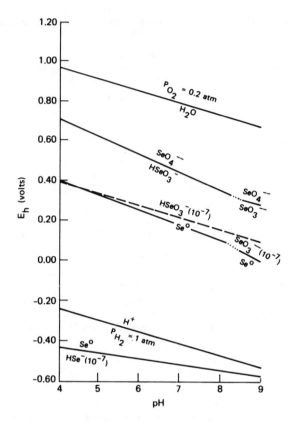

Figure 21.   Relation of oxidation–reduction potentials of some selenium compounds to pH. (The dashed line represents the dividing line between oxidized and reduced soils [186].)

Selenium is reduced to hydrogen selenide, H_2Se, or other selenides at relatively low redox potentials. Therefore, these compounds are unstable in aqueous systems. Hydrogen selenide by itself is not expected to exist in the aquatic environment since the Se^0/H_2Se couple falls even below the H^+/H_2 couple. Aqueous solutions of H_2Se are actually unstable in air due to its decomposition into elemental selenium and water. Under moderately reducing conditions, heavy metals are precipitated as the selenides, which have extremely low solubilities. The following are log K_s values of some heavy metals selenides of environmental interest [213]: -11.5 (Mn^{2+}), -26.0 (Fe^{2+}), -60.8 (Cu^+), -48.1 (Cu^{2+}), -29.4 (Zn^{2+}), -35.2 (Cd^{2+}), -64.5 (Hg^{2+}) and

−42.1 (Ph^{2+}). The precipitation of selenium as heavy metal selenides can be an important factor affecting the cycling of the element in soils and natural waters.

Selenious acid species ($HSeO_3^-$ and SeO_3^{2-}) are predominant in solution over most of the area of moderately oxidizing conditions encountered in oxygenated waters. Between pH 3.5 and 9.0 biselenite ion is the predominant ion in water. At pH values below 7.0 selenites are rapidly reduced to elemental selenium under mildly reducing conditions.

Most selenite salts are less soluble than the corresponding selenates. The extremely low solubility of ferric selenite [216] $Fe_2(SeO_3)_3 - K_S = 2.0 \pm 1.7 \times 10^{-31}$, and the basic ferric selenite [217] $Fe_2(OH)_4 SeO_3 - K_S = 10^{-61.7}$, which is found in dilute solutions of ferric chloride and sodium selenite, are important to the environmental cycling of selenium. Selenites also form stable adsorption complexes with ferric oxides of even lower solubility than the ferric selenites [215]. Under certain conditions, selenite seems to be completely adsorbed in high amounts by ferric hydroxide and, to a lesser extent, by aluminum hydroxide, while selenate is not [218]. The adsorption of the selenite was shown to follow a Langmuirian isotherm, with the specific adsorption being higher in acidic solutions and decreasing with an increase of pH value [219]. The low levels of selenium in ocean waters have been attributed to adsorption of selenite by the oxides of metals, such as iron and manganese [186].

Selenates (SeO_4^{2-}) occupy the upper area of the Eh-pH diagram and, therefore, are stable under alkaline oxidizing conditions. Most selenate salts are similar in solubility to the sulfates of the same metals. Soluble selenate would be expected in alkaline soils or alkaline weathering rocks in dry areas. Selenate appears to be the most significant form of selenium as far as environmental pollution is concerned because of its stability at alkaline pH values, its solubility and its ready availability to plants.

The stability relationships discussed in the previous section may be useful in predicting the most significant species of selenium in water under a variety of conditions. However, as previously discussed in the highly dynamic natural water systems, departure from equilibrium often occurs and rates of approach to equilibrium are not considered in the thermodynamic models. Very little information is available for the kinetics of conversion from one oxidation state of selenium to another. The rates of transformation of selenite to selenate and vice versa were shown to be very slow [168], whereas the rate of transformation of selenite to elemental selenium proceeds very rapidly. On the other hand, the oxidation of Se^0 to SeO_3^{2-} is not readily achieved. The mobility of selenium can also be affected by sorption on hydrous metal oxides in bottom sediments and microbial transformations to volatile organic forms.

CHROMIUM

Background

Chromium is widely distributed in the earth's crust. It is more abundant than cobalt, copper, zinc, molybdenum, lead, nickel and cadmium [220] and ranks fourth among the 29 elements of biological importance. Chromium, which is considered an essential trace element in man, occurs in most biological materials in the trivalent oxidation state. Chromium deficiency results in impaired glucose metabolism due to poor effectiveness of insulin. Some animal studies also indicate that chromium deficiency may induce atherosclerosis. In addition, the digestive enzyme trypsin appears to contain chromium as an integral part [221].

Current information on the toxic effects of chromium is derived entirely from occupational exposure, where it affects the skin and respiratory system. The toxicity of chromium depends on its oxidation state. Only hexavalent (+VI) chromium is known to be toxic; the trivalent (+III) state has no established toxicity [221]. Such hexavalent chromium compounds as chromic acid, sodium and potassium chromates, and dichromates have been shown to cause ulceration of the skin and perforation of nasal septum to exposed workers engaged in the manufacture of these chemicals. The toxic effects of hexavalent chromium are attributed to its oxidation potential and easy permeation of biological membranes. All the biological interactions of hexavalent chromium result in reduction to the trivalent state, the latter forming coordination complexes with organic molecules. Hexavalent chromium is reduced in the skin to the trivalent state by methionine, cystine and cysteine. The trivalent form resulting from this reduction may form protein complexes and may initiate sensitization.

Inhaled hexavalent chromium may cause cancer of the respiratory tract, and increased risk of lung cancer among those occupationally exposed to chromium (+VI) has been established. On the other hand, there is no evidence that nonoccupational exposure to chromium in the diet, water or ambient air constitutes a cancer hazard [221].

The total daily intake of chromium by man in the United States has been estimated to be 60–280 μg/day, with an average of 60 μg. This intake level is considered nutritionally marginal. It has been suggested that diets containing mostly processed foods may be chromium deficient. Therefore, marginal deficiency of chromium in the U.S. is of greater concern to a nutritionist than overexposure. Chromium (+VI) in drinking water at concentrations greater than 5 mg/l was found to accumulate in rats; however, no change in growth rate or food intake was observed [222]. Even concentrations up to 25 mg/l in the drinking water failed to produce changes in these characteristics. The maximum contaminant level for total chromium in the Primary Drinking

Water Regulations of the U.S. Environmental Protection Agency is 0.05 mg/l, which is one hundredth of the maximum "no observed adverse health effect" concentration [1].

Sources in the Environment

Natural Sources

Chromium occurs in nature mainly in chromite, a mineral of the spinel group with the formula $(Fe,Mg)O(Cr,Al,Fe)_2O_3$. The highest grade of the ore contains 52-56% chromic oxide and 10-26% ferrous oxide [223]. The theoretical end member, $FeOCr_2O_3$, would contain 68% chromic oxide, Cr_2O_3, and 32% ferrous oxide, FeO. Red lead ore, crocoite ($PbCrO_4$) is another source of chromium that is found in small quantities in limited areas of the world. Chromium also occurs in small quantities in many minerals in which it replaces Fe^{3+} or Al^{3+} (as Cr^{3+}), e.g., in chromium tourmalines, chromium garnets, chromium micas and chromium chlorites. The green color of emerald is due to the incorporation of small amounts of chromium in place of alumium in beryl (beryllium aluminum silicate) [224]. Trace amounts of chromic oxide in crystalline corundum (alumina) give ruby its distinguishing color.

Chromium concentrations in the continental crust range between 80 to 200 ppm, with an average of 125 ppm. Chromium also occurs in soils at concentrations ranging from trace to 5.23% in Puerto Rico [225]. The geometric mean of chromium concentration in the United States soils was estimated to be 37 ppm [226]. Higher concentrations of chromium are usually found in ultramafic igneous rocks (1000-3400 ppm), in shales and clays (30-590 ppm), and in phosphorites (30-3000 ppm). The phosphorites used as fertilizers are important sources of contamination of soils with chromium [221].

Anthropogenic Sources

The technological advances during the past two decades have led to the increased usage of chromium and have made the element considerably important. The United States produced all the world's supply of chromite during the first half of the 19th century; however, the discovery of a cheaper, high-grade chromite ore in foreign countries resulted in the end in 1961 of mining the ore. The metallurgy industry uses approximately 57% of the imported chromite ore, whereas 30% is used in refractory materials and 13% in chemical industry [221].

The metallurgy industry uses the highest quality chromite ore, which is usually converted to several types of ferrochromium or chromium metal that are alloyed with iron or other elements, usually nickel and cobalt. Over 60% of the chromium used in the metallurgy industry is used in making stainless steel; the remainder is used in high-speed steels, other alloy steels, high-temperature steels and nonferrous alloys.

Chromite ore containing about 34% chromic oxide and high alumina content is used by the refractory industry for the manufacture of melting-furnace linings as chromite bricks or magnesia-chrome bricks because chromite is chemically inert and has a high melting point, 2040°C. Chrome refractory materials are also used as coatings to close pores and for joining bricks within the furnace [220].

The chemical industry uses chromite ore for the manufacture of sodium chromate and sodium dichromate, which are used for the production of other chromium chemicals. Most of the chromic acid produced is used for chrome plating. Chromium chemicals are used as tanning agents, pigments, catalysts, wood preservatives and corrosion inhibitors [220].

The industrial use and applications of chromium compounds result in air emissions and wastewater discharges that could contribute to the contamination of the natural environment if not properly controlled.

Occurrence in Natural Waters

Chromium concentrations in natural waters are limited by the low solubility of chromium (+III) oxides. A survey was conducted by the U.S. Geological Survey [227] of 15 rivers of North America. It was reported that chromium levels ranged from 0.7 μg/l (Sacramento River) to 84 μg/l (Mississippi River). Most of the samples were in the range of 1 to 10 μg/l of chromium. Another study [228] of more than 1500 surface waters in the U.S. revealed the presence of chromium in only 25% of the samples, with a maximum of 110 μg/l and a mean of 10 μg/l. These data represent the dissolved chromium content of these surface waters and do not account for the chromium associated with particulate material. Review of the data of Kroner's survey [228] indicated that dissolved chromium was detected in 23% of the samples at levels ranging from 2 to 25 μg/l. However, the suspended solids fraction in 38% of the samples contained chromium in the 3 to 13 μg/l range.

In addition to direct discharge of chromium-bearing wastes from municipal and industrial wastes, considerable amounts of chromium-bearing sediments are transported to surface waters in storm runoff from urban areas and watershed drainage of eroded soils. In a study of trace element concentrations in Lake Michigan [229], it was found that the highest concentrations of chromium were found in the region traversed by the plume of Grand River, which drains agricultural and industrial areas. The chromium concentration in the lake water ranged from 0.55 to 1.1 μg/l, whereas substantially higher concentrations (77 mg/kg) were found in the sediments near the sediment–water interface. The background level of chromium in the lake sediments was in the range of 20 to 30 mg/kg [230]. Marine sediments from bays and estuaries near large industrial and urban areas have been found to contain high concentrations of heavy metals including chromium [231,232]. The clay fraction of bottom sediments near the head of New Bedford Harbor, Massachusetts contained chromium at a concentration of 2146 ppm, which gradually decreased seaward to 200 ppm in the sediments at the edge of the bay.

The Atlantic coastal river waters contain significantly higher concentrations of chromium than the Gulf and Pacific river waters [233,234]. This probably reflects the sizes of different watersheds and their composition of soils.

The chromium content of seawater is less than that of river waters, generally below 1 μg/l. The concentrations in samples taken near the coast of Japan [235] ranged from 0.04 to 0.07 μg/l, from 0.13 to 0.25 μg/l near England [236], 0.46 μg/l in the Irish Sea [237] and from 0.2 to 0.5 μg/l in the Pacific Ocean. Considerable amounts of chromium are delivered to the oceans by rivers and eventually are deposited on the ocean floor. It has been estimated that 6.7×10^6 kg of chromium are added to the oceans each year [238].

Chromium in the Air

The ambient chromium concentration in the air is generally low and is probably of crustal origin. Air quality data from the National Air Surveillance Networks [221] for the period 1960-1969 indicated that the annual mean concentration of chromium in urban stations ranged between 0.01 to 0.03 μg/m^3. For rural stations, it seldom reached 0.01 μg/m^3 (the minimum detection limit for the instruments used at that time). Additional studies [239] of chromium content of the southwestern desert atmosphere showed a level of 0.003 μg/m^3 in rural areas outside Tucson, Arizona, whereas the urban level in Tucson was 0.004 μg/m^3. This study suggested that most of the chromium in the suspended particulate matter in the air was probably due to airborne soil material, with little enrichment in the urban areas.

Coal-fired power plants can contribute significant amounts of trace metals, including chromium in the particulate emissions. The average chromium content of coal is 20 ppm, with a range of 10 to 1000 ppm, depending on its origin [221]. A study of trace metal concentrations in particulate emissions from a coal-fired power plant revealed chromium levels of 300 μg/m^3 in the inlet to the electrostatic precipitator. The concentration was reduced to 0.7 μg/m^3 in the discharge and was mostly in the 0 to 1-μm-diameter particulates [240]. A national survey of elements in fly ashes from 21 states showed a chromium content ranging from 20 to 170 ppm (dry weight) [241].

The small size of the chromium-bearing particulates suggests that they can be airborne for long distances and would probably be removed from the atmosphere by precipitation. However, the present regulations for control of particulate emissions from power plants and other industrial sources preclude the consideration of these sources as major contributors of chromium in the aquatic environment.

Effects on Aquatic Life

There is a great range of sensitivity to chromium among different species of aquatic life. Toxicity varies widely with the species, temperature, pH, valence of chromium, and synergistic and antagonistic effects, particularly that of hardness [80]. The 96-hour LC_{50} and safe concentrations for hexavalent chromium were reported [221] to be 33 and 1 mg/l, respectively, for fathead minnows in hard water; 50 and 0.6 mg/l in soft water for brook trout; and 69 and 0.3 mg/l for rainbow trout in soft water. The LC_{50} of hexavalent chromium for *Daphnia* was reported to be 0.05 mg/l, and the chronic no-effect level of trivalent chromium on reproduction was 0.33 mg/l [221]. Chromium was also reported to inhibit the growth of different species of fresh water algae in the range of 0.032 to 6.4 mg/l [242].

Chromium concentrations in sea water average about 0.04 μg/l, with concentration factors of 1600 in benthic algae, 2300 in phytoplankton, 1900 in zooplankton, 100 in crustacean muscle and 70 in fish [159]. Chromium threshold toxicity levels of 1 mg/l for polychaete, 5 mg/l for prawn and 20 mg/l for crabs have been reported [81]. A hexavalent chromium concentration of 5 mg/l chromium reduced photosynthesis in giant kelp by 50% during four days of exposure [243]. Because of the sensitivity of lower forms of aquatic life to chromium and its accumulation at all trophic levels, the maximum concentration of 0.05 mg/l has been recommended for freshwater aquatic life [81].

The toxicity of chromium toward bacteria is controlled by the valence of chromium, the type of organism and the amount of organic matter present. Trivalent chromium has been reported to lower the 5-day biochemical oxygen demand by 50%. The same effect was obtained at a level of 100 mg/l hexavalent chromium [81]. In anaerobic sludge digestion, chromium is toxic at a concentration of 500 mg/l.

Chemistry in Aqueous Systems

Chromium commonly occurs in four oxidation states: (0), (+II), (+III) and (+VI). The most important species in aqueous systems are the trivalent and hexavalent. The standard free energies of formation of some chromium species are shown in Table XXX. The stability equilibria and oxidation–

Table XXX. Standard Free Energies of Formation of Some
Chromium Species at 25°C [56]

Mineral Name	Formula[a]	ΔG_f^0 (kcal/mol)
Chromium	$Cr_{(s)}^0$	0.0
Chromium Oxide	$CrO_{(s)}$ hydr.	−83.81
Eskolaite	$Cr_2O_{3(s)}$	−250.20
Chromous Hydroxide	$Cr(OH)_{2(s)}$	−140.50
Chromic Hydroxide	$Cr(OH)_{3(s)}$	−215.30
	$Cr(OH)_3 \cdot nH_2O_{(s)}$	−205.50
Hydrated Chromium Hydroxide	$Cr(OH)_{4(s)}$	−242.38
Chromium Dioxide	$CrO_{2(s)}$	−129.00
Chromium Trioxide	$CrO_{3(s)}$	−120.00
Chromous Ion	Cr^{2+}	−42.10
Chromic Ion	Cr^{3+}	−51.50
Chromyl Ion, green	$CrOH^{2+}$	−103.00
Chromyl Ion	$Cr(OH)_2^+$	−151.21
Chromite Ion, green	CrO_2^-	−128.09
Chromite Ion, green	CrO_3^{3-}	−144.22
Chromic Acid	H_2CrO_4	−185.92
Acid Chromate Ion	$HCrO_4^-$	−184.90
Chromate Ion	CrO_4^{2-}	−176.10
Dichromate Ion	$Cr_2O_7^{2-}$	−315.40

[a]All species are aqueous unless noted otherwise.

reduction equilibria for some chromium species are shown in Tables XXXI and XXXII, respectively.

Figure 22 illustrates an Eh-pH diagram for the important species of chromium in a system containing chromium at 10^{-4} M in water at 25°C. Chromic hydroxide, $Cr(OH)_{3(s)}$, or hydrated chromic oxide, is the principal species most likely to enter into equilibria affecting the solubility of chromium in natural waters. In the presence of such complexing ions as chlorides, sulfates, cyanides, oxalates or citrates, $Cr(OH)_3 \cdot nH_2O_{(s)}$ is the solid whose solubility is low enough to occur in the system. The areas of dominance of solute species in equilibrium in the presence of chlorides for the system chromium−water at a concentration of 10^{-4} M chromium at 25°C are shown in Figure 23. The species $CrOH^{2+}$ shown in the diagram is thought to be a complex ion-containing chloride [56].

In the pH range of natural waters of 5.0 to 9.0, Eh values less than +0.5 V, and in the presence of chlorides, compounds of trivalent chromium, Cr^{3+}, are

Table XXXI. Equilibrium Constants of Chromium Species at 25°C

Reaction No.	Reaction	log K_{eq}
(143)	$Cr^{3+} + 2H_2O = CrO_2^- + 4H^+$	−26.99
(144)	$Cr^{3+} + H_2O = CrOH^{2+} + H^+$	−−3.81
(145)	$CrOH^{2+} + H_2O = Cr(OH)_2^+ + H^+$	−6.22
(146)	$Cr(OH)_2^+ = CrO_2^- + 2H^+$	−16.96
(147)	$H_2CrO_4 = HCrO_4^- + H^+$	−0.75
(148)	$2H_2CrO_4 = Cr_2O_7^{2-} + H_2O + 2H^+$	−0.18
(149)	$H_2CrO_4 = CrO_4^{2-} + 2H^+$	−7.20
(150)	$Cr_2O_7^{2-} + H_2O = 2HCrO_4^-$	−1.66
(151)	$Cr_2O_7^{2-} + H_2O = 2CrO_4^{2-} + 2H^+$	−14.59

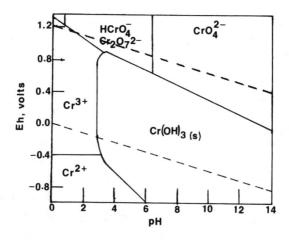

Figure 22.   Eh–pH diagram for the system chromium–water at 25°C. $[Cr]_t = 10^{-4}M$ (reproduced from Pourbaix [56] courtesy of Pergamon Press).

precipitated as chromic hydroxide, $Cr(OH)_{3(s)}$, or the hydrated chromic hydroxide, $Cr(OH)_3 \cdot nH_2O_{(s)}$. The solubility of chromic hydroxide was reported [56] to be $10^{-14.3}$ M as Cr, at pH 7.0, 25°C. Figure 24 shows the solubility of $Cr(OH)_3 \cdot nH_2O_{(s)}$ in water containing chlorides at different pH values at 25°C. The solubility of the saturated hydroxide was found to be $10^{-6.1}$ M as Cr, or 0.04 mg/l Cr at pH 7.0 [56].

Hexavalent chromium forms oxo-compounds and does not give rise to an extensive and complex series of polyacids and anions characteristic of the less

Table XXXII. Oxidation Potentials of Chromium Species at 25°C [56]

Reaction No.	Reaction	$E^{o\,a}$ (V)
(152)	$Cr^{2+} = Cr^{3+} + e$	−0.407
(153)	$Cr^{2+} + H_2O = CrOH^{2+} + H^+ + e$	−0.182
(154)	$Cr^{2+} + 2H_2O = Cr(OH)_2^+ + 2H^+ + e$	0.185
(155)	$Cr^{2+} + 2H_2O = CrO_2^- + 4H^+ + e$	1.188
(156)	$Cr^{3+} + 4H_2O = H_2CrO_4 + 6H^+ + 3e$	1.335
(157)	$Cr^{3+} + 4H_2O = HCrO_4^- + 7H^+ + 3e$	1.350
(158)	$2Cr^{3+} + 7H_2O = Cr_2O_7^{2-} + 14H^+ + 6e$	1.33
(159)	$Cr^{3+} + 4H_2O = CrO_4^{2-} + 8H^+ + 3e$	1.477
(160)	$CrOH^{2+} + 3H_2O = HCrO_4^- + 6H^+ + 3e$	1.275
(161)	$2CrOH^{2+} + 5H_2O = Cr_2O_7^{2-} + 12H^+ + 6e$	1.258
(162)	$CrOH^{2+} + 3H_2O = CrO_4^{2-} + 7H^+ + 3e$	1.402
(163)	$Cr(OH)_2^+ + 2H_2O = HCrO_4^- + 5H^+ + 3e$	1.152
(164)	$2Cr(OH)_2^+ + 3H_2O = Cr_2O_7^{2-} + 10H^+ + 6e$	1.135
(165)	$Cr(OH)_2^+ + 2H_2O = CrO_4^{2-} + 6H^+ + 3e$	1.279
(166)	$CrO_2^- + 2H_2O = CrO_4^{2-} + 4H^+ + 3e$	0.945
(167)	$2CrO_2^- + 3H_2O = Cr_2O_7^{2-} + 6H^+ + 6e$	0.801
(168)	$CrO_3^{3-} + H_2O = CrO_4^{2-} + 2H^+ + 2e$	0.359

[a]Signs of the half-cell potentials are in accord with the IUPAC convention.

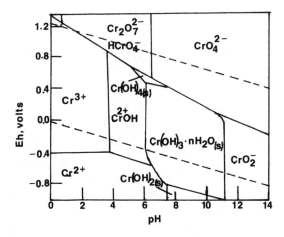

Figure 23.   Eh–pH diagram for the system chromium–water at 25°C in solutions containing chlorides. $[Cr]_t = 10^{-4}M$ (reproduced from Pourbaix [56] courtesy of Pergamon Press).

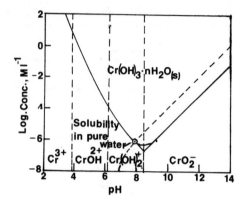

Figure 24.   Effect of pH on solubility of $Cr(OH)_3 \cdot nH_2O_{(s)}$ in solutions containing chlorides (reproduced from Pourbaix [56] courtesy of Pergamon Press).

acidic oxides. This is probably due to the greater extent of multiple bonding of the smaller chromium ion. The main oxy-acids or anions of hexavalent chromium are chromate, CrO_4^{2-}, and dichromate, $Cr_2O_7^{2-}$. Figures 22 and 23 show the domains of relative predominance of the dissolved species of hexavalent chromium, $HCrO_4^-$, CrO_4^{2-} and $Cr_2O_7^{2-}$, as a function of pH and Eh. Both the chromate and dichromate ions are strong oxidizing agents, especially in acid solutions. The powerful oxidizing action of dichromates in very acidic solutions forms the basis for the determination of chemically oxidizable organic matter in the chemical oxygen demand test. The equilibrium of CrO_4^{2-}, $HCrO_4^-$ and $Cr_2O_7^{2-}$ with $Cr(OH)_{3(s)}$ lies below the line where water is oxidized to oxygen at atmospheric pressure. This indicates the extreme stability of dichromate solutions, a fact that is well recognized in analytical chemistry.

At low Eh values or in the presence of oxidizable compounds, hexavalent chromium is reduced to chromic hydroxide, $Cr(OH)_{3(s)}$, in neutral or alkaline solutions or to chromic salt, Cr^{3+}, in acid solutions. Chromous salts, Cr^{2+}, are formed in acid solutions under strongly reducing conditions.

The equilibria considerations discussed previously should be considered as an attempt to understand the boundary conditions toward which the chromium–water system may proceed. In natural waters, kinetic considerations, complex formation, and biological interaction may influence the achievement of equilibrium conditions and, therefore, the precise levels of chromium species may not agree with the thermodynamic predictions.

Trivalent chromium has a strong tendency to form complexes whose ligand rates of exchange are low. Complexes of Cr^{3+} with ammonia, urea, ethylenediamine, halides, sulfates, organic acids, proteins and peptides are known [220,244]. Many of these complexes persist for relatively long periods in solution, even in conditions in which they are thermodynamically unstable, because of their kinetic inertness. The extent of such interactions of chromium in natural waters is unknown; however, in the pH range prevailing in natural waters and with the strong tendency of trivalent chromium to olate, most of these complexes could be converted ultimately to the more stable chromic hydroxide.

Hexavalent chromium can be reduced in natural waters containing organic matter to Cr^{3+}, which would be precipitated as the hydroxide. Both hexavalent and trivalent chromium have been shown to exist in sea water. Fukai [245] reported that an increased proportion of chromium in the trivalent form was found in deeper sea waters.

Both Cr^{6+} and Cr^{3+} and their organic and inorganic complexes can be adsorbed on suspended particulate materials in natural waters, which would be precipitated ultimately at the bottom of lakes or oceans. Several studies on the distribution of trace elements in Lake Michigan [229,230] have indicated that higher concentrations of trace elements including chromium were found near the sediment-water interface than in underlying sediments, as in the water column above them. Leland et al. [230] showed that the average concentration of chromium in the uppermost layer of sediments near the sediment interface was 77 ppm, compared to a background level of 20–40 ppm. Stoffers [232] reported that the clay fraction of sediments collected from New Bedford Harbor contained 2146 ppm, compared to a background of 100 ppm. Schroeder and Lee [246] reported that Cr^{6+} was reduced by Fe^{2+}, dissolved sulfides, cysteine and mercaptosuccinic acid to the trivalent form. They also showed that sand, bentonite and iron oxide sorbed 90–99% of the Cr^{3+} from solution after seven days. This study indicates that under the reducing condition in bottom sediments, Cr^{3+} would be the prevailing species that may be precipitated as the hydroxide or sorbed on the available particulate materials.

The above discussion indicates that a complex model would have to be developed to accurately evaluate the behavior and cycling of chromium in natural waters. Such a model would have to consider the speciation of chromium, complex formation, sorption on available surfaces and biological interactions.

REFERENCES

1. U.S. Environmental Protection Agency, "National Interim Primary Drinking Water Regulations," EPA-570/9-76-003, Washington, DC (1976).
2. Kurland, L. T., et al. *World Neurol.* 1:320 (1960).

3. Irukayama, K. *Proc. Third Int. Conf. on Adv. in Water Pol. Res.* 3: 153 (1967).
4. Fujiki, M. *Kumanato Igk. Z.* 37:10 (1963).
5. Borg, K., et al. *Appl. Ecol.* 3 (Suppl):171 (1966).
6. Johnels, A. G., and T. Westermark. In: *Chemical Fallout*, M. Miller and G. Berg, Eds. (Springfield, IL: Charles C. Thomas, Publisher, 1969).
7. Wobester, G., et al. *J. Fish. Res. Bd. Can.* 27:830 (1970).
8. Klein, D. H., and E. D. Goldberg. *Environ. Sci. Technol.* 4:765 (1970).
9. Bache, C. A., et al. *Science* 172:951 (1971).
10. Fimreite, N., et al. *Can. Field Naturalist* 85:211 (1971).
11. Hammond, A. L. *Science* 171:788 (1971).
12. Nelson, N. *Environ. Res.* 4:1 (1971).
13. Uhte, J. F., and E. G. Bligh. *J. Fish. Res. Bd. Can.* 28:786 (1971).
14. Knauer, G. A., and J. H. Martin. *Limnol. Oceanog.* 17:868 (1972).
15. Peakall, D. B., and R. J. Lovett. *Bioscience* 22:20 (1972).
16. Evans, R. J., et al. *Environ. Sci. Technol.* 6:901 (1972).
17. Walter, C. M., et al. *J. Water Poll. Control Fed.* 45:2203 (1973).
18. Langley, D. G. *J. Water Poll. Control Fed.* 45:45 (1973).
19. Fujita, M., and K. Hashizume. *Water Res.* 9:889 (1975).
20. Westöö, G. In: *Chemical Fallout*, M. Miller and G. Berg, Eds. (Springfield, IL: Charles C. Thomas, Publisher, 1969).
21. Jernelöv, A. In: *Chemical Fallout*, M. Miller and G. Berg, Eds. (Springfield, IL: Charles C. Thomas, Publisher, 1969).
22. Kamps, L. R., et al. *Bull. Environ. Contam. Toxicol.* 8:273 (1972).
23. Ableson, P. H. *Science* 169: 3 (1970).
24. Jonasson, I. R., *Geol. Survey of Canada Paper, 70-57* (1970).
25. *Man's Impact on the Global Environment*, Massachusetts Institute of Technology, Cambridge, MA (1971), p. 137.
26. Plunkett, E. R. *Handbook of Industrial Toxicology*, (Bronx, NY: Chemical Publishing Co., Inc., 1966).
27. Fleischer, M. *U.S. Geol. Survey Prof. Paper 713*, 6 (1970).
28. McCarthy, J. H., Jr., et al. *U.S. Geol. Survey Prof. Paper 713*, 37 (1970).
29. Johnson, D. L., and R. S. Braman. *Environ. Sci. Technol.* 8:1003 (1974).
30. Kothny, E. L. "Trace Elements in the Environment," Advances in Chemistry Series 123, (Washington, DC: American Chemical Society, 1973).
31. Jenne, E. A. *U.S. Geol. Survey Prof. Paper 713*, 40 (1970).
32. Wershaw, R. L. *U.S. Geol. Survey Prof. Paper 713*, 29 (1970).
33. Joensuu, D. I. *Science* 172:1027 (1971).
34. Klein, D. H. "Mercury in the Environment," Environmental Protection Technology Series, EPA-600/2-73-008, U.S. Environmental Protection Agency, Washington, DC (1973).
35. Jensen, S., and A. Jernelöv. *Nature* 223:753 (1969).
36. Wood, J. M., et al. *Nature* 220:173 (1968).
37. Hannertz, L. *Rept. Inst. Freshwater Res. (Dottiningholm)* 48:121 (1968).
38. Bishop, P. L., and E. J. Kirsch. *Proc. 27th Purdue Ind. Waste Conf.*, Purdue University, Lafayette, IN (1972).
39. Tonomura, K., et al. *Environ. Toxicol. Pestic. Proc. U.S.-Japan*, 115 (1972).

40. Langley, D. G. *J. Water Poll. Control Fed.* 45:44 (1973).
41. Spangler, W. J., et al. *Science* 180:192 (1973).
42. Jacobs, L. W., and D. R. Keeney. *J. Environ. Qual.* 3:121 (1974).
43. Bisogni, J. J., and A. W. Lawrence. *J. Water Poll. Control Fed.* 47:135 (1975).
44. Fagerström, T., and A. Jernelöv. *Water Res.* 5:121 (1971).
45. Jernelöv, A., and H. Lann. *Environ. Sci. Technol.* 7:712 (1973).
46. Wood, J. M. *Science* 183:1049 (1974).
47. Billen, G., et al. *Water Res.* 8:219 (1974).
48. Jensen, S., and A. Jernelöv. *Int. Atomic Energy Agency Tech. Rep. Ser. No. 173* 43 (1972).
49. Wood, J. M. *Environment* 14:33 (1972).
50. Hem, J. D. *U.S. Geol. Survey Prof. Paper 713*, 19 (1970).
51. Simpson, R. B. *J. Am. Chem. Soc.* 83:4711 (1961).
52. Schwarzenbach, G., and M. Schellenberg. *Helv. Chim. Acta* 48:28 (1965).
53. Waugh, T. D., et al. *J. Phys. Chem.* 59:395 (1955).
54. Parikh, S. S., and T. R. Sweet. *J. Phys. Chem.* 65:1909 (1961).
55. Sytsma, L. F. *Diss. Abstr. Int. B* 32:6311 (1972).
56. Pourbaix, M. *Atlas of Electrochemical Equilibria in Aqueous Solutions*, (Elmsford, NY: Pergamon Press, Inc., 1966).
57. Wagman, D. D., et al. "Selected Values of Chemical Thermodynamic Properties," NBS Technical Note No. 270-3, National Bureau of Standards, Washington, DC (1968).
58. Cotton, F. A., and G. Wilkinson. *Advanced Inorganic Chemistry* (New York: Wiley-Interscience, 1966), p. 616.
59. Hietanen, S., and L. D. Sillen. *Acta Chem. Scand.* 6:747 (1952).
60. Baughman, G. L., et al. "Chemistry of Organomercurials in Aquatic Systems," Ecological Research Series, EPA-660/3-73-012, U.S. Environmental Protection Agency, Athens, GA (1973).
61. Boyer, P. D. *J. Am. Chem. Soc.* 76:4331 (1954).
62. Simpson, R. B. *J. Chem. Phys.* 46:4775 (1967).
63. Eigen, M., et al. "Essays in Coordination Chemistry," *Experientia* Suppl. 9:164 (1964).
64. Moser, H. C., and A. F. Voigt. *J. Am. Chem. Soc.* 79:1837 (1957).
65. Sidgewick, N. V. *The Chemical Elements and Their Compounds* (London: Oxford University Press, 1950), p. 324.
66. Jensen, F. R., and B. Rickborn. *Electrophilic Substitution of Organomercurials* (New York: McGraw-Hill Book Co., 1968).
67. Reutov, O. A., and I. P. Beletskaya. *Reaction Mechanisms of Organometallic Compounds* (Amsterdam: North Holland Publishing Co., 1968).
68. Shibko, S. I., and N. Nelson. *Environ. Res.* 4:23 (1971).
69. Kreevoy, M. M. *J. Am. Chem. Soc.* 79:5927 (1957).
70. Brown, R. D., et al. *Aust. J. Chem.* 18:1507 (1965).
71. Fagerström, T., and A. Jernelöv. *Water Res.* 6:1195 (1972).
72. Ouellette, R. J. Ph.D. Thesis, University of California, Berkeley, CA (1969).
73. Talmi, Y., and R. E. Mesmer. *Water Res.* 9:547 (1975).
74. Tsivoglou, E. C. "Tracer Measurement of Stream Reaeration," USDI, Federal Water Pollution Control Administration, Washington, DC (1967).

75. Harvey, S. C. In: *The Pharmacological Basis of Therapeutics*, L. S. Goodman, and A. Gilman, Eds. (New York: Macmillan Publishing Co., Inc., 1965).
76. "Arsenicals as Growth Promoters," *Nutrit. Rev.* 14:206 (1956).
77. Schroeder, H. A., and J. Balassa. *J. Chron. Dis.* 19:85 (1966).
78. Vallee, B. L., et al. *M. M. A. Arch. Ind. Health* 21:132 (1960).
79. U.S. Environmental Protection Agency. "Water Quality Criteria Data Book," Vol. 3, *Effects of Chemicals on Aquatic Life*, Washington, DC, Pub. No. 18050GWV05/71 (May, 1971).
80. National Academies of Sciences and Engineering. "Water Quality Criteria, 1972," Washington, DC (1972).
81. McKee, J. E., and H. W. Wolf. *Water Quality Criteria*, The Resources Agency of California, State Water Quality Control Board, Sacramento, CA (1963).
82. Yeh, S., et al. *Cancer* 21:312 (1968).
83. Chen, K. P., and H. Y. Wu. *J. Formosan Med. Assoc.* 61:611 (1962).
84. Tseng, W. P., et al. *J. Nat. Cancer Inst.* 40:453 (1968).
85. Neubauer, O. *Brit. J. Cancer* 1:192 (1947).
86. Baroni, C., et al. *Arch. Environ. Health* 7:688 (1963).
87. Fleischer, L. "Data of Geochemistry," 6th ed., *U.S. Geol. Survey Prof. Paper 440* (1963).
88. Vinogradov, A. P. *The Geochemistry of Rare and Dispersed Chemical Elements in Soils*, 2nd ed. Consultants Bureau, Inc., New York (1959).
89. Golschmidt, V. M. *Geochemistry* A. Muir, Ed. (Oxford: Clarendon Press, Publisher, 1958), p. 646.
90. Ferguson, J. F., and J. Gavis. *Water Res.* 6:1259 (1972).
91. Chamberlain, W., and J. Shapiro. *J. Limnol. Oceanog.* 14:921 (1969).
92. Wedepohl, K. H., In: *Handbook of Geochemistry*, Vol. II (New York: Springer-Verlag New York, Inc., 1969).
93. Durum, W. H., et al. "Reconnaissance of Selected Minor Elements in Surface Water of the United States," *Geol. Survey Circ. 643*, U.S. Department of the Interior, Washington, DC (1971).
94. Veal, D. J. *Anal. Chem.* 38:1080 (1966).
95. Chapman, A. *Analyst* 51:548 (1926).
96. Ellis, M. M., et al. *Ind. Eng. Chem.* 33:1331 (1941).
97. Coulson, E. J., et al. *J. Nutrition* 10:225 (1935).
98. Ullmann, W. W., et al. *J. Water Poll. Control Fed.* 33:416 (1961).
99. McBride, B. C., and R. S. Wolfe. *Biochemistry* 10:4312 (1971).
100. Turner, A. W. *Aust. J. Biol. Sci.* 7:452 (1954).
101. Turner, A. W., and J. W. Legge. *Aust. J. Biol. Sci.* 7:479 (1954).
102. Gosio, B. *Ber.* 30:1024 (1897).
103. Challenger, F., et al. *J. Chem. Soc. (London)* 95 (1933).
104. Cox, D. P., and M. Alexander. *Appl. Microbiol.* 25:408 (1973).
105. Woolson, E. A., and P. E. Kearney. *Environ. Sci. Technol.* 7:47 (1973).
106. Schuth, C. K., et al. *J. Agric. Food Chem.* 22:999 (1974).
107. Braman, R. S., and C. C. Foreback. *Science* 182:1247 (1973).
108. Bjerrum, N. "Stability Constants of Metal-Ion Complexes," *Special Publ. No. 17*, The Chemical Society, London (1964).
109. Browning, E. *Toxicity of Industrial Metals* (London: Butterworths, 1961).
110. Schroeder, H. A., and I. H. Tipton. *Arch. Environ. Health.* 17:965 (1968).

111. Fchess, H., et al. *J. Am. Water Works Assoc.* 30:1425 (1938).
112. Kohoe, R. A., *Med. Clin. N. Am.* 26:126 (1942).
113. Lewis, K. H. "Symposium on Environmental Lead Contamination," *Public Health Service Publication 1440*, Washington, DC (1966).
114. Blauth, J., and W. Duhl. *Roczn. ZaKl. Hig. Warsz.* 437 (1953); *Chem. Abstr.* 48:7824 (1954).
115. Huff, L. C. *Econ. Geol.* 47:517 (1952).
116. Wright, J. R., et al. *Soil Sci. Soc. Am. Proc.* 19:340 (1955).
117. Chow, T. J., and C. C. Patterson. *Geochim. Cosmochim. Acta* 26:263 (1962).
118. Lagerwerff, J. V., and A. W. Specht. *Environ. Sci. Technol.* 4:583 (1970).
119. Patterson, C. C. *Arch. Environ. Health* 11:344 (1965).
120. Blanchard, R. L. *Proc. Int. Symp. on Radioecological Concentration Process.*, B. Aberg and F. P. Hungate, Eds. (Elmsford, NY: Pergamon Press, Inc., 1966), p. 281.
121. Murozumi, M., et al. *Geochim. Cosmochim. Acta* 33:1247 (1969).
122. Chow, T. J. *J. Water Poll. Control. Fed.* 40:399 (1968).
123. Livingstone, D. A. "Data of Geochemistry," *U.S. Geol. Survey Prof. Paper 440-G*, (1967).
124. "Biological Effects of Atmospheric Pollutants: Lead, Airborne Lead in Perspective," (Washington, DC: National Academy of Sciences, 1972).
125. Abernethy, R. G., et al. *Bureau of Mines Report RI-7281*, U.S. Department of Interior, Washington, DC (1969).
126. First, M. W. In: *Air Pollution*, Vol. 3, A. C. Stern, Ed. (New York: Academic Press Inc., 1968), p. 298.
127. Atkins, P. R. *J. Air Poll. Control Fed.* 19:591 (1969).
128. U.S. Department of Health, Education, and Welfare. "Survey of Lead in the Atmosphere of Three Urban Communities," Public Health Service Publication 999-AP-12, Cincinnati, OH (1965).
129. McCaldin, R. O. "Symposium on Environmental Lead Contamination," Public Health Service Publication 1440, Washington, DC (1966), p. 7.
130. U.S. Department of Health, Education, and Welfare. "Air Quality Data from the National Air Sampling Networks and Contributing State and Local Networks," NAPCA Publication APTD 68-9 (1968).
131. McMullen, T. B., et al. *J. Air Poll. Control Assoc.* 20:369 (1970).
132. Page, A. L., and T. J. Ganje. *Environ. Sci. Technol.* 4:140 (1970).
133. Hem, J. D., and W. H. Durum. *J. Am. Water Works Assoc.* 65:562 (1973); Also, "Lead in the Environment," *U.S. Geol. Survey Prof. Paper 957*, Washington, DC (1976).
134. Hunt, W. F., Jr., et al. *Trace Substances in Environmental Health IV*, D. D. Hemphill, Ed. (Columbia, MO: University of Missouri Press, 1971), p. 56.
135. Lazrus, A. L., et al. *Environ. Sci. Technol.* 4:55 (1970).
136. Ter Haar, G. L., and M. A. Bayard. *Nature* 232:553 (1971).
137. Burton, W. M., and N. G. Steward. *Nature* 186:584 (1960).
138. Francis, C. W., et al. *Environ. Sci. Technol.* 4:586 (1970).
139. Leland, H. V., et al. In: *Trace Metals and Metal-Organic Interactions in Natural Waters*, P. C. Singer, Ed. (Ann Arbor, MI: Ann Arbor Science Publishers, Inc., 1973).
140. Jones, J. R. G. *Fish and River Pollution* (London: Butterworths, 1964), p. 53.

390 CHEMISTRY OF NATURAL WATERS

141. Carpenter, K. E. *Ann. Appl. Biol.* 12:1 (1925).
142. Wetterberg, L. *Lancet* 1:498 (1966).
143. Goldberg, E. D. *Composition of Sea Water, Comparative and Descriptive Oceanography, The Sea,* Vol. 2 (New York: Wiley-Interscience, 1963), p. 3.
144. Taisumoto, M., and C. C. Patterson. *Nature* 199:350 (1963).
145. Schroeder, H. A., et al. *J. Chron. Dis.* 14:408 (1961).
146. Pringle, B. H. et al. *J. Sanit. Eng. Div., ASCE* 94:455 (1968).
147. Bilinski, H., and W. Stumm. "Pb(II)-Species in Natural Waters," EAWAG No. 1, Swiss Federal Institutes of Technology, Federal Institute of Water Resources and Water Pollution Control (1973).
148. Clarkson, T. W. "International Conference on Environmental Sensing and Assessment," Las Vegas, NE. 1:1 (1975).
149. Lieber, M., and W. F. Welsch. *J. Am. Water Works Assoc.* 46:541 (1954).
150. Kobayashi, J. *Nippon Kagaku Zasshi* 39:286, 369, 424, (1969) (Japanese).
151. Hem. J. D. *Water Resources Res.* 8:661 (1972).
152. Bowen, H. J. M. *Trace Elements in Biochemistry* (New York: Academic Press, Inc., 1966), pp. 164, 179, 209.
153. MacGregor, A. *Environ. Health Pers.* 12:137 (1975).
154. Kobayashi, J. "Fifth Annual Conference on Trace Substances in Environmental Health," University of Missouri, Columbia, MO (June 1971).
155. Emerson, R. E., et al. "International Conference on Environmental Sensing and Assessment," Las Vegas, NE 1:6 (1975).
156. Bruland, K. W., et al. *Limnol. Oceanog.* 23:618 (1978).
157. "Deliberations of the International Decade of Ocean Exploration (IDOE) Baseline Conference," IDOE, Washington, DC (May 1972).
158. Mount, D. I. *J. Wildlife Management,* 31:168 (1967).
159. Lowman, F. G. "Radioactivity in the Marine Environment," (Washington, DC: National Academy of Sciences, 1971), p. 161.
160. Kerfoot, W. B., and S. A. Jacobs. *Environ. Sci. Technol.* 10:662 (1976).
161. Houblou, W. F., et al. *Oregon Fish Comm. Briefs* 5:1 (1954).
162. Brooks, R. R., and M. G. Rumbsby. *Limnol. Oceanog.* 10:521 (1965).
163. Gardiner, J. *Water Res.* 8:23 (1974).
164. Lake, P. E., and J. M. Goodings. *Can. J. Chem.* 36:1089 (1968).
165. Zirino, A., and S. Yamamoto. *Limnol. Oceanog.* 17:661 (1972).
166. Gardiner, J. *Water Res.* 8:157 (1974).
167. Moxon, A. L. "Alkali Disease or Selenium Poisoning," South Dakota Agricultural Experiment Station Bulletin No. 311, Brookings, South Dakota State College of Agriculture and Mechanic Arts (1937).
168. Rosenfeld, I., and O. A. Beath. *Selenium, Geobotany, Biochemistry, Toxicity, and Nutrition,* (New York: Academic Press, Inc., 1964).
169. Byard, J. L. *Arch. Biochem. Biophys.* 130:556 (1969).
170. McConnell, K. P., and O. W. Portman. *J. Biol. Chem.* 195:277 (1952).
171. Flohe, L., et al. *Fed. Eur. Biochem. Soc. Lett.* 32:132 (1973).
172. Levander, O. A., et al. *Toxicol. Appl. Pharmacol.* 16:79 (1970).
173. Allaway, W. H., et al. *Arch. Environ. Health* 16:342 (1968).
174. Parizek, J., et al. *Newer Trace Elements in Nutrition,* (New York: Marcel Dekker, Inc., 1971), p. 85.

175. Schroeder, H. A., et al. *J. Chronic Dis.* 23:227 (1970).
176. Luttrell, G. W. "Annotated Bibliography on the Geology of Selenium," *U.S. Geological Survey Bull. No. 1019M*, Washington, DC (1959).
177. Goldschmidt, V. M., and L. W. Strock. *Ges. Wiss. Gottingen Math. Physik. Klasse* 1:123 (1935).
178. Lakin, H. W. In: *Trace Elements in the Environment*, Advances in Chemistry Series 123 (Washington, DC: American Chemical Society, 1973), p. 96.
179. Byers, H. G., et al. *Ind. Eng. Chem. Ind. Ed.* 28:821 (1936).
180. Beath, O. A., et al. "Some Rocks of High Selenium Content," *Wyoming Geol. Survey Bull. No. 36*, University of Wyoming (1946).
181. Wells, N. *N.Z. Geol. Geophys.* 10:198 (1967).
182. Fleming, G. A. *Soil Sci.* 94:28 (1962).
183. Byers, H. G. *U.S. Dept. Agric. Tech. Bull.* 482:48 (1935).
184. Pillay, K. K. S., et al. *Nucl. Appl. Technol.* 7:478 (1969).
185. Hashimoto, Y. *Environ. Sci. Technol.* 4:157 (1970).
186. "Selenium," In: *Medical and Biologic Effects of Environmental Pollutants* (Washington, DC: National Academy of Sciences, 1976), p. 23.
187. Lewis, B. G., et al. *Biochem. Biophys. Acta* 237:603 (1971).
188. Evans, C. S., et al. *Austral. J. Biol. Sci.* 21:13 (1968).
189. Abu-Erreish, G. M., et al. *Soil Sci.* 106:415 (1968).
190. Selyankina, K. P. *Hyg. Sanit.* 35:431 (1970).
191. Lakin, H. W., and H. G. Byers. "Selenium Occurrence in Certain Soils in the United States, with a Discussion of Related Topics," Sixth Report, *U.S. Dept. Agric. Tech. Bull.* (1941), p. 783.
192. Hasimoto, Y., and J. W. Winchester. *Environ. Sci. Technol.* 1:338 (1967).
193. Stahl, Q. R. "Preliminary Air Pollution Survey of Selenium and Its Compounds, A Literature Review." Air Pollution Control Administration Publication No. APTD69-47, Raleigh, NC (1969).
194. Klarkar, D. P., et al. *Geochem. Cosmochim. Acta* 32:285 (1968).
195. Scott, R. C., and P. T. Voegeli, Sr. "Radiochemical Analyses of Ground and Surface Water in Colorado, 1954-1961," Colorado Water Conservation Board Basic Data Report No. 7 (1961).
196. Edmond, C. R. *Austral. Min. Develop. Lab. Bull.* 4:17 (1967).
197. Oelsch Läger, W., and K. M. Menke. *Z. Ernährungswissenschaft* 9:216 (1969).
198. Lakin, H. W., and D. F. Davidson. *Symposium, Selenium in Biomedicine*, First International Symposium, Oregon State University (Westport, CT: AVI Publishing Co., 1967), p. 27.
199. Schutz, D. F., and K. K. Turekian. *Geochim. Cosmochim. Acta* 29:259 (1965).
200. Russell, F. C. "Minerals in Pasture, Deficiencies and Excesses in relation to Animal Health," Imperial Bur. of Animal Nutrition, Aberdeen, Scotland, Tech. Communication 15 (1944).
201. Lunde, G. *J. Sci. Food Agric.* 21:242 (1970).
202. Kifer, R. R. *Feedstuffs* 41:24 (1969).
203. Morris, V. C., and O. A. Levander. *J. Nutr.* 100:1383 (1970).
204. Woolfolk, C. A., and H. R. Whiteley. *J. Bacteriol.* 84:647 (1962).
205. Nickerson, W. J., and G. Falcone. *J. Bacteriol.* 85:763 (1963).
206. McCready, R. G. L., et al. *Can. J. Microbiol.* 12:703 (1966).

207. Tilton, R. C., et al. *Can. J. Microbiol.* 13:1175 (1967).
208. Shrift, A., et al. *Plant Physiol.* 36:502 (1961).
209. Letunova, S. V. "Trace Element Metabolism in Animals," *Proc. WAAP/IBP Int. Symp.*, Aberdeen, Scotland, 1969, E. & S. Livingstone, Ltd., Edinburgh (1970), p. 432.
210. Challenger, F. *Adv. Enzymol.* 12:429 (1951).
211. Fleming, G. A., and M. Alexander. *Appl. Microbiol.* 24:424 (1972).
212. Chau, Y. K., et al. *Science* 192:1131 (1976).
213. Sillen, L. G., and A. E. Martell. *Stability Constants of Metal-Ion Complexes*, Special Publication No. 17 (2nd ed.), The Chemical Society, London (1964), p. 754.
214. Geering, H. R., et al. *Soil Sci. Soc. Am. Proc.* 32:35 (1968).
215. Johnson, C. M. *Residue Rev.* 38:101 (1976).
216. Chukhlanstev, V. G., and G. P. Tomashevsky. *Russ. J. Inorg. Chem.* 1:303 (1956).
217. Williams, K. T., and H. G. Byers. *Ind. Eng. Chem.* 28:912 (1936).
218. Plotnikov, V. I. *Zhur. Neorg. Klim.* 3:1761 (1958).
219. Hingston, F. J., et al. *Adsorption from Aqueous Solutions*, Advances in Chemistry Series 79 (Washington, DC: American Chemical Society, 1968), p. 82.
220. Schroeder, H. A. *Chromium*, Air Quality Monograph #70-15, American Petroleum Institute, Washington, DC (1970).

221. "Chromium, Medical and Biologic Effects of Environmental Pollutants," (Washington, DC: National Academy of Sciences, 1974).
222. Byerrum, R. U. *Proc. Fifteenth Ind. Waste Conf.* Engineering Extension Series No. 106, Purdue University Lafayette, IN (1961).
223. Rollinson, C. L. In: *Comprehensive Inorganic Chemistry* (Elmsford, NY: Pergamon Press, Inc., 1973), p. 623.
224. Weeks, M. E., and H. E. Leicester. *Discovery of the Elements*, 7th Ed., (Easton, PA: Mack Printing Co., 1968).
225. Robinson, W. O., et al. U.S. Department of Agriculture Technical Bulletin No. 471, U.S. Department of Agriculture, Washington, DC (1935).
226. Shacklette, H. T., et al. "Elemental Composition of Surficial Materials in Conterminous United States," *U.S. Geol. Survey Prof. Paper 574-D.*, Washington, DC (1971).
227. Haffty, J. "Residue Method for Common Minor Elements," *U.S. Geol. Survey Water Supply Paper 1540-A.*, Washington, DC (1960).
228. Kroner, R. G., "Traces of Heavy Metals in Water-Removal Processes and Monitoring," EPA-90219-74-001, U.S. Environmental Protection Agency, Washington, DC (1973).
229. Copeland, R. A., and J. C. Ayers. "Trace Element Distribution in Water, Sediment, Phytoplankton, Zooplankton and Benthos of Lake Michigan," (Ann Arbor: Environmental Research Group, Inc., 1972).
230. Leland, H. V., et al. In: *Trace Metals and Metal-Organic Interactions in Natural Waters*, P. C. Singer, Ed. (Ann Arbor, MI: Ann Arbor Science Publishers, Inc., 1973).
231. Schell, W. R., and A. Nevissi. *Environ. Sci. Technol.* 11:887 (1977).
232. Stoffers, P., et al. *Environ. Sci. Technol.* 11:819 (1977).
233. Durum, W. H., and J. Haffty. *Geochim. Cosmochim. Acta* 27:1 (1963).

234. "Investigation of Geothermal Waters in the Long Valley Area," State of California, Department of Water Resources, Mono County (1967).
235. Ishibashi, M., and T. Skigematsu. *Bull. Inst. Chem. Res. (Kyoto Univ.)* 23:59 (1950).
236. Loveridge, B. A., et al. "Atomic Research Establishment Report R-3323 (England)," Atomic Energy Research Establishment, Harwell, England (1960).
237. Chuecas, L., and J. P. Riley. *Anal. Chim. Acta* 35:240 (1966).
238. Bowen, H. J. M. *Trace Elements in Biochemistry* (New York: Academic Press, Inc., 1966).
239. Moyers, J. L., et al. *Environ. Sci. Technol.* 11:789 (1977).
240. Lee, R. E., et al. *Environ. Sci. Technol.* 9:643 (1975).
241. Furr, A. K. *Environ. Sci. Technol.* 11:1194 (1977).
242. Harvey, R. J. *Botan. Gaz.* 111:1 (1949).
243. Clendennig, K. A., and W. J. North. In: *Proc. First Int. Conf. on Waste Disposal in the Marine Environment*, E. A. Pearson, Ed., (Elmsford, NY: Pergamon Press, Inc., 1960).
244. Naismith, W. E. F. *Arch. Biochem. Biophys.* 73:255 (1958).
245. Fukai, R. *Nature* 213:901 (1967).
246. Schroeder, D. C., and C. F. Lee. *Water Air Soil Poll.* 4:355 (1975).